Econophysics and Data Driven Modelling of Market Dynamics

New Economic Windows

More information about this series at http://www.springer.com/series/6901

Frédéric Abergel · Hideaki Aoyama
Bikas K. Chakrabarti · Anirban Chakraborti
Asim Ghosh

Editors

Econophysics and Data Driven Modelling of Market Dynamics

 Springer

Editors

Frédéric Abergel
Laboratory of Mathematics Applied to
  System
CentraleSupélec
Châtenay-Malabry
France

Anirban Chakraborti
School of Computational and Integrative
  Sciences
Jawaharlal Nehru University
New Delhi
India

Hideaki Aoyama
Department of Physics
Kyoto University
Kyoto
Japan

Asim Ghosh
Saha Institute of Nuclear Physics
Kolkata
India

Bikas K. Chakrabarti
Saha Institute of Nuclear Physics
Kolkata
India

ISSN 2039-411X
New Economic Windows
ISBN 978-3-319-38393-4
DOI 10.1007/978-3-319-08473-2

ISSN 2039-4128   (electronic)

ISBN 978-3-319-08473-2   (eBook)

Springer Cham Heidelberg New York Dordrecht London

Printed on acid-free paper

Springer International Publishing AG Switzerland is part of Springer Science+Business Media
(www.springer.com)

Preface

This proceedings volume is based on the conference entitled 'Econophysics and Data Driven Modelling of Market Dynamics' that was held at Saha Institute of Nuclear Physics, Kolkata during 14–17 March 2014. This was the eighth event of the 'Econophys-Kolkata' series of conferences, and was organized jointly by Saha Institute of Nuclear Physics, École Centrale Paris, Jawaharlal Nehru University and Kyoto University.

During the past decades, the financial market landscape has been dramatically changing: deregulation of markets, growing complexity of products, etc. The ever-rising speed and decreasing costs of computational power and networks have led to the emergence of huge databases. We chose this particular theme for the conference, as we thought that it would be most appropriate with the availability of these data. Econophysicists, along with many others, have been relying primarily on empirical observations in order to construct models and validate them, or study models that are better empirically founded. Thus, a major part of the efforts of econophysicists have been the study of empirical data and financial time series analyses. Often, the empirics have guided researchers to design more realistic and practical models. The recent turmoil on financial markets and the 2008 crash seem to plead for new models or approaches, and the econophysics community indeed has an important role to play in market modelling in the future years to come.

This proceedings volume contains papers by distinguished experts from all over the world, mostly based on the talks and seminars delivered at the meeting and accepted after refereeing. For completeness, a few articles by experts who could not participate in the meeting due to unavoidable reasons were also invited and these too have been incorporated in this volume. This volume is organized as follows: A first part dedicated to 'Market Analysis and Modelling'. A second part entitled 'Miscellaneous' presents other ongoing studies in related areas on econophysics and sociophysics. We have included in the third part, 'Reviews', two reviews which address recent developments in econophysics and sociophysics. We have included in the fourth part, 'Discussions and Commentary', an extensive note on the impact of econophysics researches (obtained from responses of leading researchers to questionnaire). Another write-up in this part discusses the influence of econo-physics research on contemporary researches in social sciences.

We are grateful to all the participants at the meeting and for their contributions. We are also grateful to Mauro Gallegati and the Editorial Board of the 'New Economic Windows' series of Springer-Verlag (Italy) for their continuing support in getting this proceedings volume published in their esteemed series.

The conveners (editors) also express their thanks to Saha Institute of Nuclear Physics, École Centrale Paris, Jawaharlal Nehru University and Kyoto University for their support in organizing this conference. The support from J.C. Bose project fund (DST, India) of Bikas K. Chakrabarti is gratefully acknowledged.

Châtenay-Malabry, France   Frédéric Abergel
Kyoto, Japan   Hideaki Aoyama
Kolkata, India   Bikas K. Chakrabarti
New Delhi, India   Anirban Chakraborti
Kolkata, India   Asim Ghosh
October 2014

Contents

Part I   Market Analysis and Modelling

Empirical Evidence of Market Inefficiency:
Predicting Single-Stock Returns 3
Marouane Anane and Frédéric Abergel
1   Introduction 3
2   Data, Methodology and Performance Measures 4
   2.1   Data 4
   2.2   Methodology 5
   2.3   Performance Measures 6
3   Conditional Probability Matrices 6
   3.1   Binary Method 8
   3.2   Four-Class Method 11
4   Linear Regression 12
   4.1   Ordinary Least Squares (OLS) 14
   4.2   Ridge Regression 15
   4.3   Least Absolute Shrinkage and Selection
        Operator (LASSO) 20
   4.4   Elastic Net (EN) 22
5   Conclusions 24
References 66

Calibration of a Stock's Beta Using Option Prices 67
Sofiene El Aoud and Frédéric Abergel
1   Introduction 67
2   Model with Constant Idiosyncratic Volatility 69
   2.1   Presentation of the Model 69
   2.2   Calibration of Implied Beta 70
   2.3   Limits of the Model 71

3   Model with Stochastic Idiosyncratic Volatility   71
     3.1   Presentation of the Model   72
     3.2   Pricing Options on the Index and the Stock.   73
     3.3   Calibration of Implied Beta Using Options Prices.   80
4   Applications for the Estimation of the Parameter β.   83
     4.1   Prediction of Forward Beta   87
     4.2   Hedging of Options on the Stock by Instruments
            on the Index   89
5   Conclusion   94
References   103

Long-Term Evolution of the Topological Structure
of Interactions Among Stocks in the New York Stock
Exchange 1925–2012   105
Chandrashekar Kuyyamudi, Anindya S. Chakrabarti
and Sitabhra Sinha
1   Introduction.   106
2   The NYSE   107
3   Results   109
     3.1   Return Cross-Correlation Matrix.   109
     3.2   Eigenvalue Analysis of the Correlation Matrix   110
     3.3   The Largest Eigenvalue.   111
     3.4   Filtering the Correlation Matrix   112
     3.5   Time Evolution of the Market Mode.   113
     3.6   Time Evolution of the Random Modes   114
     3.7   Time Evolution of the Intra-group Interaction Modes   115
     3.8   Percolation Inspired Approach to Clustering   117
References   120

Bitcoin Dynamics: The *Inverse Square Law* of Price Fluctuations
and Other Stylized Facts   121
Soumya Easwaran, Manu Dixit and Sitabhra Sinha
1   Introduction.   121
2   The Bitcoin.   122
3   Results   123
4   Discussion.   127
References   127

Are Firms that Are Awarded More Patents More Productive?   129
Shouji Fujimoto, Atushi Ishikawa, Takayuki Mizuno
and Tsutomu Watanabe
1   Introduction.   129
2   Databases   130
3   Distribution of Firm Patents.   131

4   Cobb-Douglas Production Function and Total
     Factor Productivity... 135
5   Correlation Between C and A 136
6   Summary 140
References 141

Nonlinear Dynamics of Stock Markets During Critical Periods 143
Kousik Guhathakurta
1   Introduction... 143
2   The Stock Market Crashes Under Study 145
     2.1   Black Monday (Oct 19, 1987) Crash.................... 145
     2.2   Japanese Bubble (1986–1991) 146
     2.3   The Friday the 13th Mini-crash (Oct 13, 1989) 146
     2.4   October 27, 1997 Mini-crash 146
     2.5   11 Sept NYSE, 2001 Crash............................. 146
     2.6   Stock Market Downturn of 2002, DJIA.................. 147
     2.7   2010 Flash Crash...................................... 147
     2.8   China 2007 Crash...................................... 147
3   Data & Software 148
4   Theoretical Background of Empirical Tools................... 149
     4.1   Recurrence Plot 149
     4.2   Quantification of Recurrence Plots (Recurrence
            Quantification Analysis) With Confidence Intervals 150
5   Analysis of Empirical Results 152
6   Conclusions... 163
References 164

**Probabilistic Flows of Inhabitants in Urban Areas
and Self-organization in Housing Markets** 167
Takao Hishikawa and Jun-ichi Inoue
1   Introduction... 167
2   The Model System... 170
     2.1   A City—Working Space— 170
     2.2   Agents 171
     2.3   Attractiveness of Locations 172
     2.4   Probabilistic Search of Locations by Buyers 174
     2.5   Offering Prices by Sellers 175
     2.6   The Condition on Which the Transaction is Approved 176
     2.7   Computer Simulations: A Preliminary.................... 176
3   Empirical Data in City of Sapporo 178
4   An Extension to a City Having Multiple Centers 179
5   Computer Simulations... 180

5.1   Spatial Structure in the Distribution
        of Visiting Times   180
5.2   The Rent Distribution   182
5.3   On the Locations of Offices   183
5.4   On the Effective Time-Scale of Update Rule   185
6   Summary and Discussion   185
6.1   The 'Quasi-One-dimensional' Model   186
6.2   Probabilistic Search Depending on the Location
        of Office   186
References   188

Part II   Miscellaneous

Performance of Inequality Indices   191
Anurag Chaplot and Rituparna Sen
1   Introduction   191
2   Different Measures of Inequality   192
3   Estimation of Inequality Indices   193
3.1   Point Estimates   194
3.2   Normal Approximation Confidence Intervals   195
3.3   Bootstrap Percentile Confidence Interval   196
3.4   Bootstrap-t Confidence Interval   196
4   Simulation Study   196
5   Real Data Examples   204
6   Conclusions   209
References   211

**A Dynamical View of Different Solution Paradigms
in Two-Person Symmetric Games: Nash Versus
Co-action Equilibria**   213
V. Sasidevan and Sitabhra Sinha
1   Introduction   213
2   A Dynamical Framework for Analysing 2-Person Games   215
3   Examples   218
3.1   Prisoner's Dilemma   218
3.2   Chicken   219
3.3   Stag-Hunt   222
4   Conclusions   223
References   223

Spatio-Temporal Patterns of Development in India:
Identifying Geographical Features of Growth and Urbanisation. 225
S. Sridhar, Tara Thiagarajan and Sitabhra Sinha
1   Introduction. 226
2   Temporal Patterns 227
3   Patterns in Settlement Size Scale 229
4   Spatiotemporal Patterns. 231
5   Conclusions. 233
References 233

Part III   Reviews

Physicists' Approaches to a Few Economic Problems 237
Anirban Chakraborti, Yoshi Fujiwara, Asim Ghosh,
Jun-ichi Inoue and Sitabhra Sinha
1   Background and Motivation. 238
2   Income and Wealth Distributions: Kinetic Exchange Models 239
     2.1   Introduction. 239
     2.2   Model with Uniform Savings. 243
     2.3   Model with Distributed Savings 244
     2.4   Summary and Discussions 246
3   Market Mechanism: Agent-Based Models 247
     3.1   Introduction. 247
     3.2   El Farol Bar Problem 248
     3.3   Minority Game. 248
     3.4   Kolkata Paise Restaurant Problem. 250
     3.5   Summary and Discussions 253
4   Economic Success and Failures: Analyses and Modelling 254
     4.1   Introduction. 254
     4.2   The Economic Fate of Movies: A Case Study 255
     4.3   Log-Normal Nature of Economic Performance 258
     4.4   Bimodality of Success and Failure 259
     4.5   Power Law Decay of Income with Time 260
     4.6   The Stylized Facts of "Popularity" 261
5   Inter-firms and Banks-firms Network Structures:
     Empirical Studies. 261
     5.1   Introduction. 261
     5.2   Banks-firms Credit Network 263
     5.3   Production Network Among Firms 266
     5.4   Summary. 270

6   Financial Time-Series Analyses: Wiener Processes
     and Beyond.   271
     6.1   Introduction.   271
     6.2   Empirical Evidence and Stable Distributions   271
     6.3   Time-Dependent Volatility and the Prediction Models.   273
     6.4   Duration Between Price Changes: First-Passage Process   274
     6.5   Microscopic Reconstruction of Prices   276
     6.6   Summary.   279
7   Outlook   279
References   280

Socio-Economic Inequalities: A Statistical Physics Perspective   287
Arnab Chatterjee
1   Introduction.   287
2   Evolutionary Perspective to Socio-economic Inequality   290
3   Why Statistical Physics?   290
4   Processes Leading to Inequality and Broad Distributions   291
     4.1   Random Walk   291
     4.2   Combination of Exponentials   292
     4.3   Self-organized Criticality   292
     4.4   Multiplicative Processes   293
     4.5   Preferential Attachment   294
5   Income, Wealth & Energy.   297
     5.1   Modelling Income and Wealth Distributions   298
     5.2   Is Wealth and Income Inequality Natural?   300
6   Urban and Similar Systems   300
     6.1   City Size.   301
     6.2   Scaling of Urban Measures   302
     6.3   Firms   302
7   Consensus.   303
     7.1   Voting   303
     7.2   Religion   304
     7.3   Modelling Opinion and Its Dynamics   306
8   Bibliometrics   309
     8.1   Annual Citation Indices.   310
     8.2   Universality in Citation Distributions   314
9   Networks   315
10   Measuring Inequality   315
11   How to Deal with Inequality?   317
12   Discussions   319
References   320

Part IV   Discussions and Commentary

**Judging the Impact of 'Econophysics' Through Response
to Questionnaire**   327
Kishore C. Dash
1   Introduction.   327
2   Response to Questionnaire.   331
     2.1   Questions   331
     2.2   Answers   332
3   Conclusion   347
References   348

**Commentary: Influence of Econophysics Research
on Contemporary Researches in Social Sciences**   349
Asim Ghosh
1   Introduction.   349
2   Result.   350
3   Conclusion   351
References   351

Photo   353

Part I
Market Analysis and Modelling

Empirical Evidence of Market Inefficiency: Predicting Single-Stock Returns

Marouane Anane and Frédéric Abergel

Abstract Although it is widely assumed that the stock market is efficient, some empirical studies have already tried to address the issue of forecasting stock returns. As far as is known, it is hard to find a paper involving not only the forecasting statistics but also the forecasting profitability. This paper aims to provide an empirical evidence of the market inefficiency and to present some simple realistic strategies based on forecasting stocks returns. In order to achieve this study, some linear and non linear algorithms are used to prove the predictability of returns. Many regularization methods are introduced to enhance the linear regression model. In particular, the RIDGE method is used to address the colinearity problem and the LASSO method is used to perform variable selection. The different obtained results show that the stock market is inefficient and that profitable strategies can be computed based on forecasting returns. Empirical tests also show that simple forecasting methods perform almost as well as more complicated methods.

1 Introduction

Forecasting the market has been one of the most exciting financial subjects for over a century. In 1900, Bachelier[1] admitted, "Undoubtedly, the Theory of Probability will never be applicable to the movements of quoted prices and the dynamics of the Stock Exchange will never be an exact science. However, it is possible to study mathematically the static state of the market at a given instant to establish the probability law for the price fluctuations that the market admits at this instant". Seventy years later, Fama[2] proposed some formal definitions of the market efficiency; "A market in which prices always fully reflect available information is called efficient". Opinions

M. Anane
Chair of Quantitative Finance, MAS Laboratory, Ecole Centrale Paris,
BNP Paribas 20 Boulevard des Italiens, 75009 Paris, France
e-mail: marouane.anane@gmail.com

F. Abergel (✉)
Chair of Quantitative Finance, MAS Laboratory, Ecole Centrale Paris,
Grande voie des vignes, 92290 Chatenay Malabry, France
e-mail: frederic.abergel@ecp.fr

© Springer International Publishing Switzerland 2015
F. Abergel et al. (eds.), *Econophysics and Data Driven Modelling of Market Dynamics*,
New Economic Windows, DOI 10.1007/978-3-319-08473-2_1

have been always divergent about the market efficiency. Malkiel [3] concluded that most investors trying to predict stocks' returns always ended up with profits inferior to passive strategies. In his famous book, *Fooled by Randomness*, Taleb [4] argued that even the best performances can be explained by luck and randomness. On the other hand, finance professionals demonstrated, in real life, that they can always make money beating the market; see Warren Buffett's response to efficient market claims [5].

The recent rise in electronic markets lead to big available financial data. The attempt to discover some predictable, and hopefully profitable, signal in the middle of those millions of numbers has never been as high as today.

In the academic world, Chakraborti et al. [6] studied in detail the statistical properties of the intraday returns, and came to the conclusion that there is no evidence of correlation between successive returns. Similarly, Lillo and Farmer [7] concluded that stock returns contain negligible temporal autocorrelation. Fortunately, Zheng and Abergel [8] found some promising results, in particular the liquidity imbalance on the best bid/ask seems to be informative to predict the next trade sign.

In the professional world, many books present hundred of strategies predicting the market and always earning money; see [9, 10] for example. When testing those strategies in other samples, results are so different and the strategies are no longer profitable. It is possible that the overfit of such methods played a key role in the good performances published in those books.

This study was performed from both an academic and a professional perspective. For each prediction method, not only are statistical results presented, but also presented are the performances of the correspondent strategies. The aim is to give another point of view of a good prediction and of an efficient market.

This work is organized as follows: In the first section, the data and the test methodology are presented. In the second section a non linear method, based on conditional probability matrices, is used to test the predictive power of each indicator. In the last section, the linear regression is introduced to combine the different indicators and many regularization ideas are tested in order to enhance the performances of the strategies.

2 Data, Methodology and Performance Measures

2.1 Data

This paper focuses on the EURO STOXX 50 European liquid stocks. One year (2013) of full daily order book data provided by BNP Paribas are used to achieve the study. For a stock with a mid price S_t at time t, the return to be predicted over a period dt is $Ln(\frac{S_{t+dt}}{S_t})$. At the time t, one can use all the available data for any time $s \leq t$ to perform the prediction.

In Sects. 2 and 3, the focus is on predicting the stocks' returns over a fixed period dt using some order book indicators. Once the returns and the indicators are computed, the data are sampled on a fixed time grid from 10 to 17 h with a resolution dt. Three different resolutions are tested; 1, 5 and 30 min.

Below are the definitions of the studied indicators and the rationale behind using them to predict the returns:

Past return: The past return is defined as $Ln(\frac{S_t}{S_{t-dt}})$. Two effects justify the use of the past return indicator to predict the next return; the mean-reversion effect and the momentum effect. If a stock, suddenly, shows an abnormal return that, significantly, deviates the stock's price from its historical mean value, the mean reversion effect is observed when an opposite return occurs rapidly to put the stock back in its usual average price range. On the other hand, if the stock shows, progressively, an important and continuous deviation; the momentum effect occurs when more market participants are convinced of the move and trade in the same sense increasing the deviation even more.

Order book imbalance: The liquidity on the bid (respectively ask) side is defined as $Liq_{bid} = \sum_{i=1}^{5} w_i b_i bq_i$ (respectively $Liq_{ask} = \sum_{i=1}^{5} w_i a_i aq_i$), where b_i (respectively a_i) is the price at the limit i on the bid (respectively ask) side, bq_i (respectively aq_i) is the corresponding available quantity, and w_i is a decreasing function on i used to give more importance to the best limits. Those indicators give an idea about the instantaneous money available for trading on each side of the order book. Finally, the order book imbalance is defined as $Ln(\frac{Liq_{bid}}{Liq_{ask}})$. This indicator summarizes the order book static state and gives an idea about the buy-sell instantaneous equilibrium. When this indicator is significantly higher (respectively lower) than 0, the available quantity at the bid side is significantly higher (respectively lower) than the one at the ask side; only few participants are willing to sell (respectively buy) the stock, which might reflect a market consensus that the stock will move up (respectively down).

Flow quantity: This indicator summarizes the order book dynamic over the last period dt. Q_b (respectively Q_s) is denoted as the sum of the bought (respectively sold) quantities, over the last period dt and the flow quantity is defined as $Ln(\frac{Q_b}{Q_s})$. This indicator is close to the order flow and shows a high positive autocorrelation. The rationale behind using the flow quantity is to verify if the persistence of the flow is informative about the next return.

EMA: For a process $(X)_{t_i}$ observed on discrete times (t_i), the Exponential Moving Average $EMA(d, X)$ of delay d is defined as $EMA(d, X)_{t_0} = X_{t_0}$ and for $t_{1 \leq i}$, $EMA(d, X)_{t_i} = \omega X_{t_i} + (1 - \omega)EMA(d, X)_{t_{i-1}}$, where $\omega = min(1, \frac{t_i - t_{i-1}}{d})$. The EMA is a weighted average of the process with an exponential decay. The smaller d is, the shorter the EMA memory is.

2.2 Methodology

The aim of this study is to prove, empirically, the market inefficiency by predicting the stocks' returns for three different periods: 1, 5 and 30 min. In Sect. 2, the used indicators are the past returns, the order book imbalance and the flow quantity. A simple method based on historical conditional probabilities is used to prove, separately, the informative effect of each indicator. In Sect. 3, the three indicators and their $EMA(X, d)$ for $d \in (1, 2, 4, 8, 16, 32, 64, 128, 256)$ are combined in order to perform a better prediction than the mono indicator case. Different methods, based on the linear regression, are tested. In particular, the statistical and the numerical stability problems of the linear regression are addressed. In the different sections, the predictions are tested statistically, then used to design a simple trading strategy. The goal is to verify, whether or not one can find a profitable strategy covering 0.5 basis point trading costs. This trading cost is realistic and corresponds to many funds, brokers, and banks trading costs. The possibility of computing, if it exists, a strategy, profitable, after paying the costs, would be an empirical argument of the market inefficiency. Notice that, in all the sections, the learning samples are sliding windows containing sufficient number of days, and the testing samples are the next days. The models parameters are fitted on the learning sample (called in-sample) and the strategies are tested on the testing sample (called out of sample). The sliding training avoids any overfit problem since performances are only computed out of sample.

2.3 Performance Measures

In the most of the studies addressing the market efficiency, the results are summarized in the linear correlation. However, this measure is not enough to conclude about the returns predictability or the market efficiency. Results interpretation should depend on the predicted signal and the trading strategy. A 1 % correlation is high if the signal is supposed to be totally random, and 99 % correlation is insufficient if the signal is supposed to be perfectly predictable. Moreover, a trader making 1 euro each time trading a stock with 50.01 % probability and losing 1 euro with 49.99 % probability, might be considered as a noise trader. However, if this strategy can be run, over 500 stocks, one time a second, for 8 hours a day, at the end of the day the gain will be the sum S_n of $n = 14.4$ million realisations. Using the central limit theorem, $\frac{S_n}{n}$ has a normal law $N(E, \frac{\sigma}{\sqrt{n}})$. Thus the probability of having a negative trading day is $\Phi(\frac{-E\sqrt{n}}{\sigma}) = \Phi(-0.62) = 26.5\,\%$, so much lower than the one of a noise trader. In this paper, returns are considered predictable and thus the market is considered inefficient, if one can run a profitable strategy covering the trading costs.

3 Conditional Probability Matrices

The conditional probability matrices method uses observed frequencies as an estimation of the conditional probability law. To apply this method, data need to be descritized in a small number of classes. Denote the explanatory variable as X, the return as Y and the frequencies matrix as M. Denote the classes of X (respectively Y) as $C^X = \{C_i^X : i \in \mathbb{N}_+ \cap \{i \leq S_X\}\}$ (respectively $C^Y = \{C_i^Y : i \in \mathbb{N}_+ \cap \{i \leq S_Y\}\}$).

Table 1 Historical frequencies matrix for Deutsh Telecom over 2013

	$A = $ "$Y < 0$"	$B = $ "$Y > 0$"
$A = $ "$X < 0$"	19,950	21,597
$B = $ "$X > 0$"	21,597	20,448

S_X (respectively S_Y) denotes the total number of classes for X (respectively Y). For a given learning period $[0, T]$ containing N observations, the frequencies matrix at the time T is constructed as:

$$M_T^{i,j} = card(\{(X_{t_n} \in C_i^X, Y_{t_n} \in C_j^Y)\})$$

where $n \in \mathbb{N}_+ \cap \{n \leq N\}$, and X_{t_n} (respectively Y_{t_n}) is the nth observed value of X (respectively Y), observed at the time t_n. Note that the return Y_{t_n} is backshifted for one instant (namely $Y_{t_n} = Ln(\frac{S_{t_{n+1}}}{S_{t_n}})$). Finally, the prediction of the next Y conditional to the last observed X_T can be computed using the matrix M_T.

The idea of this method is a simple application of the statistical independence test. If some events $A = $ "$X_{t_n} \in C_i^X$" and $B = $ "$Y_{t_n} \in C_j^Y$" are statistically independent then $P(A|B) = P(A)$. For example, to check if the past returns (denoted X in this example) can help predicting the future returns (denoted Y in this example), the returns are classified into two classes, then the empirical historical frequencies matrix is computed. Table 1 shows the results for the 1-min returns of Deutsh Telecom over the year 2013.

In probabilistic terms, the historical probability to observe a negative return is $P(A) = 49.70\%$ and to observe a positive return is $P(B) = 50.30\%$. Thus a trader always buying the stock would have a success rate of 50.30%. Notice that: $P(A/A) = 48.02\%, P(B/A) = 51.98\%, P(A/B) = 51.37\%, P(B/B) = 48.63\%$. Thus, a trader playing the mean-reversion (buy when the past return is negative and sell when the past return is positive), would have a success rate of 51.67%. Notice that the same approach as **1.3** gives a success rate, when trading the strategy over 500 stocks, of 54.38% for the buy strategy and of 72.91% for the mean reversion strategy.

This simple test shows that the smallest statistical bias can be profitable and useful for designing a trading strategy. However the previous strategy is not realistic; the conditional probabilities are computed in sample and the full sample data of Deutsh Telecom was used for the computation. In reality, predictions have to be computed using only the past data. It is, thus, important to have stationary probabilities. Table 2 shows that the monthly observed frequencies are quite stable, and thus can be used

Table 2 Monthly historical conditional probabilities: in the most cases, $P(A/A)$ and $P(B/B)$ are lower than 50% where $P(B/A)$ and $P(A/B)$ are higher than 50%

Month	Jan	Feb	Mar	Apr	May	Jun	Jul	Aug	Sep	Oct	Nov	Dec
P(A/A)	0.49	0.48	0.47	0.48	0.47	0.50	0.48	0.51	0.51	0.47	0.46	0.44
P(B/A)	0.51	0.52	0.53	0.52	0.53	0.50	0.52	0.49	0.49	0.53	0.54	0.56
P(A/B)	0.50	0.52	0.51	0.53	0.52	0.49	0.51	0.50	0.51	0.50	0.51	0.55
P(B/B)	0.50	0.48	0.49	0.47	0.48	0.51	0.49	0.50	0.49	0.50	0.49	0.45

to estimate out of sample probabilities. Each month, one can use the observed frequencies of the previous month as an estimator of current month probabilities.

In the following paragraphs, frequencies matrices are computed on sliding windows for the different indicators. Several classification and prediction methods are presented.

3.1 Binary Method

In the binary case, returns are classified into positive and negative as the previous example and explanatory variables are classified relatively to their historical mean. A typical constructed matrix is shown in Table 1. Denote, in the Table 1 example, $C_1^X = \{X < \overline{X} = 0\}$, $C_2^X = \{X > \overline{X} = 0\}$, $C_1^Y = \{Y < 0\}$, $C_2^Y = \{Y > 0\}$. Y can be predicted using different formula based on the frequency matrix. Below some estimators examples:

$\widehat{Y}_1$: The sign of the most likely next return conditionally to the current state.
$\widehat{Y}_2$: The expectation of the most likely next return conditionally to the current state.
$\widehat{Y}_3$: The expectation of next return conditionally to the current state.

$$\widehat{Y}_1 = \begin{cases} +1 & \text{if } X_T \in C_1^X \\ -1 & \text{if } X_T \in C_2^X \end{cases}$$

$$\widehat{Y}_2 = \begin{cases} E(Y | Y \in C_2^Y \cap X \in C_1^X) & \text{if } X_T \in C_1^X \\ E(Y | Y \in C_1^Y \cap X \in C_2^X) & \text{if } X_T \in C_2^X \end{cases}$$

$$\widehat{Y}_3 = \begin{cases} E(Y | X \in C_1^X) & \text{if } X_T \in C_1^X \\ E(Y | X \in C_2^X) & \text{if } X_T \in C_2^X \end{cases}$$

In this study, only results based on the estimator $\widehat{Y}_3$ (denoted $\widehat{Y}$ in the rest of the paper) are presented. Results computed using different other estimators are equivalent and the differences do not impact the conclusions. To measure the quality of the prediction, four tests are applied:

AUC: (Area under the curve) combines the true positive rate and the false positive rate to give an idea about the classification quality.

Accuracy: defined as the ratio of the correct predictions (Y and $\widehat{Y}$ have the same sign).

Gain: computed on a simple strategy to measure the prediction performance. Predictions are used to run a strategy that buys when the predicted return is positive and sells when it is negative. At each time, for each stock the strategy's position is in $\{-100,000, 0, +100,000\}$.

Profitability: defined as the gain divided by the traded notional of the strategy presented above. This measure is useful to estimate the gain with different transaction costs. Figure 1 summarizes the statistical results of predicting the 1-min returns

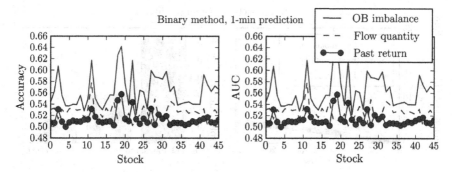

Fig. 1 The quality of the binary prediction: the AUC and the accuracy are higher than 50 %. The three predictors are better than random guessing and are significantly informative

using the three indicators. For each predictor, the AUC and the accuracy are computed over all the stocks. Notice that for each stock, results are computed over more than 100,000 observations and the amplitude of the 95 % confidence interval is around 0.6 %. For the three indicators, the accuracy and the AUC are significantly higher than the 50 % random guessing threshold. The graph shows also that the order book imbalance gives the best results and that the past resturn is the least successful predictor. Detailed results per stock are given in Table 3 of Appendix 1.

In Fig. 2, the performances of the trading strategies based on the prediction of the 1-min returns are presented. The strategies are profitable and the results confirm the predictability of the returns (see the details in Table 4 of Appendix 1).

In Fig. 3, the cumulative gains of the strategies based on the 3 indicators over 2013 are represented. When trading without costs, predicting the 1-min return using the past return and betting 100,000 euros at each time, would make a 5-million Euro profit. Even better, predicting using the order book imbalance would make more than 20 million Euros profit. The results confirm the predictability of the returns,

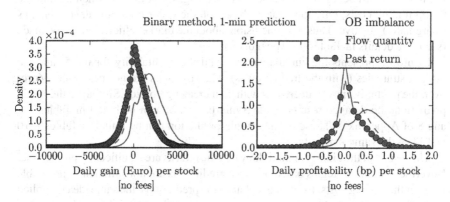

Fig. 2 The quality of the binary prediction: for the 3 predictors, the densities of the gain and the profitability are positively biased, confirming the predictability of the returns

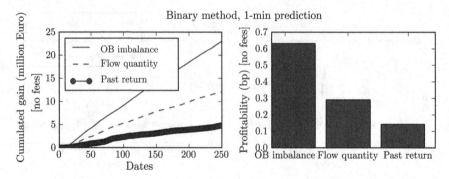

Fig. 3 The quality of the binary prediction: the graphs confirm that the 3 indicators are informative and that the order book imbalance indicator is the most profitable

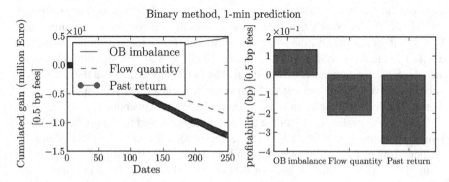

Fig. 4 The quality of the binary prediction: when adding the 0.5 bp trading costs, the strategies are no longer very profitable

but not the inefficiency of the market. In fact, Fig. 4 shows that, when adding the 0.5 bp trading costs, only the strategy based on the order book imbalance remains (marginally) positive. Thus, no conclusion, about the market efficiency, can be made (see more details in Table 5 of Appendix 1).

Figure 5 represents the cumulative gain and the profitability for the 5- and the 30-min strategies (with the trading costs). The strategies are not profitable. Moreover, the predictive power decreases with an increasing horizon.Similar as the 1-min prediction, the detailed results of the 5-min prediction can be found in Tables 6, 7 and 8 of Appendix 1. Those of the 30-min prediction can be found in Tables 9, 10 and 11 of the same Appendix.

The results of the binary method show that the returns are significantly predictable. Nevertheless, the strategies based on those predictions are not sufficiently profitable to cover the trading costs. In order to enhance the predictions, the same idea is applied to the four-class case. Moreover, a new strategy based on a minimum threshold of the expected return is tested.

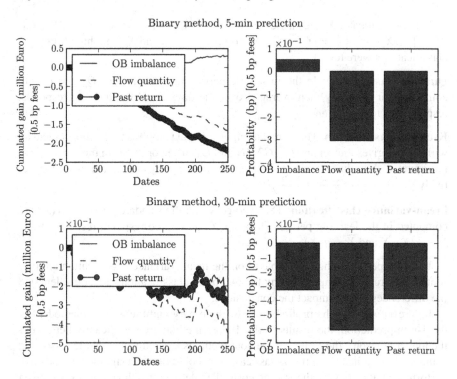

Fig. 5 The quality of the binary prediction: the strategies are not profitable. Moreover, the performances decreases significantly compared to the 1-min horizon

3.2 Four-Class Method

The indicator X is now classified into four classes; "very low values" C_1^X, "low values" C_2^X, "high values" C_3^X and "very high values" C_4^X. At each time t_n, Y is predicted as $\widehat{Y} = E(Y|X \in C_i^X)$, where C_i^X is the class of the current observation X_{t_n}. As the previous case, the expectation is estimated from the historical frequencies matrix.

Finally, a new trading strategy is tested. The strategy is to buy (respectively sell) 100,000 euros when $\widehat{Y}$ is positive (respectively negative) and $|\widehat{Y}| > \theta$, where θ is a minimum threshold (1 bp in this paper). Notice that the case $\theta = 0$ corresponds to the strategy tested in the binary case.

The idea of choosing $\theta > 0$ aims to avoid trading the stock when the signal is noisy. In particular, when analyzing the expectations of Y relative to the different classes of X, it is always observed that the absolute value of the expectation is high when X is in one of its extreme classes (C_1^X or C_4^X). On the other hand, when X is in one of the intermediary classes (C_2^X or C_3^X) the expectation of Y is close to 0 reflecting a noisy signal.

For each indicator X, the classes are defined as $C_1^X =]-\infty, X_a[$, $C_2^X =]X_a, X_b[$, $C_3^X =]X_b, X_c[$ and $C_4^X =]X_c, +\infty[$. To compute X_a, X_b and X_c, the three following classifications were tested:

Quartile classification: In the in-sample period, the quartile Q_1, Q_2 and Q_3 are computed for each day then averaged over the days. X_a, X_b and X_c corresponds, respectively, to $\overline{Q_1}, \overline{Q_2}$ and $\overline{Q_3}$.

K-means classification: The K-means algorithm [11], applied to the in-sample data with $k = 4$, gives the centers G_1, $G2$, G_3 and G_4 of the optimal (in the sense of the minimum within-cluster sum of squares) clusters. X_a, X_b and X_c are given respectively by $\frac{G_1+G_2}{2}$, $\frac{G_2+G_3}{2}$ and $\frac{G_3+G_4}{2}$.

Mean-variance classification: The average $\overline{X}$ and the standard deviation $\sigma(X)$ are computed in the learning period. Then, X_a, X_b and X_c correspond, respectively, to $\overline{X} - \sigma(X), \overline{X}$ and $\overline{X} + \sigma(X)$.

In this paper, only the results based on the mean-variance classification are presented. The results computed using the two other classifications are equivalent and the differences do not impact the conclusions.

Figure 6 compares the profitabilities of the binary and the 4-class methods. For the 1-min prediction, the results of the 4-class method are significantly better. For the longer horizons, the results of the both methods are equivalent. Notice also that, using the best indicator, in the 4-class case, one could obtain a significantly positive performance after paying the trading costs. The detailed results per stock are given in Tables 12, 13, 14, 15, 16, 17, 18, 19 and 20, of Appendix 2.

The interesting result of this first section is that even when using the simplest statistical learning method, the used indicators are informative and provide a better prediction than random guessing. However, in most cases, the obtained performances are too low to conclude about the market inefficiency. In order to enhance the performances, the 3 indicators and their exponential moving average are combined using some classic linear methods in the next section.

4 Linear Regression

In this section, the matrix X denotes a 30-column matrix containing the 3 indicators and their $EMA(d)$ for $d \in (1, 2, 4, 8, 16, 32, 64, 128, 256)$. The vector Y denotes the target to be predicted. Results of the previous section proved that the used indicators are informative and thus can be used to predict the target. In general, one can calibrate, on the learning sample, a function f such that $f(X)$ is "the closet possible" to Y and hope that, for some period after the learning sample, the relation between X and Y is still close enough to the function f. Hence $f(X)$ would be a "good" estimator of Y. Due to the finite number of observations in the learning sample, one can always find $f(X)$ arbitrarily close to Y by increasing the number of the freedom degree. However,

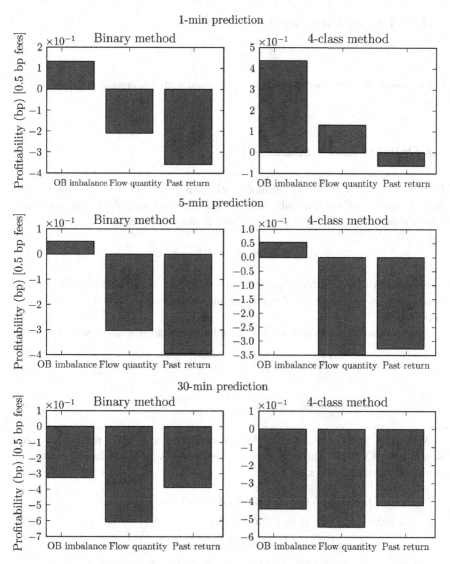

Fig. 6 The quality of the 4-class prediction: for the 1-min prediction, the results of the 4-class method are significantly better than the results of the binary one. For longer horizons, both strategies are not profitable when adding the trading costs

such perfect in-sample calibration overfits the data and the out of sample results are always irrelevant.

In the linear case, f is supposed to be linear and the model errors are supposed to be independent and identically distributed [12] (Gaussian in the standard textbook model). A more mathematical view of linear regression is that it is a probabilistic model of Y given X that assumes:

$$Y = X\beta + \varepsilon, \quad \varepsilon \sim N(0, \sigma^2)$$

For technical reasons, the computations are done with z-scored data (use $\frac{X_i - \overline{X_i}}{\sigma(X_i)}$ in stead of X_i).

4.1 Ordinary Least Squares (OLS)

OLS method consists of estimating the unknown parameter β by minimizing the sum of squares of the residuals between the observed variable Y and the linear approximation $X\beta$. The estimator is denoted $\widehat{\beta}$ and is defined as

$$\widehat{\beta} = argmin_\beta (J_\beta = ||Y - X\beta||_2^2)$$

This criterion is reasonable if at each time i the row X_i of the matrix X and the observation Y_i of the vector Y represent independent random sample from their populations.

The cost function J_β is quadratic on β and differentiating with respect to β gives:

$$\frac{\delta J_\beta}{\delta \beta} = 2t_X X\beta - 2t_X Y$$

$$\frac{\delta^2 J_\beta}{\delta \beta \delta \beta} = 2t_X X$$

When $t_X X$ is invertible, setting the first derivative to 0, gives the unique solution $\widehat{\beta} = (t_X X)^{-1} t_X Y$. The statistical properties of this estimator can be calculated straightforward as follows:

$$E(\widehat{\beta}|X) = (t_X X)^{-1} t_X E(Y|X) = (t_X X)^{-1} t_X X\beta = \beta$$

$$Var(\widehat{\beta}|X) = (t_X X)^{-1} t_X Var(Y|X) X (t_X X)^{-1}$$

$$= (t_X X)^{-1} t_X \sigma^2 IX (t_X X)^{-1} = \sigma^2 (t_X X)^{-1}$$

$$E(||\widehat{\beta}||_2^2|X) = E(t_Y X (t_{XX})^{-2} t_X Y|X)) = Trace(X(t_{XX})^{-2} t_X \sigma^2 I) + ||\beta||_2^2$$

$$= \sigma^2 Trace((t_{XX})^{-1}) + ||\beta||_2^2$$

$$MSE(\widehat{\beta}) = E(||\widehat{\beta} - \beta||_2^2|X) = E(||\widehat{\beta}||_2^2|X) - ||\beta||_2^2$$

$$= \sigma^2 Trace((t_{XX})^{-1}) = \sigma^2 \sum \frac{1}{\lambda_i}$$

where MSE denotes the mean squared error and $(\lambda)_i$ denote the eigen values of $t_X X$. Notice that the OLS estimator is unbiased, but can show an arbitrary high MSE when the matrix $t_X X$ has close to 0 eigen values. In the out of sample period, $\widehat{Y} = X\widehat{\beta}$ is used to predict the target. As seen in Sect. 2, the corresponding trading strategy

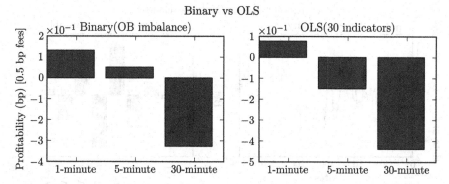

Fig. 7 The quality of the OLS prediction: the results of the OLS method are not better than those of the binary one

is to buy (respectively sell) 100,000 euros when $\widehat{Y} > 0$ (respectively $\widehat{Y} < 0$). To measure the quality of the predictions, the binary method based on the order book imbalance indicator is taken as a benchmark. The linear regression is computed using 30 indicators, including the order book imbalance, thus it should perform at least as well as the binary method. Figure 7 compares the profitabilities of the two strategies. The detailed statistics per stock are given in Tables 21, 22 and 23 of Appendix 3. Similar to the binary method, the performance of the OLS method decrease with an increasing horizon. Moreover, the surprising result is that when combining all the 30 indicators, the results are not better than just applying the binary method to the order book imbalance indicator. This leads to questioning the quality of the regression.

Figure 8 gives some example of the OLS regression coefficients. It is observed that the coefficients are not stable over the time. For example, for some period, the regression coefficient of the order book imbalance indicator is negative. This does not make any financial sense. In fact, when the imbalance is high, the order book shows more liquidity on the bid side (participants willing to buy) than the ask side (participants willing to sell). This state of the order book is observed on average before an up move—i.e. a positive return. The regression coefficient should, thus, be always positive. It is also observed that, for highly correlated indicators, the regression coefficients might be so different. This result also does not make sense, since one would expect to have close coefficients for similar indicators.

From a statistical view, this is explained by the high MSE caused by the high colinearity between the variables. In the following paragraphs, the numerical view is also addressed and some popular solutions to the OLS estimation problems are tested.

4.2 Ridge Regression

When solving a linear system $AX = B$, with A invertible, if a small change in the coefficient matrix (A) or a small change in the right hand side (B) results in a

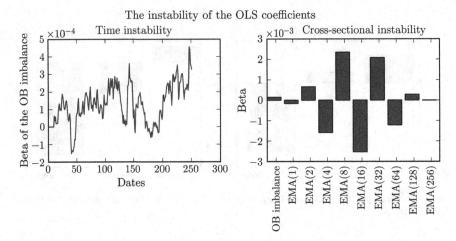

Fig. 8 The quality of the OLS prediction: the graph on the *left* shows the instability of the regression coefficient of the order book imbalance indicator over the year 2013 for the stock Deutsh Telecom. The graph on the *right* shows, for a random day, a very different coefficients for similar indicators; the order book imbalance and its exponential moving averages

large change in the solution vector (X) the system is considered ill-conditioned. The resolution of the system might give a non reliable solution which seems to satisfy the system very well.

An example of an ill-conditioned system is given bellow:

$$\begin{bmatrix} 1.000 & 2.000 \\ 3.000 & 5.999 \end{bmatrix} \times \begin{bmatrix} x \\ y \end{bmatrix} = \begin{bmatrix} 4.000 \\ 11.999 \end{bmatrix} => \begin{bmatrix} x \\ y \end{bmatrix} = \begin{bmatrix} 2.000 \\ 1.000 \end{bmatrix}$$

When making a small change in the matrix A:

$$\begin{bmatrix} 1.001 & 2.000 \\ 3.000 & 5.999 \end{bmatrix} \times \begin{bmatrix} x \\ y \end{bmatrix} = \begin{bmatrix} 4.000 \\ 11.999 \end{bmatrix} => \begin{bmatrix} x \\ y \end{bmatrix} = \begin{bmatrix} -0.400 \\ 2.200 \end{bmatrix}$$

When making a small change in the vector B:

$$\begin{bmatrix} 1.000 & 2.000 \\ 3.000 & 5.999 \end{bmatrix} \times \begin{bmatrix} x \\ y \end{bmatrix} = \begin{bmatrix} 4.001 \\ 11.999 \end{bmatrix} => \begin{bmatrix} x \\ y \end{bmatrix} = \begin{bmatrix} -3.999 \\ 4.000 \end{bmatrix}$$

When dealing with experimental data, it is not reliable to have a completely different calibration because of a small change in the observations. Hence, it is mandatory to take into consideration such effects before achieving any computation.

In literature, various measures of the ill-conditioning of a matrix have been proposed [13], perhaps the most popular one [14] is $K(A) = ||A||_2 ||A^{-1}||_2$, where $||.||_2$ denotes the l_2-norm defined for a vector X as $||X||_2 = \sqrt{t_X X}$ and for a matrix A as $||A||_2 = \max_{||X||_2 \neq 0} \frac{||AX||_2}{||X||_2}$. The larger is $K(A)$, the more ill-conditioned is A.

The rationale behind defining the condition number $K(A)$ is to measure the sensitivity of the solution X relative to a perturbation of the matrix A or the vector B. More precisely, it is proved (see the Appendix 4) that:

- If $AX = B$ and $A(X + \delta X) = B + \delta B$ then $\frac{||\delta X||_2}{||X||_2} \leq K(A)\frac{||\delta B||_2}{||B||_2}$

- If $AX = B$ and $(A + \delta A)(X + \delta X) = B$ then $\frac{||\delta X||_2}{||X + \delta X||_2} \leq K(A)\frac{||\delta A||_2}{||A||_2}$

Notice that $K(A)$ can be easily computed as the maximum singular value of A. For example, in the system above, $K(A) = 49,988$. The small perturbations can, thus, be amplified by almost 50,000, causing the previous observations.

Figure 9 represents the singular values of $t_X X$ used to compute the regression of the right graph of Fig. 8. The graph shows a hard decreasing singular values. In particular, the condition number is higher than 80,000.

This finding explains the instability observed on the previous section. Moreover that the OLS estimator is statistically not satisfactory, the numerical problems due to the ill-conditioning of the matrix makes the result numerically unreliable.

One popular solution to enhance the stability of the estimation of the regression coefficients is the Ridge method. This method was introduced independently by A. Tikhonov, in the context of solving ill-posed problems, around the middle of the 20th century, and by A.E. Hoerl in the context of addressing the linear regression problems by the sixteeth. The Ridge regression consists of adding a regularization term to the original OLS problem:

$$\widehat{\beta_\Gamma} = argmin_\beta(||Y - X\beta||_2^2 + ||\Gamma\beta||_2^2)$$

The new term gives preference to a particular solution with desirable properties. Γ is called the Tikhonov matrix and chosen usually as a multiple of the identity matrix; $\lambda_R I$, where $\lambda_R \geq 0$. The new estimator of the linear regression coefficients is called the Ridge estimator, denoted $\widehat{\beta_R}$, and defined as follows:

$$\widehat{\beta_R} = argmin_\beta(||Y - X\beta||_2^2 + \lambda_R||\beta||_2^2)$$

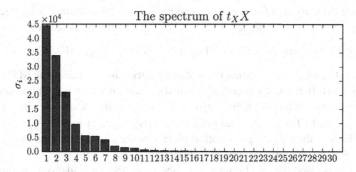

Fig. 9 The quality of the OLS prediction: the graph shows that the matrix inverted when computing the OLS coefficient is ill-conditioned

Similar to the OLS case, by straightforward calculation:

$$\widehat{\beta}_R = (t_XX + \lambda_R I)^{-1} t_X Y = Z\widehat{\beta} \quad \text{where} \qquad Z = (I + \lambda_R(t_XX)^{-1})^{-1} = W^{-1}$$

and

$$E(\widehat{\beta}_R|X) = E(Z_R\widehat{\beta}|X) = Z\beta$$
$$Var(\widehat{\beta}_R|X) = Var(Z\widehat{\beta}|X) = \sigma^2 Z(t_XX)^{-1} t_Z$$
$$MSE(\widehat{\beta}_R) = E(t_{(Z\widehat{\beta}-\beta)}(Z\widehat{\beta} - \beta)|X) = E(t_\beta t_Z Z\beta|X) - 2t_\beta Z\beta + t_\beta\beta$$
$$= Trace(t_Z Z\sigma^2(t_XX)^{-1}) + t_\beta t_Z Z\beta - 2t_\beta Z\beta + t_\beta\beta$$

Notice that:

$$(t_XX)^{-1} = \frac{Z^{-1} - I}{\lambda_R}$$
$$I - Z = (I - Z)WW^{-1} = (W - I)W^{-1} = \lambda_R(t_XX)^{-1}W^{-1}$$
$$= \lambda_R(Wt_XX)^{-1} = \lambda_R(t_XX + \lambda_R I)^{-1}$$

Thus:

$$MSE(\widehat{\beta}_R) = Trace(\frac{\sigma^2 Z}{\lambda_R Z}) - Trace(\frac{\sigma^2 Z}{\lambda_R Z^2}) + t_\beta(I - Z)^2\beta$$
$$= \sigma^2 \sum \frac{\lambda_i}{(\lambda_i + \lambda_R)^2} + \lambda_R^2 t_\beta(t_XX + \lambda_R I)^{-2}\beta$$

The first element of the MSE corresponds exactly to the trace of the covariance matrix of $\widehat{\beta}_R$, i.e. the total variance of the parameters estimations. The second element is the squared distance from $\widehat{\beta}_R$ to β and corresponds to the square of the bias introduced when adding the ridge penalty. Notice that, when increasing the λ_R, the bias increases and the variance decreases. On the other hand, when decreasing the λ_R, the bias decreases and the variance increases converging to their OLS values. To enhance the stability of the linear regression, one should compute a λ_R, such that $MSE(\widehat{\beta}_R) \leq MSE(\widehat{\beta})$. As proved by Hoerl [15], this is always possible.

Theorem: There always exist $\lambda_R \geq 0$ such that $MSE(\widehat{\beta}_R) \leq MSE(\widehat{\beta})$.

From a statistical view, adding the Ridge penalty aims to reduce the MSE of the estimator, and is particularly necessary when the covariance matrix is ill-conditioned. From a numerical view, the new matrix to be inverted is $t_XX + \lambda_R I$ with as eigen values $(\lambda_i + \lambda_R)_i$. The conditional number is $K(t_XX + \lambda_R I) = \frac{\lambda_{max} + \lambda_R}{\lambda_{min} + \lambda_R} \leq \frac{\lambda_{max}}{\lambda_{min}} = K(t_XX)$. Hence, the ridge regularization enhances the conditioning of the problem and improves the numerical reliability of the result.

From the previous, it can be seen that increasing the λ_R leads to numerical stability and reduces the variance of the estimator, however it increases the bias of the estimator. One has to chose the λ_R as a tradeoff between those two effects. Next, two estimators of λ_R are tested; the Hoerl-Kennard-Baldwin (HKB) estimator [16] and the Lawless-Wang (LW) estimator [17].

In order to compare the stability of the Ridge and the OLS coefficients, Figs. 10 and 11 represent the same test of Fig. 8, applied, respectively, to the Ridge HKB and

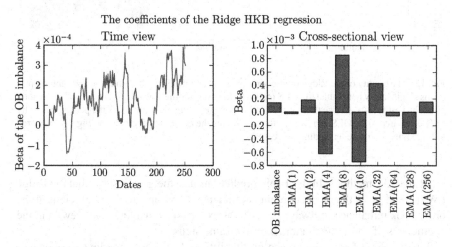

Fig. 10 The quality of the Ridge HKB prediction: the graphs show that the results of the Ridge HKB method are not significantly different from those of the OLS method (Fig. 8). In this case, the λ_R is close to 0 and the effect of the regularization is limited

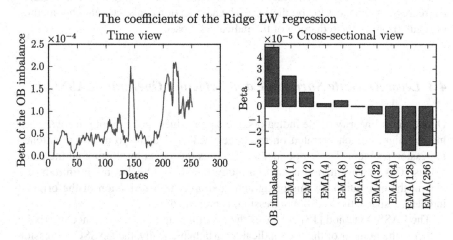

Fig. 11 The quality of the Ridge LW prediction: the graph on the *left* shows the stability of the regression coefficient of the order book imbalance over the year 2013 for Deutsh Telecom. The coefficient is positive during all the period, in line with the financial view. The graph on the *right* shows, for a random day, a positive coefficients for the order book imbalance and its short term EMAs. The coefficients decreases with the time; i.e. the state of the order book "long time ago" has a smaller effect than its current state. More over, for longer than a 10-sec horizon, the coefficients become negative confirming the mean-reversion effect

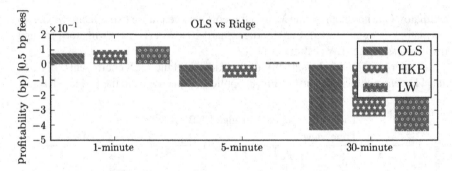

Fig. 12 The quality of the Ridge prediction: for the 1- and the 5-min horizons the LW method performs significantly better than the OLS method. However, for the 30-min horizon, the HKB method gives the best results. Notice that for the 1-min case, the LW method improves the performances by 58 % compared to the OLS, confirming that stabilizing the regression coefficients (Fig. 11 compared to Fig. 8), leads to a better trading strategies

the Ridge LW methods. In the 1-min prediction case, the graphs show that the Ridge LW method gives the most coherent coefficients. In particular, the coefficient of the order book imbalance is always positive (as expected from a financial view) and the coefficients of similar indicators have the same signs.

Finally, Fig. 12 summarizes the profitabilities of the corresponding strategies of the two methods. Tables 24, 25, 26, 27, 28 and 29 of Appendix 4 detail the results per stock.

From the previous results, it can be concluded that adding a regularization term to the regression enhances the predictions. The next section deals with an other method of regularization: the reduction of the indicators' space.

4.3 Least Absolute Shrinkage and Selection Operator (LASSO)

Due to the colinearity of the indicators, the eigen values spectrum of the covariance matrix might be concentrated on the largest values, leading to an ill-conditioned regression problem. The Ridge method, reduces this effect by shifting all the eigen values. This transformation leads to a more reliable results, but might introduce a bias in the estimation. In this paragraph, a simpler transformation of the original indicators' space, the LASSO regression, is presented.

The LASSO method [18] enhances the conditioning of the covariance matrix by reducing the number of the used indicators. Mathematically, the LASSO regression aims to produce a sparse regression coefficients—i.e. with some coefficients exactly equal to 0. This is possible thanks to the l_1-penalization. More precisely, the LASSO regression is to estimate the linear regression coefficient as:

$$\widehat{\beta_L} = argmin_\beta(||Y - X\beta||_2^2 + \lambda_L||\beta||_1)$$

where $||.||_1$ denotes the l_1-norm, defined as the sum of the coordinates' absolute values. Writing $|\beta_i| = \beta_{i+} - \beta_{i-}$ and $\beta_i = \beta_{i+} + \beta_{i-}$, with $\beta_{i+} \geq 0$ and $\beta_{i-} \leq 0$, a classic quadratic problem, with a linear constraints, is obtained and can be solved by a classic solver. As far as known, there is no estimator for λ_L. In this study, the cross-validation [11] method is applied to select λ_L out of a set of parameters; $T10^{-k}$, where $k \in (2, 3, 4, 5, 6)$ and T denotes the number of the observations.

Figure 13 compares, graphically, the Ridge and the LASSO regularization, Fig. 14 addresses the instability problems observed in Figs. 8 and 15 summarizes the results of the strategies corresponding to the LASSO method. The detailed results per stock

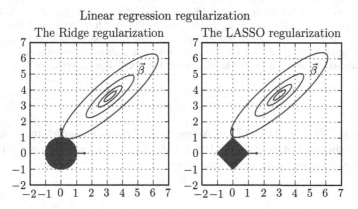

Fig. 13 The quality of the LASSO prediction: the estimation graphs for the Ridge (on the *left*) and the LASSO regression (on the *right*). Notice that the l_1-norm leads to 0 coefficients on the less important axis

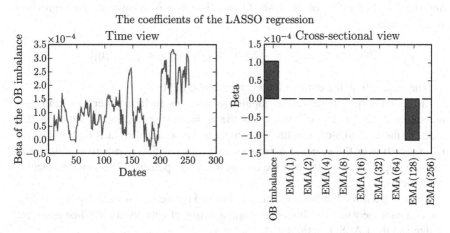

Fig. 14 The quality of the LASSO prediction: the graphs show that the LASSO regression gives a regression coefficients in line with the financial view (similarly to Fig. 11). Moreover, the coefficients are sparse and simple for the interpretation

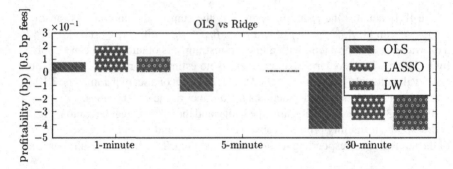

Fig. 15 The quality of the LASSO prediction: similar as the Ridge regression, the LASSO regression gives a better profitability than the OLS one. Notice that for the 1-min case, the LASSO method improves the performances by 165 % compared to the OLS. Eventhough the LASSO metho is using less regressors than the OLS method, (and thus less signal), the out of sample results are significantly better in the LASSO case. This result confirms the importance of the signal by noise ratio and highlights the importance of the regularization when addressing an ill-conditioned problem

are given in Tables 30, 31 and 32 of Appendix 5. The next paragraph introduces the natural combination of the Ridge and the LASSO regression and presents this paper's conclusions concerning the market inefficiency.

4.4 Elastic Net (EN)

The EN regression aims to combine the regularization effect of the Ridge method and the selection effect of the LASSO one. The idea is to estimate the regression coefficients as:

$$\widehat{\beta}_{EN} = argmin_\beta(||Y - X\beta||_2^2 + \lambda_{EN_1}||\beta||_1 + \lambda_{EN_2}||\beta||_2^2)$$

The detail about the computation can be found in [19].

In this study, the estimation is computed in two steps. In the first step λ_{EN_1} and λ_{EN_2} are selected via the crossvalidation and the problem is solved same as the LASSO case. In the second step, the final coefficients are obtained by a Ridge regression ($\lambda_{EN_1} = 0$) over the selected indicators (indicators with a non-zero coefficient in the first step). The two step method avoids useless l_1-penalty effects on the selected coefficients.

Figure 16 shows that the coefficients obtained by the EN method are in line with the financial view and combine both regularization effects observed when using the Ridge and the LASSO methods.

Finally, the strategy presented in **2.2** (trading only if $\widehat{Y} \geq |\theta|$) is applied to the different regression methods. Figure 17 summarizes the obtained results.

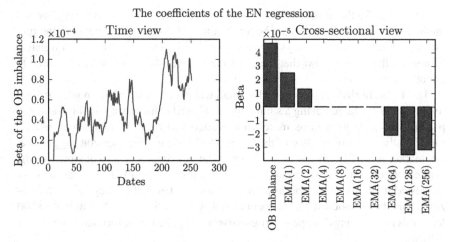

Fig. 16 The quality of the EN prediction: the graphs show that the EN regression gives a regression coefficients in line with the financial view (similarly to Figs. 11 and 14)

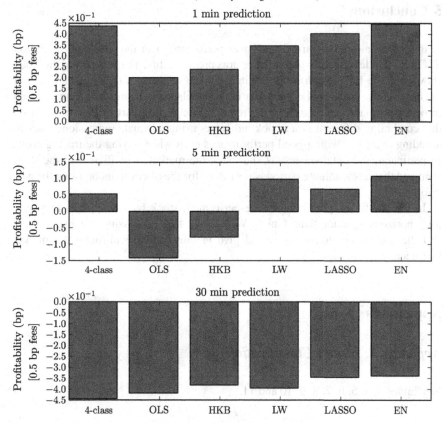

Fig. 17 The quality of the EN prediction: the EN method gives the best results compared to the other regressions

The results for the three horizons confirm that the predictions of all the regularized method (Ridge, LASSO, EN) are better than the OLS ones. As detailed in the previous paragraphs, this is always the case when the indicators are highly correlated. Moreover, the graphs show that the EN method gives the best results compared to the other regressions.

The 1-min horizon results underline that, when an indicator has an obvious correlation with the target, using a simple method based exhaustively on this indicator, performs as least as well as more sophisticated methods including more indicators. Finally, the performance of the EN method for the 1-min horizon suggest that the market is inefficient for such horizon. The conclusion is less obvious for the 5-min horizon. On the other hand, the 30-min horizon results show that, none of the tested methods could find any proof of the market inefficiency for such horizon. From the previous, it can be concluded that the market is inefficient in the short term, this inefficiency disappears progressively when the new information are widely diffused.

5 Conclusions

In this paper, a large empirical study was performed, over the stocks of the EURO STOXX 50 index, in order to test the returns predictability. The first part of the study shows that the future returns are not independent of the past dynamic and state of the order book. In particular, the order book imbalance indicator is informative and provides a reliable prediction of the returns. The second part of the study shows that combining different order book indicators using adequate regressions leads to a trading strategies with a good performances even when paying the trading costs. In particular, the obtained results show that the market is inefficient in the short term and that a few-minute period is necessary for the prices to incorporate the new information.

In this study the predictions were performed, stock by stock, for a relatively long horizons (greater than 1 min). We believe that addressing the multi-stock, and the shorter horizon cases would give in line results confirming the market inefficiency.

Appendices

Appendix 1: Binary Classification

See Tables 3, 4, 5, 6, 7, 8, 9, 10 and 11.

Table 3 The quality of the binary prediction: 1-min prediction AUC and accuracy per stock

Stock	Order book imbalance		Flow quantity		Past return	
	AUC	Accuracy	AUC	Accuracy	AUC	Accuracy
INTERBREW	0.54	0.54	0.53	0.53	0.51	0.51
Air LIQUIDE	0.56	0.56	0.53	0.53	0.51	0.51
ALLIANZ	0.61	0.61	0.51	0.51	0.53	0.53
ASML Holding NV	0.56	0.56	0.53	0.53	0.51	0.51
BASF AG	0.54	0.54	0.52	0.52	0.50	0.50
BAYER AG	0.54	0.54	0.53	0.53	0.51	0.51
BBVARGENTARIA	0.54	0.54	0.53	0.53	0.51	0.51
BAY MOT WERKE	0.54	0.54	0.53	0.53	0.51	0.51
DANONE	0.56	0.56	0.53	0.53	0.51	0.51
BNP PARIBAS	0.53	0.53	0.52	0.52	0.51	0.51
CARREFOUR	0.55	0.55	0.53	0.53	0.51	0.51
CRH PLC IRLANDE	0.62	0.62	0.58	0.58	0.53	0.53
AXA	0.55	0.55	0.51	0.51	0.52	0.52
DAIMLER CHRYSLER	0.54	0.54	0.53	0.53	0.51	0.51
DEUTSCHE BANK AG	0.53	0.53	0.52	0.52	0.51	0.51
VINCI	0.54	0.54	0.53	0.53	0.51	0.51
DEUTSCHE TELEKOM	0.56	0.56	0.52	0.52	0.51	0.51
ESSILOR INTERNATIONAL	0.56	0.56	0.54	0.54	0.50	0.50
ENEL	0.63	0.63	0.51	0.51	0.55	0.55
ENI	0.64	0.64	0.51	0.51	0.56	0.56
E.ON AG	0.58	0.58	0.51	0.51	0.51	0.51
TOTAL	0.54	0.54	0.52	0.52	0.51	0.51
GENERALI ASSIC	0.62	0.62	0.50	0.50	0.54	0.54
SOCIETE GENERALE	0.52	0.52	0.51	0.51	0.51	0.51
GDF SUEZ	0.56	0.56	0.52	0.52	0.50	0.50
IBERDROLA I	0.56	0.56	0.54	0.54	0.51	0.51
ING	0.53	0.53	0.53	0.53	0.51	0.51
INTESABCI	0.60	0.60	0.51	0.51	0.53	0.53
INDITEX	0.59	0.59	0.55	0.55	0.50	0.50
LVMH	0.59	0.59	0.52	0.52	0.52	0.52
MUNICH RE	0.58	0.58	0.52	0.52	0.51	0.51
LOREAL	0.60	0.60	0.53	0.53	0.52	0.52
PHILIPS ELECTR	0.56	0.56	0.55	0.55	0.50	0.50
REPSOL	0.57	0.57	0.54	0.54	0.51	0.51
RWE ST	0.54	0.54	0.53	0.53	0.51	0.51
BANCO SAN CENTRAL HISPANO	0.54	0.54	0.53	0.53	0.51	0.51
SANOFI	0.54	0.54	0.53	0.53	0.50	0.50

(continued)

Table 3 (continued)

Stock	Order book imbalance		Flow quantity		Past return	
	AUC	Accuracy	AUC	Accuracy	AUC	Accuracy
SAP AG	0.54	0.54	0.52	0.52	0.51	0.51
SAINT GOBAIN	0.54	0.54	0.53	0.53	0.51	0.51
SIEMENS AG	0.54	0.54	0.53	0.53	0.51	0.51
SCHNEIDER ELECTRIC SA	0.54	0.54	0.52	0.52	0.51	0.51
TELEFONICA	0.59	0.59	0.53	0.53	0.51	0.51
UNICREDIT SPA	0.57	0.57	0.50	0.50	0.52	0.52
UNILEVER CERT	0.56	0.56	0.52	0.52	0.51	0.51
VIVENDI UNIVERSAL	0.57	0.57	0.53	0.53	0.51	0.51
VOLKSWAGEN	0.57	0.57	0.52	0.52	0.51	0.51

Table 4 The quality of the binary prediction: the daily gain average and standard deviation for the 1-min prediction (without trading costs)

Stock	Order book imbalance		Flow quantity		Past return	
	$\overline{Gain}$	$\sigma(Gain)$	$\overline{Gain}$	$\sigma(Gain)$	$\overline{Gain}$	$\sigma(Gain)$
INTERBREW	1,388	1,201	1,107	1,308	174	1,264
AIR LIQUIDE	1,603	1,112	996	1,005	169	936
ALLIANZ	2,775	1,219	221	1,107	638	1,175
ASML Holding NV	1,969	1,278	1,244	1,316	190	1,419
BASF AG	1,156	1,102	921	1,311	2	1,185
BAYER AG	1,269	1,055	1,142	1,251	289	1,296
BBVARGENTARIA	1,954	1,537	1,866	1,700	595	1,934
BAY MOT WERKE	1,330	1,219	1,240	1,325	347	1,394
DANONE	1,591	993	958	1,143	231	1,196
BNP PARIBAS	1,120	1,608	831	1,620	526	1,911
CARREFOUR	1,878	1,572	1,461	1,601	600	1,665
CRH PLC IRLANDE	4,144	1,881	2,853	1,691	1,496	1,542
AXA	2,003	1,373	674	1,428	582	1,603
DAIMLER CHRYSLER	1,380	1,275	1,130	1,228	208	1,390
DEUTSCHE BANK AG	1,251	1,372	905	1,405	310	1,672
VINCI	1,410	1,113	1,252	1,211	376	1,113
DEUTSCHE TELEKOM	1,586	1,416	848	1,196	308	1,298
ESSILOR INTERNATIONAL	1,762	1,315	1,523	1,295	12	1,281
ENEL	3,723	1,655	295	1,384	1,219	1,307
ENI	2,996	1,185	321	1,161	1,109	1,201
E.ON AG	2,245	1,193	481	1,722	323	1,445
TOTAL	1,256	956	831	977	326	950
GENERALI ASSIC	3,977	1,764	177	1,324	1,210	1,577
SOCIETE GENERALE	1,195	1,763	853	1,896	643	2,060

(continued)

Table 4 (continued)

Stock	Order book imbalance		Flow quantity		Past return	
	$\overline{Gain}$	$\sigma(Gain)$	$\overline{Gain}$	$\sigma(Gain)$	$\overline{Gain}$	$\sigma(Gain)$
GDF SUEZ	2,031	1,227	934	1,389	156	1,355
IBERDROLA I	2,220	1,433	1,626	1,514	566	1,403
ING	1,511	1,564	1,493	1,491	217	1,720
INTESABCI	4,019	1,911	153	1,787	1,048	1,954
INDITEX	2,481	1,452	1,742	1,525	145	1,344
LVMH	2,445	1,220	533	1,148	613	1,267
MUNICH RE	1,895	1,107	791	1,485	194	1,006
LOREAL	2,367	1,109	894	1,242	438	1,220
PHILIPS ELECTR	1,978	1,173	1,670	1,565	182	1,251
REPSOL	2,694	1,451	1,700	1,607	292	1,558
RWE ST	1,323	1,348	1,475	1,880	307	1,747
BANCO SAN CENTRAL HISPANO	1,717	1,535	1,393	1,577	383	1,684
SANOFI	1,368	1,040	1,118	1,123	107	1,190
SAP AG	1,225	1,022	939	1,071	117	1,084
SAINT GOBAIN	1,612	1,359	1,209	1,449	455	1,607
SIEMENS AG	1,108	983	967	1,196	164	1,124
SCHNEIDER ELECTRIC SA	1,419	1,294	1,014	1,275	379	1,436
TELEFONICA	2,694	1,267	1,156	1,341	290	1,194
UNICREDIT SPA	3,039	2,025	382	1,850	683	2,002
UNILEVER CERT	1,402	766	551	860	222	949
VIVENDI UNIVERSAL	2,142	1,223	1,114	1,391	244	1,326
VOLKSWAGEN	2,044	1,440	1,165	1,397	225	1,359

Table 5 The quality of the binary prediction: the daily gain average and standard deviation for the 1-min prediction (with trading costs)

Stock	Order book imbalance		Flow quantity		Past return	
	$\overline{Gain}$	$\sigma(Gain)$	$\overline{Gain}$	$\sigma(Gain)$	$\overline{Gain}$	$\sigma(Gain)$
INTERBREW	−191	1,189	−788	1,325	−1,222	1,531
AIR LIQUIDE	81	1,112	−980	1,057	−1,211	1,164
ALLIANZ	1,141	1,063	−1,199	1,309	−952	1,162
ASML Holding NV	370	1,179	−697	1,335	−1,301	1,574
BASF AG	−422	1,064	−955	1,338	−1,298	1,558
BAYER AG	−363	1,002	−734	1,249	−1,122	1,503
BBVARGENTARIA	303	1,477	−58	1,681	−910	2,027
BAY MOT WERKE	−260	1,176	−530	1,263	−1,256	1,510
DANONE	−40	963	−906	1,164	−1,246	1,369

(continued)

Table 5 (continued)

Stock	Order book imbalance		Flow quantity		Past return	
	$\overline{Gain}$	$\sigma(Gain)$	$\overline{Gain}$	$\sigma(Gain)$	$\overline{Gain}$	$\sigma(Gain)$
BNP PARIBAS	−402	1,596	−1,022	1,618	−1,115	1,998
CARREFOUR	251	1,486	−492	1,606	−975	1,690
CRH PLC IRLANDE	2,971	1,714	934	1,612	−27	1,549
AXA	313	1,299	−1,064	1,488	−1,152	1,560
DAIMLER CHRYSLER	−231	1,243	−748	1,235	−1,206	1,529
DEUTSCHE BANK AG	−394	1,368	−959	1,423	−1,277	1,819
VINCI	−170	1,072	−656	1,224	−1,093	1,324
DEUTSCHE TELEKOM	50	1,407	−949	1,225	−1,128	1,516
ESSILOR INTERNATIONAL	185	1,265	−389	1,296	−1,104	1,575
ENEL	2,151	1,456	−1,069	1,610	−329	1,198
ENI	1,513	971	−1,136	1,375	−281	1,046
E.ON AG	583	1,096	−1,108	1,887	−1,047	1,592
TOTAL	−362	934	−1,058	1,024	−1,278	1,206
GENERALI ASSIC	2,369	1,565	−1,403	1,539	−484	1,490
SOCIETE GENERALE	−405	1,718	−846	1,901	−968	2,002
GDF SUEZ	402	1,140	−951	1,438	−1,249	1,513
IBERDROLA I	762	1,332	−312	1,503	−1,094	1,475
ING	−186	1,519	−450	1,470	−1,186	1,890
INTESABCI	2,333	1,715	−1,081	1,822	−517	1,820
INDITEX	1,110	1,375	−195	1,535	−1,155	1,457
LVMH	831	1,119	−1,183	1,296	−928	1,235
MUNICH RE	366	1,011	−1,019	1,490	−1,260	1,177
LOREAL	816	985	−797	1,274	−982	1,236
PHILIPS ELECTR	377	1,113	−272	1,575	−1,255	1,490
REPSOL	1,233	1,308	−184	1,585	−1,188	1,713
RWE ST	−182	1,251	−399	1,864	−1,122	1,960
BANCO SAN CENTRAL HISPANO	205	1,431	−492	1,566	−1,064	1,822
SANOFI	−279	998	−720	1,127	−1,382	1,454
SAP AG	−340	1,000	−944	1,093	−1,428	1,277
SAINT GOBAIN	−48	1,326	−694	1,463	−1,060	1,655
SIEMENS AG	−472	966	−898	1,209	−1,353	1,363
SCHNEIDER ELECTRIC SA	−162	1,263	−872	1,296	−1,339	1,493
TELEFONICA	1,124	1,130	−686	1,342	−1,044	1,257
UNICREDIT SPA	1,434	1,940	−896	1,953	−738	2,067
UNILEVER CERT	−253	730	−1,246	938	−1,344	1,142
VIVENDI UNIVERSAL	547	1,113	−804	1,386	−1,186	1,452
VOLKSWAGEN	446	1,373	−785	1,408	−979	1,584

Table 6 The quality of the binary prediction: 5-min prediction AUC and accuracy per stock

Stock	Order book imbalance		Flow quantity		Past return	
	AUC	Accuracy	AUC	Accuracy	AUC	Accuracy
INTERBREW	0.50	0.50	0.50	0.50	0.51	0.51
AIR LIQUIDE	0.53	0.53	0.50	0.50	0.50	0.51
ALLIANZ	0.54	0.54	0.51	0.51	0.52	0.52
ASML Holding NV	0.51	0.51	0.51	0.51	0.51	0.51
BASF AG	0.50	0.50	0.50	0.50	0.50	0.50
BAYER AG	0.51	0.51	0.51	0.51	0.51	0.51
BBVARGENTARIA	0.50	0.50	0.50	0.50	0.50	0.50
BAY MOT WERKE	0.50	0.50	0.51	0.51	0.50	0.50
DANONE	0.51	0.51	0.51	0.51	0.50	0.50
BNP PARIBAS	0.50	0.50	0.50	0.50	0.50	0.50
CARREFOUR	0.51	0.51	0.50	0.51	0.51	0.51
CRH PLC IRLANDE	0.56	0.56	0.53	0.53	0.51	0.51
AXA	0.50	0.50	0.49	0.49	0.50	0.50
DAIMLER CHRYSLER	0.51	0.51	0.50	0.50	0.51	0.51
DEUTSCHE BANK AG	0.50	0.50	0.50	0.50	0.50	0.50
VINCI	0.51	0.51	0.51	0.51	0.50	0.51
DEUTSCHE TELEKOM	0.52	0.52	0.50	0.51	0.50	0.50
ESSILOR INTERNATIONAL	0.51	0.51	0.51	0.51	0.50	0.50
ENEL	0.55	0.55	0.51	0.51	0.51	0.51
ENI	0.56	0.56	0.51	0.51	0.51	0.51
E.ON AG	0.52	0.52	0.50	0.50	0.50	0.50
TOTAL	0.51	0.51	0.50	0.51	0.51	0.51
GENERALI ASSIC	0.56	0.56	0.50	0.50	0.51	0.51
SOCIETE GENERALE	0.50	0.50	0.50	0.50	0.50	0.50
GDF SUEZ	0.52	0.52	0.50	0.50	0.50	0.50
IBERDROLA I	0.52	0.52	0.51	0.51	0.51	0.51
ING	0.50	0.50	0.50	0.50	0.50	0.50
INTESABCI	0.54	0.54	0.49	0.49	0.50	0.50
INDITEX	0.53	0.53	0.50	0.50	0.52	0.52
LVMH	0.53	0.53	0.50	0.50	0.51	0.51
MUNICH RE	0.53	0.53	0.51	0.51	0.51	0.51
LOREAL	0.53	0.53	0.50	0.50	0.51	0.51
PHILIPS ELECTR	0.52	0.52	0.51	0.51	0.50	0.50
REPSOL	0.53	0.53	0.51	0.51	0.50	0.50
RWE ST	0.51	0.51	0.50	0.50	0.50	0.50
BANCO SAN CENTRAL HISPANO	0.51	0.51	0.51	0.51	0.51	0.51
SANOFI	0.51	0.51	0.50	0.50	0.50	0.50

(continued)

Table 6 (continued)

Stock	Order book imbalance		Flow quantity		Past return	
	AUC	Accuracy	AUC	Accuracy	AUC	Accuracy
SAP AG	0.51	0.51	0.51	0.51	0.50	0.50
SAINT GOBAIN	0.50	0.50	0.50	0.50	0.51	0.51
SIEMENS AG	0.51	0.51	0.50	0.50	0.51	0.51
SCHNEIDER ELECTRIC SA	0.50	0.50	0.50	0.50	0.50	0.50
TELEFONICA	0.53	0.53	0.51	0.51	0.51	0.51
UNICREDIT SPA	0.52	0.52	0.51	0.51	0.50	0.50
UNILEVER CERT	0.52	0.52	0.50	0.50	0.50	0.50
VIVENDI UNIVERSAL	0.52	0.52	0.50	0.50	0.50	0.50
VOLKSWAGEN	0.52	0.52	0.50	0.50	0.51	0.51

Table 7 The quality of the binary prediction: the daily gain average and standard deviation for the 5-min prediction (without trading costs)

Stock	Order book imbalance		Flow quantity		Past return	
	$Gain$	$\sigma(Gain)$	$Gain$	$\sigma(Gain)$	$Gain$	$\sigma(Gain)$
INTERBREW	45	978	112	958	44	1,010
AIR LIQUIDE	308	752	40	741	42	798
ALLIANZ	479	1,073	74	871	182	906
ASML Holding NV	146	1,027	83	1,029	−39	1,143
BASF AG	−7	976	107	969	22	987
BAYER AG	195	972	161	1016	50	963
BBVARGENTARIA	129	1,529	107	1,307	83	1,332
BAY MOT WERKE	67	1,005	196	1,010	43	969
DANONE	203	1,008	52	938	−65	845
BNP PARIBAS	65	1,327	1	1,350	−61	1,376
CARREFOUR	134	1,238	193	1,214	32	1,267
CRH PLC IRLANDE	1,167	1,433	567	1,378	310	1,523
AXA	112	1,230	−68	1,254	−120	1,246
DAIMLER CHRYSLER	79	1,037	−72	1,088	27	1,059
DEUTSCHE BANK AG	−13	1,362	−35	1,287	6	1,277
VINCI	226	877	195	927	147	892
DEUTSCHE TELEKOM	319	857	195	837	31	980
ESSILOR INTERNATIONAL	103	990	114	977	−14	968
ENEL	700	1,227	−4	1,183	108	1,117
ENI	556	822	39	841	71	815
E.ON AG	279	1,005	78	1,158	23	1,022
TOTAL	139	738	71	842	150	845
GENERALI ASSIC	853	1,233	−22	1,126	−30	1,257
SOCIETE GENERALE	121	1,523	−72	1,542	−75	1,587

(continued)

Table 7 (continued)

Stock	Order book imbalance		Flow quantity		Past return	
	$\overline{Gain}$	$\sigma(Gain)$	$\overline{Gain}$	$\sigma(Gain)$	$\overline{Gain}$	$\sigma(Gain)$
GDF SUEZ	328	993	61	1,105	105	964
IBERDROLA I	443	1,173	169	1,165	66	1,085
ING	49	1,342	250	1,521	−18	1,341
INTESABCI	757	1,549	−102	1,540	−75	1,536
INDITEX	333	1,108	160	1,099	138	1,078
LVMH	367	915	1	882	71	927
MUNICH RE	362	917	100	930	135	903
LOREAL	345	920	5	860	124	955
PHILIPS ELECTR	308	1,053	268	1,087	52	980
REPSOL	548	1,138	182	1,190	41	1,175
RWE ST	209	1,229	252	1,668	104	1,627
BANCO SAN CENTRAL HISPANO	246	1,309	190	1,289	58	1,136
SANOFI	171	891	78	951	−26	860
SAP AG	45	799	76	787	−0	846
SAINT GOBAIN	134	1,135	26	1,106	153	1,149
SIEMENS AG	161	927	42	755	84	896
SCHNEIDER ELECTRIC SA	140	1,015	83	993	109	1,075
TELEFONICA	443	924	192	1,028	141	927
UNICREDIT SPA	383	1,738	156	1,594	43	1,697
UNILEVER CERT	169	734	−11	677	−13	704
VIVENDI UNIVERSAL	324	1,000	5	971	14	1,018
VOLKSWAGEN	219	1,185	38	1,002	46	1,087

Table 8 The quality of the binary prediction: the daily gain average and standard deviation for the 1-min prediction (with 0.5 bp trading costs)

Stock	Order book imbalance		Flow quantity		Past return	
	$\overline{Gain}$	$\sigma(Gain)$	$\overline{Gain}$	$\sigma(Gain)$	$\overline{Gain}$	$\sigma(Gain)$
INTERBREW	−182	973	−128	974	−212	1,028
AIR LIQUIDE	54	746	−214	764	−184	800
ALLIANZ	156	1,059	−172	894	−113	878
ASML Holding NV	−115	1,022	−167	1,032	−278	1,138
BASF AG	−190	997	−180	980	−228	992
BAYER AG	−46	960	−73	1,032	−202	967
BBVARGENTARIA	−95	1,528	−152	1,314	−137	1,330
BAY MOT WERKE	−110	997	−71	1,004	−183	952
DANONE	−62	995	−225	962	−309	867

(continued)

Table 8 (continued)

Stock	Order book imbalance		Flow quantity		Past return	
	$\overline{Gain}$	$\sigma(Gain)$	$\overline{Gain}$	$\sigma(Gain)$	$\overline{Gain}$	$\sigma(Gain)$
BNP PARIBAS	−154	1,327	−240	1,348	−309	1,371
CARREFOUR	−119	1,229	−50	1,213	−228	1,270
CRH PLC IRLANDE	852	1,407	235	1,360	−7	1,477
AXA	−149	1,203	−310	1,269	−346	1,261
DAIMLER CHRYSLER	−119	1,032	−288	1,107	−184	1,078
DEUTSCHE BANK AG	−246	1,368	−249	1,293	−249	1,279
VINCI	35	887	9	937	−41	904
DEUTSCHE TELEKOM	52	857	−53	844	−205	1,000
ESSILOR INTERNATIONAL	−116	1,002	−134	967	−251	948
ENEL	395	1,196	−201	1,208	−97	1,111
ENI	226	783	−181	837	−138	807
E.ON AG	19	996	−157	1,173	−166	1,051
TOTAL	−68	766	−162	860	−122	856
GENERALI ASSIC	496	1,205	−251	1,155	−323	1,241
SOCIETE GENERALE	−82	1,526	−282	1,562	−291	1,591
GDF SUEZ	64	977	−182	1,114	−134	978
IBERDROLA I	175	1,155	−96	1,165	−205	1,073
ING	−186	1,359	−12	1,506	−287	1,349
INTESABCI	419	1,505	−302	1,560	−311	1,520
INDITEX	70	1,074	−142	1,097	−119	1,060
LVMH	71	900	−206	912	−166	911
MUNICH RE	74	888	−133	935	−145	880
LOREAL	83	901	−184	889	−134	957
PHILIPS ELECTR	−1	1,058	−30	1,083	−222	967
REPSOL	256	1,116	−93	1,183	−173	1,177
RWE ST	12	1,237	19	1,668	−114	1,635
BANCO SAN CENTRAL HISPANO	6	1,306	−79	1,281	−184	1,150
SANOFI	−59	892	−165	955	−207	901
SAP AG	−188	794	−184	794	−255	850
SAINT GOBAIN	−51	1,152	−206	1,112	−81	1,155
SIEMENS AG	−65	927	−144	782	−162	889
SCHNEIDER ELECTRIC SA	−107	1,007	−155	1,002	−127	1,065
TELEFONICA	134	906	−29	1,041	−79	930
UNICREDIT SPA	120	1,703	−50	1,590	−197	1,701
UNILEVER CERT	−88	715	−229	713	−215	730
VIVENDI UNIVERSAL	38	989	−222	983	−266	1,029
VOLKSWAGEN	−20	1,178	−175	1,021	−173	1,088

Table 9 The quality of the binary prediction: 30-min prediction AUC and accuracy per stock

Stock	Order book imbalance		Flow quantity		Past return	
	AUC	Accuracy	AUC	Accuracy	AUC	Accuracy
INTERBREW	0.50	0.50	0.50	0.50	0.49	0.49
AIR LIQUIDE	0.50	0.50	0.50	0.50	0.49	0.49
ALLIANZ	0.51	0.51	0.51	0.51	0.49	0.49
ASML Holding NV	0.49	0.49	0.50	0.50	0.51	0.51
BASF AG	0.49	0.49	0.50	0.50	0.49	0.49
BAYER AG	0.51	0.51	0.49	0.49	0.49	0.50
BBVARGENTARIA	0.50	0.50	0.51	0.51	0.51	0.51
BAY MOT WERKE	0.50	0.50	0.51	0.51	0.50	0.50
DANONE	0.50	0.50	0.51	0.51	0.51	0.51
BNP PARIBAS	0.48	0.49	0.49	0.49	0.51	0.51
CARREFOUR	0.52	0.52	0.51	0.51	0.50	0.50
CRH PLC IRLANDE	0.51	0.51	0.51	0.51	0.50	0.50
AXA	0.49	0.49	0.49	0.49	0.49	0.49
DAIMLER CHRYSLER	0.51	0.52	0.50	0.51	0.50	0.50
DEUTSCHE BANK AG	0.52	0.52	0.49	0.49	0.51	0.51
VINCI	0.51	0.52	0.50	0.51	0.52	0.52
DEUTSCHE TELEKOM	0.49	0.50	0.50	0.51	0.51	0.51
ESSILOR INTERNATIONAL	0.50	0.50	0.50	0.50	0.51	0.51
ENEL	0.50	0.50	0.50	0.50	0.51	0.51
ENI	0.51	0.51	0.50	0.50	0.51	0.51
E.ON AG	0.49	0.49	0.50	0.50	0.51	0.51
TOTAL	0.51	0.51	0.52	0.52	0.51	0.51
GENERALI ASSIC	0.50	0.50	0.51	0.51	0.50	0.50
SOCIETE GENERALE	0.50	0.50	0.51	0.51	0.51	0.51
GDF SUEZ	0.51	0.51	0.50	0.50	0.52	0.52
IBERDROLA I	0.52	0.52	0.50	0.50	0.52	0.52
ING	0.50	0.50	0.50	0.50	0.49	0.49
INTESABCI	0.50	0.50	0.50	0.50	0.49	0.49
INDITEX	0.50	0.50	0.49	0.49	0.51	0.51
LVMH	0.50	0.50	0.50	0.50	0.49	0.49
MUNICH RE	0.50	0.50	0.51	0.51	0.50	0.50
LOREAL	0.50	0.51	0.51	0.51	0.50	0.50
PHILIPS ELECTR	0.49	0.49	0.50	0.50	0.51	0.51
REPSOL	0.49	0.49	0.50	0.50	0.51	0.51
RWE ST	0.50	0.51	0.50	0.51	0.51	0.51
BANCO SAN CENTRAL HISPANO	0.50	0.50	0.48	0.48	0.50	0.50
SANOFI	0.50	0.50	0.51	0.51	0.49	0.49

(continued)

Table 9 (continued)

Stock	Order book imbalance		Flow quantity		Past return	
	AUC	Accuracy	AUC	Accuracy	AUC	Accuracy
SAP AG	0.51	0.51	0.50	0.50	0.51	0.51
SAINT GOBAIN	0.50	0.50	0.51	0.52	0.50	0.50
SIEMENS AG	0.52	0.52	0.50	0.50	0.50	0.50
SCHNEIDER ELECTRIC SA	0.50	0.50	0.48	0.48	0.51	0.51
TELEFONICA	0.52	0.52	0.49	0.49	0.50	0.50
UNICREDIT SPA	0.52	0.52	0.50	0.50	0.51	0.51
UNILEVER CERT	0.52	0.52	0.50	0.50	0.50	0.50
VIVENDI UNIVERSAL	0.50	0.50	0.50	0.50	0.50	0.50
VOLKSWAGEN	0.52	0.51	0.50	0.50	0.50	0.50

Table 10 The quality of the binary prediction: the daily gain average and standard deviation for the 30-min prediction (without trading costs)

Stock	Order book imbalance		Flow quantity		Past return	
	$Gain$	$\sigma(Gain)$	$Gain$	$\sigma(Gain)$	$Gain$	$\sigma(Gain)$
INTERBREW	−24	952	−9	832	−54	936
AIR LIQUIDE	−1	676	−15	682	−68	737
ALLIANZ	102	866	15	911	−9	855
ASML Holding NV	−69	1,009	31	934	−27	1,046
BASF AG	−88	855	57	905	−20	859
BAYER AG	7	989	−65	917	−68	884
BBVARGENTARIA	−76	1,271	19	1,185	86	1,214
BAY MOT WERKE	−87	941	57	919	−50	867
DANONE	−85	813	−46	806	−11	812
BNP PARIBAS	−87	1,273	−63	1,252	54	1,257
CARREFOUR	56	1,153	−45	1,208	−86	1,068
CRH PLC IRLANDE	79	1,125	84	1,292	−26	1,263
AXA	−128	1,067	−94	1,095	−44	1,188
DAIMLER CHRYSLER	106	906	−53	964	−17	950
DEUTSCHE BANK AG	168	1,090	−137	1,120	50	1,121
VINCI	123	837	36	821	109	801
DEUTSCHE TELEKOM	10	855	47	866	13	796
ESSILOR INTERNATIONAL	−26	932	29	922	11	976
ENEL	19	1,044	10	1,039	72	1,015
ENI	38	746	−59	767	−2	775
E.ON AG	−9	968	−29	979	57	971
TOTAL	72	752	106	707	47	743
GENERALI ASSIC	−107	1,067	−8	1,124	−57	1,207
SOCIETE GENERALE	1	1,554	38	1,454	35	1,466

(continued)

Table 10 (continued)

Stock	Order book imbalance		Flow quantity		Past return	
	$\overline{Gain}$	$\sigma(Gain)$	$\overline{Gain}$	$\sigma(Gain)$	$\overline{Gain}$	$\sigma(Gain)$
GDF SUEZ	61	908	−3	891	49	881
IBERDROLA I	78	1,114	−28	1,033	111	1,075
ING	−48	1,348	−52	1,258	−34	1,324
INTESABCI	−77	1,457	−18	1,431	17	1,437
INDITEX	8	975	−62	984	57	900
LVMH	−5	857	−20	873	−81	807
MUNICH RE	−17	787	−17	754	−31	744
LOREAL	14	842	72	804	45	877
PHILIPS ELECTR	−69	845	−25	844	14	903
REPSOL	−66	1,011	−32	1,022	47	999
RWE ST	60	1,242	113	1,259	63	1,228
BANCO SAN CENTRAL HISPANO	−26	1,227	−109	1,205	10	1,180
SANOFI	−60	924	34	952	−34	890
SAP AG	48	776	−16	863	25	725
SAINT GOBAIN	12	1,016	93	1,072	−40	1,075
SIEMENS AG	137	912	−39	909	−41	893
SCHNEIDER ELECTRIC SA	17	985	−122	917	72	940
TELEFONICA	135	927	38	906	56	879
UNICREDIT SPA	188	1,709	−3	1,592	41	1,625
UNILEVER CERT	29	605	−54	639	−14	661
VIVENDI UNIVERSAL	−18	945	43	935	−27	933
VOLKSWAGEN	100	1,110	−6	1,113	43	1,135

Table 11 The quality of the binary prediction: the daily gain average and standard deviation for the 30-min prediction (with 0.5 bp trading costs)

Stock	Order book imbalance		Flow quantity		Past return	
	$\overline{Gain}$	$\sigma(Gain)$	$\overline{Gain}$	$\sigma(Gain)$	$\overline{Gain}$	$\sigma(Gain)$
INTERBREW	−56	952	−33	835	−88	937
AIR LIQUIDE	−30	679	−46	686	−99	735
ALLIANZ	67	866	−22	910	−38	858
ASML Holding NV	−107	1,008	−9	934	−63	1,045
BASF AG	−117	857	28	903	−51	862
BAYER AG	−29	991	−95	919	−95	886
BBVARGENTARIA	−104	1,273	−8	1,186	53	1,212
BAY MOT WERKE	−115	942	22	918	−78	868
DANONE	−117	811	−82	806	−50	810

(continued)

Table 11 (continued)

Stock	Order book imbalance		Flow quantity		Past return	
	$\overline{Gain}$	$\sigma(Gain)$	$\overline{Gain}$	$\sigma(Gain)$	$\overline{Gain}$	$\sigma(Gain)$
BNP PARIBAS	−113	1,274	−96	1,254	15	1,253
CARREFOUR	20	1,153	−79	1,209	−122	1,065
CRH PLC IRLANDE	48	1,121	48	1,290	−58	1,263
AXA	−159	1,067	−120	1,093	−84	1,185
DAIMLER CHRYSLER	74	906	−79	966	−47	950
DEUTSCHE BANK AG	129	1,091	−171	1,120	17	1,120
VINCI	95	839	8	827	86	802
DEUTSCHE TELEKOM	−18	856	16	866	−15	794
ESSILOR INTERNATIONAL	−51	931	−1	922	−26	973
ENEL	−11	1,044	−16	1,041	46	1,015
ENI	5	745	−88	767	−27	774
E.ON AG	−36	968	−57	983	23	967
TOTAL	41	756	78	706	13	744
GENERALI ASSIC	−139	1066	−39	1,124	−92	1,208
SOCIETE GENERALE	−28	1,556	4	1,453	−2	1,464
GDF SUEZ	25	909	−34	888	17	880
IBERDROLA I	47	1,112	−60	1,035	75	1,073
ING	−82	1,348	−86	1,257	−69	1,325
INTESABCI	−103	1,459	−48	1,429	−15	1,438
INDITEX	−20	976	−88	986	24	899
LVMH	−38	856	−55	871	−112	806
MUNICH RE	−49	787	−50	758	−61	746
LOREAL	−14	842	43	805	17	879
PHILIPS ELECTR	−97	847	−57	844	−20	898
REPSOL	−97	1,012	−65	1,023	8	998
RWE ST	34	1,243	81	1,263	33	1,227
BANCO SAN CENTRAL HISPANO	−53	1,228	−137	1,208	−23	1,177
SANOFI	−93	928	1	953	−63	890
SAP AG	15	776	−54	864	−20	716
SAINT GOBAIN	−14	1,020	57	1,075	−73	1,074
SIEMENS AG	101	912	−68	911	−71	893
SCHNEIDER ELECTRIC SA	−15	984	−155	917	37	938
TELEFONICA	107	926	8	910	24	881
UNICREDIT SPA	154	1,708	−33	1,593	10	1,628
UNILEVER CERT	−8	606	−89	639	−51	659
VIVENDI UNIVERSAL	−53	947	3	935	−66	932
VOLKSWAGEN	75	1,111	−32	1,116	10	1,138

Appendix 2: Four-Class Classification

See Tables 12, 13, 14, 15, 16, 17, 18 and 19.

Notice that the nans on the tables of the Appendix 2 correspond to the cases where $|\widehat{Y}|$ is always lower than θ thus no positions are taken.

Table 12 The quality of the 4-class prediction: 1-min prediction AUC and accuracy per stock

Stock	Order book imbalance		Flow quantity		Past return	
	AUC	Accuracy	AUC	Accuracy	AUC	Accuracy
INTERBREW	0.58	0.59	0.50	0.42	0.50	0.50
AIR LIQUIDE	0.71	0.72	Nan	Nan	0.50	0.58
ALLIANZ	0.69	0.69	0.50	0.54	0.61	0.61
ASML Holding NV	0.60	0.60	0.50	0.54	0.48	0.48
BASF AG	0.60	0.60	Nan	Nan	0.49	0.50
BAYER AG	0.53	0.55	0.50	0.59	0.50	0.56
BBVARGENTARIA	0.57	0.57	0.55	0.55	0.55	0.56
BAY MOT WERKE	0.57	0.58	0.55	0.55	0.54	0.55
DANONE	0.60	0.60	Nan	Nan	0.58	0.58
BNP PARIBAS	0.58	0.59	0.50	0.50	0.52	0.53
CARREFOUR	0.59	0.60	0.50	0.56	0.56	0.56
CRH PLC IRLANDE	0.70	0.70	0.64	0.64	0.55	0.56
AXA	0.58	0.60	Nan	Nan	0.56	0.56
DAIMLER CHRYSLER	0.57	0.57	0.50	0.51	0.54	0.54
DEUTSCHE BANK AG	0.55	0.55	0.54	0.56	0.52	0.52
VINCI	0.60	0.60	0.55	0.56	0.56	0.56
DEUTSCHE TELEKOM	0.71	0.72	Nan	Nan	0.51	0.51
ESSILOR INTERNATIONAL	0.60	0.60	0.50	0.55	0.52	0.56
ENEL	0.73	0.73	Nan	Nan	0.57	0.60
ENI	0.76	0.76	Nan	Nan	0.61	0.61
E.ON AG	0.64	0.64	Nan	Nan	0.53	0.53
TOTAL	0.54	0.59	Nan	Nan	0.50	0.46
GENERALI ASSIC	0.68	0.68	Nan	Nan	0.60	0.60
SOCIETE GENERALE	0.55	0.56	0.50	0.54	0.52	0.54
GDF SUEZ	0.62	0.62	Nan	Nan	0.53	0.53
IBERDROLA I	0.63	0.63	0.56	0.56	0.57	0.57
ING	0.55	0.55	0.54	0.55	0.52	0.55
INTESABCI	0.67	0.67	Nan	Nan	0.58	0.58
INDITEX	0.68	0.68	0.58	0.58	0.55	0.55
LVMH	0.65	0.66	Nan	Nan	0.58	0.58
MUNICH RE	0.66	0.66	0.55	0.55	0.54	0.54
LOREAL	0.67	0.67	Nan	Nan	0.58	0.58

(continued)

Table 12 (continued)

Stock	Order book imbalance		Flow quantity		Past return	
	AUC	Accuracy	AUC	Accuracy	AUC	Accuracy
PHILIPS ELECTR	0.61	0.62	0.50	0.51	0.52	0.54
REPSOL	0.63	0.63	0.53	0.58	0.57	0.57
RWE ST	0.58	0.58	0.53	0.55	0.52	0.52
BANCO SAN CENTRAL HISPANO	0.57	0.56	0.52	0.51	0.58	0.58
SANOFI	0.60	0.60	Nan	Nan	0.50	0.60
SAP AG	0.52	0.61	0.50	0.56	0.52	0.54
SAINT GOBAIN	0.56	0.58	0.54	0.58	0.54	0.55
SIEMENS AG	0.56	0.61	0.55	0.56	0.59	0.59
SCHNEIDER ELECTRIC SA	0.57	0.58	Nan	Nan	0.56	0.57
TELEFONICA	0.68	0.68	0.53	0.57	0.56	0.56
UNICREDIT SPA	0.64	0.65	0.50	0.54	0.57	0.57
UNILEVER CERT	0.50	0.63	Nan	Nan	Nan	Nan
VIVENDI UNIVERSAL	0.63	0.63	Nan	Nan	0.51	0.52
VOLKSWAGEN	0.62	0.62	0.49	0.49	0.52	0.53

Table 13 The quality of the 4-class prediction: the daily gain average and standard deviation for the 1-min prediction (without trading costs)

Stock	Order book imbalance		Flow quantity		Past return	
	$\overline{Gain}$	$\sigma(Gain)$	$\overline{Gain}$	$\sigma(Gain)$	$\overline{Gain}$	$\sigma(Gain)$
INTERBREW	137	388	−6	98	4	131
AIR LIQUIDE	306	577	0	0	3	42
ALLIANZ	1,363	779	4	47	68	276
ASML Holding NV	440	651	5	63	−2	132
BASF AG	87	287	0	0	−2	48
BAYER AG	21	128	14	137	14	99
BBVARGENTARIA	390	665	273	669	208	582
BAY MOT WERKE	107	281	47	276	52	238
DANONE	168	366	0	0	4	47
BNP PARIBAS	171	428	3	66	44	453
CARREFOUR	486	715	11	139	136	469
CRH PLC IRLANDE	2,534	1,240	1,364	1,077	202	560
AXA	594	786	0	0	55	320
DAIMLER CHRYSLER	93	289	2	24	16	191
DEUTSCHE BANK AG	34	224	38	212	12	291
VINCI	154	451	13	111	27	147
DEUTSCHE TELEKOM	488	827	0	0	3	66
ESSILOR INTERNATIONAL	351	596	17	164	10	106

(continued)

Table 13 (continued)

Stock	Order book imbalance		Flow quantity		Past return	
	$\overline{Gain}$	$\sigma(Gain)$	$\overline{Gain}$	$\sigma(Gain)$	$\overline{Gain}$	$\sigma(Gain)$
ENEL	2,219	1,056	0	0	193	503
ENI	2,000	773	0	0	110	300
E.ON AG	651	680	0	0	10	168
TOTAL	10	93	0	0	1	38
GENERALI ASSIC	2,520	1,420	0	0	249	756
SOCIETE GENERALE	184	503	2	25	56	410
GDF SUEZ	504	692	0	0	21	171
IBERDROLA I	738	951	155	512	115	409
ING	109	373	59	296	7	138
INTESABCI	2,512	1,248	0	0	185	731
INDITEX	1,039	914	151	587	44	223
LVMH	930	847	0	0	64	277
MUNICH RE	370	533	26	145	3	50
LOREAL	800	674	0	0	22	112
PHILIPS ELECTR	440	613	6	94	11	116
REPSOL	1,234	1,013	142	445	110	555
RWE ST	192	556	85	380	29	364
BANCO SAN CENTRAL HISPANO	228	501	4	158	168	635
SANOFI	26	127	0	0	6	90
SAP AG	50	196	24	187	6	200
SAINT GOBAIN	210	519	30	186	88	362
SIEMENS AG	26	139	31	198	28	162
SCHNEIDER ELECTRIC SA	123	434	0	0	37	214
TELEFONICA	1,402	825	36	232	34	205
UNICREDIT SPA	1,316	1,393	17	197	247	835
UNILEVER CERT	16	104	0	0	0	0
VIVENDI UNIVERSAL	583	826	0	0	5	141
VOLKSWAGEN	530	745	−0	78	1	215

Table 14 The quality of the 4-class prediction: the daily gain average and standard deviation for the 1-min prediction (with trading costs)

Stock	Order book imbalance		Flow quantity		Past return	
	$\overline{Gain}$	$\sigma(Gain)$	$\overline{Gain}$	$\sigma(Gain)$	$\overline{Gain}$	$\sigma(Gain)$
INTERBREW	22	263	−9	150	−38	183
AIR LIQUIDE	128	329	0	0	1	31
ALLIANZ	586	559	−1	16	8	194
ASML Holding NV	125	408	−0	32	−25	168
BASF AG	15	189	0	0	−7	51
BAYER AG	−14	105	1	86	−2	77

(continued)

Table 14 (continued)

Stock	Order book imbalance		Flow quantity		Past return	
	$\overline{Gain}$	$\sigma(Gain)$	$\overline{Gain}$	$\sigma(Gain)$	$\overline{Gain}$	$\sigma(Gain)$
BBVARGENTARIA	107	507	31	465	16	474
BAY MOT WERKE	−1	193	1	199	−12	184
DANONE	21	210	0	0	−4	42
BNP PARIBAS	34	271	−12	126	−65	481
CARREFOUR	116	506	−8	131	18	362
CRH PLC IRLANDE	1,848	1,102	518	844	18	442
AXA	174	550	0	0	−23	274
DAIMLER CHRYSLER	−7	245	−1	14	−32	199
DEUTSCHE BANK AG	−23	204	8	122	−33	311
VINCI	38	281	−3	73	−5	111
DEUTSCHE TELEKOM	241	526	0	0	−10	79
ESSILOR INTERNATIONAL	88	388	−14	157	−4	91
ENEL	1,338	881	0	0	−18	443
ENI	1,082	613	0	0	−25	211
E.ON AG	185	475	0	0	−22	173
TOTAL	−5	72	0	0	−3	49
GENERALI ASSIC	1,518	1,179	0	0	58	636
SOCIETE GENERALE	2	412	−2	24	−41	394
GDF SUEZ	142	464	0	0	−8	126
IBERDROLA I	340	722	28	331	18	292
ING	−13	329	6	209	−12	147
INTESABCI	1,514	1,096	0	0	−20	658
INDITEX	547	702	3	400	−10	198
LVMH	372	581	0	0	−11	169
MUNICH RE	111	322	−3	62	−6	46
LOREAL	285	443	0	0	−5	85
PHILIPS ELECTR	113	417	−6	96	−8	105
REPSOL	611	809	40	254	27	437
RWE ST	38	450	−2	299	−42	372
BANCO SAN CENTRAL HISPANO	20	392	−31	203	49	463
SANOFI	1	69	0	0	−0	79
SAP AG	2	120	−4	137	−30	207
SAINT GOBAIN	25	403	−1	114	−7	289
SIEMENS AG	2	74	−2	89	−3	141
SCHNEIDER ELECTRIC SA	16	317	0	0	−14	195
TELEFONICA	656	663	6	139	−7	183
UNICREDIT SPA	693	1,159	−5	173	19	628
UNILEVER CERT	1	56	0	0	0	0
VIVENDI UNIVERSAL	214	617	0	0	−27	175
VOLKSWAGEN	171	545	−7	115	−45	246

Table 15 The quality of the 4-class prediction: 5-min prediction AUC and accuracy per stock

Stock	Order book imbalance		Flow quantity		Past return	
	AUC	Accuracy	AUC	Accuracy	AUC	Accuracy
INTERBREW	0.51	0.51	0.50	0.50	0.51	0.51
AIR LIQUIDE	0.55	0.55	0.50	0.50	0.51	0.51
ALLIANZ	0.56	0.56	0.51	0.51	0.54	0.54
ASML Holding NV	0.52	0.52	0.51	0.51	0.52	0.52
BASF AG	0.51	0.51	0.51	0.51	0.52	0.52
BAYER AG	0.51	0.52	0.50	0.51	0.50	0.50
BBVARGENTARIA	0.52	0.52	0.51	0.51	0.52	0.52
BAY MOT WERKE	0.51	0.51	0.50	0.50	0.51	0.51
DANONE	0.52	0.52	0.51	0.51	0.50	0.50
BNP PARIBAS	0.49	0.49	0.51	0.51	0.50	0.50
CARREFOUR	0.52	0.52	0.50	0.50	0.51	0.51
CRH PLC IRLANDE	0.57	0.57	0.54	0.54	0.52	0.52
AXA	0.51	0.50	0.49	0.49	0.50	0.50
DAIMLER CHRYSLER	0.52	0.52	0.49	0.50	0.49	0.50
DEUTSCHE BANK AG	0.49	0.49	0.49	0.49	0.50	0.50
VINCI	0.51	0.52	0.50	0.51	0.52	0.53
DEUTSCHE TELEKOM	0.55	0.55	0.51	0.51	0.49	0.50
ESSILOR INTERNATIONAL	0.52	0.52	0.52	0.52	0.51	0.51
ENEL	0.58	0.58	0.51	0.51	0.52	0.53
ENI	0.60	0.60	0.52	0.52	0.52	0.52
E.ON AG	0.55	0.55	0.51	0.51	0.51	0.51
TOTAL	0.51	0.52	0.50	0.50	0.52	0.53
GENERALI ASSIC	0.58	0.58	0.49	0.49	0.53	0.53
SOCIETE GENERALE	0.50	0.50	0.49	0.49	0.50	0.50
GDF SUEZ	0.52	0.52	0.50	0.50	0.51	0.51
IBERDROLA I	0.54	0.54	0.51	0.51	0.51	0.51
ING	0.50	0.50	0.50	0.50	0.49	0.49
INTESABCI	0.55	0.55	0.50	0.50	0.51	0.51
INDITEX	0.56	0.56	0.51	0.51	0.53	0.53
LVMH	0.55	0.55	0.49	0.49	0.52	0.52
MUNICH RE	0.57	0.57	0.53	0.53	0.54	0.54
LOREAL	0.57	0.57	0.50	0.50	0.52	0.52
PHILIPS ELECTR	0.53	0.53	0.50	0.51	0.51	0.52
REPSOL	0.54	0.54	0.52	0.52	0.51	0.51
RWE ST	0.51	0.51	0.52	0.52	0.51	0.51
BANCO SAN CENTRAL HISPANO	0.51	0.51	0.51	0.51	0.51	0.51
SANOFI	0.50	0.50	0.49	0.49	0.49	0.49

(continued)

Table 15 (continued)

Stock	Order book imbalance		Flow quantity		Past return	
	AUC	Accuracy	AUC	Accuracy	AUC	Accuracy
SAP AG	0.52	0.51	0.52	0.52	0.48	0.48
SAINT GOBAIN	0.51	0.51	0.51	0.51	0.50	0.50
SIEMENS AG	0.51	0.52	0.50	0.50	0.52	0.52
SCHNEIDER ELECTRIC SA	0.52	0.52	0.51	0.51	0.52	0.52
TELEFONICA	0.56	0.56	0.52	0.52	0.51	0.51
UNICREDIT SPA	0.54	0.54	0.50	0.50	0.51	0.51
UNILEVER CERT	0.54	0.54	0.51	0.51	0.52	0.52
VIVENDI UNIVERSAL	0.53	0.53	0.50	0.50	0.49	0.49
VOLKSWAGEN	0.53	0.53	0.50	0.50	0.51	0.50

Table 16 The quality of the 4-class prediction: the daily gain average and standard deviation for the 5-min prediction (without trading costs)

Stock	Order book imbalance		Flow quantity		Past return	
	$Gain$	$\sigma(Gain)$	$Gain$	$\sigma(Gain)$	$Gain$	$\sigma(Gain)$
INTERBREW	56	392	7	362	−5	557
AIR LIQUIDE	73	317	−5	214	4	364
ALLIANZ	298	629	22	388	68	379
ASML Holding NV	112	573	26	538	39	547
BASF AG	−1	328	55	506	71	441
BAYER AG	81	501	50	446	19	388
BBVARGENTARIA	86	993	117	868	123	856
BAY MOT WERKE	5	404	28	516	34	452
DANONE	83	522	16	344	7	282
BNP PARIBAS	−46	691	26	823	6	642
CARREFOUR	87	614	59	740	5	655
CRH PLC IRLANDE	982	1,141	366	998	264	987
AXA	26	685	−58	618	−8	708
DAIMLER CHRYSLER	87	480	−7	509	−20	588
DEUTSCHE BANK AG	−26	715	−24	696	31	651
VINCI	30	432	22	409	57	494
DEUTSCHE TELEKOM	132	410	27	394	3	421
ESSILOR INTERNATIONAL	50	513	74	518	61	517
ENEL	515	877	−0	625	58	641
ENI	296	471	29	313	19	354
E.ON AG	217	667	29	698	8	570
TOTAL	52	311	8	353	100	389
GENERALI ASSIC	528	951	−64	575	89	677
SOCIETE GENERALE	100	1,026	−66	928	35	863

(continued)

Table 16 (continued)

Stock	Order book imbalance		Flow quantity		Past return	
	$\overline{Gain}$	$\sigma(Gain)$	$\overline{Gain}$	$\sigma(Gain)$	$\overline{Gain}$	$\sigma(Gain)$
GDF SUEZ	133	604	25	566	−3	425
IBERDROLA I	209	620	84	603	52	561
ING	64	819	96	948	−8	734
INTESABCI	504	1,016	−35	916	−17	962
INDITEX	264	661	85	539	86	612
LVMH	152	503	−21	380	41	375
MUNICH RE	183	416	74	373	63	399
LOREAL	221	527	29	462	40	416
PHILIPS ELECTR	85	472	84	560	71	480
REPSOL	254	701	88	733	44	607
RWE ST	21	661	139	785	108	990
BANCO SAN CENTRAL HISPANO	68	791	77	840	23	760
SANOFI	27	418	−12	369	−32	406
SAP AG	14	342	25	292	−21	312
SAINT GOBAIN	23	532	102	609	47	736
SIEMENS AG	9	338	19	358	65	337
SCHNEIDER ELECTRIC SA	121	618	24	428	64	560
TELEFONICA	314	632	72	517	65	498
UNICREDIT SPA	331	1,222	30	1,032	29	1,113
UNILEVER CERT	31	274	5	193	13	191
VIVENDI UNIVERSAL	171	575	0	406	−17	449
VOLKSWAGEN	115	645	24	616	−4	608

Table 17 The quality of the 4-class prediction: the daily gain average and standard deviation for the 1-min prediction (with 0.5 bp trading costs)

Stock	Order book imbalance		Flow quantity		Past return	
	$\overline{Gain}$	$\sigma(Gain)$	$\overline{Gain}$	$\sigma(Gain)$	$\overline{Gain}$	$\sigma(Gain)$
INTERBREW	−42	387	−86	378	−90	576
AIR LIQUIDE	−3	306	−56	238	−69	360
ALLIANZ	117	610	−72	400	−29	389
ASML Holding NV	−29	566	−95	544	−58	537
BASF AG	−64	341	−56	511	−41	438
BAYER AG	−46	511	−57	438	−74	389
BBVARGENTARIA	−65	994	−69	864	−49	842
BAY MOT WERKE	−87	415	−99	526	−72	453
DANONE	−20	506	−69	349	−44	287

(continued)

Table 17 (continued)

Stock	Order book imbalance		Flowquantity		Past return	
	$\overline{Gain}$	$\sigma(Gain)$	$\overline{Gain}$	$\sigma(Gain)$	$\overline{Gain}$	$\sigma(Gain)$
BNP PARIBAS	−180	708	−131	838	−120	645
CARREFOUR	−60	612	−104	733	−133	667
CRH PLC IRLANDE	730	1,111	146	974	88	962
AXA	−122	695	−206	643	−131	722
DAIMLER CHRYSLER	−27	475	−133	524	−131	596
DEUTSCHE BANK AG	−163	723	−188	713	−99	657
VINCI	−75	440	−82	419	−39	479
DEUTSCHE TELEKOM	33	398	−51	396	−86	439
ESSILOR INTERNATIONAL	−64	521	−48	510	−49	521
ENEL	298	845	−132	639	−70	642
ENI	126	441	−54	330	−58	360
E.ON AG	56	660	−114	712	−110	582
TOTAL	−25	310	−80	366	−1	380
GENERALI ASSIC	291	918	−178	603	−44	678
SOCIETE GENERALE	−80	1,015	−222	947	−125	859
GDF SUEZ	−16	593	−101	576	−109	438
IBERDROLA I	57	605	−70	595	−56	545
ING	−94	815	−109	948	−141	738
INTESABCI	262	1,000	−210	923	−164	970
INDITEX	113	629	−38	533	−34	613
LVMH	8	482	−100	397	−54	372
MUNICH RE	56	397	4	343	−30	400
LOREAL	81	512	−69	462	−62	420
PHILIPS ELECTR	−40	482	−78	564	−52	478
REPSOL	86	684	−76	722	−72	605
RWE ST	−119	679	−48	783	−17	988
BANCO SAN CENTRAL HISPANO	−83	785	−102	842	−108	767
SANOFI	−73	429	−106	386	−102	422
SAP AG	−77	352	−54	296	−90	324
SAINT GOBAIN	−106	545	−66	605	−101	725
SIEMENS AG	−75	349	−44	358	−25	331
SCHNEIDER ELECTRIC SA	−5	616	−102	435	−54	560
TELEFONICA	134	608	−58	513	−55	496
UNICREDIT SPA	113	1,207	−149	1,042	−140	1,112
UNILEVER CERT	−38	275	−27	201	−24	198
VIVENDI UNIVERSAL	24	562	−104	423	−109	457
VOLKSWAGEN	−17	641	−95	621	−124	615

Table 18 The quality of the 4-class prediction: 30-min prediction AUC and accuracy per stock

Stock	Order book imbalance		Flow quantity		Past return	
	AUC	Accuracy	AUC	Accuracy	AUC	Accuracy
INTERBREW	0.50	0.50	0.50	0.50	0.49	0.49
AIR LIQUIDE	0.48	0.48	0.50	0.50	0.51	0.51
ALLIANZ	0.50	0.50	0.51	0.51	0.50	0.50
ASML Holding NV	0.48	0.48	0.51	0.51	0.49	0.49
BASF AG	0.51	0.51	0.50	0.49	0.48	0.48
BAYER AG	0.50	0.50	0.49	0.50	0.50	0.50
BBVARGENTARIA	0.50	0.50	0.51	0.51	0.50	0.51
BAY MOT WERKE	0.51	0.51	0.50	0.50	0.51	0.51
DANONE	0.50	0.50	0.52	0.52	0.51	0.51
BNP PARIBAS	0.48	0.48	0.50	0.50	0.51	0.51
CARREFOUR	0.52	0.52	0.51	0.51	0.50	0.50
CRH PLC IRLANDE	0.50	0.51	0.49	0.49	0.50	0.50
AXA	0.50	0.50	0.48	0.48	0.51	0.51
DAIMLER CHRYSLER	0.52	0.52	0.50	0.50	0.51	0.52
DEUTSCHE BANK AG	0.51	0.51	0.50	0.50	0.49	0.49
VINCI	0.52	0.53	0.50	0.51	0.51	0.52
DEUTSCHE TELEKOM	0.51	0.51	0.51	0.52	0.50	0.51
ESSILOR INTERNATIONAL	0.51	0.51	0.52	0.52	0.51	0.51
ENEL	0.50	0.50	0.49	0.49	0.51	0.51
ENI	0.51	0.51	0.51	0.51	0.50	0.50
E.ON AG	0.48	0.48	0.50	0.50	0.51	0.52
TOTAL	0.52	0.52	0.51	0.51	0.52	0.52
GENERALI ASSIC	0.51	0.51	0.51	0.51	0.51	0.51
SOCIETE GENERALE	0.51	0.51	0.51	0.51	0.50	0.50
GDF SUEZ	0.50	0.50	0.50	0.50	0.51	0.51
IBERDROLA I	0.50	0.50	0.49	0.49	0.50	0.50
ING	0.49	0.49	0.51	0.51	0.49	0.49
INTESABCI	0.51	0.51	0.49	0.49	0.50	0.50
INDITEX	0.50	0.50	0.48	0.48	0.51	0.51
LVMH	0.50	0.50	0.51	0.51	0.49	0.48
MUNICH RE	0.51	0.51	0.50	0.50	0.51	0.51
LOREAL	0.50	0.50	0.51	0.51	0.50	0.50
PHILIPS ELECTR	0.49	0.49	0.51	0.52	0.50	0.50
REPSOL	0.49	0.49	0.51	0.51	0.50	0.50
RWE ST	0.50	0.50	0.50	0.50	0.50	0.50
BANCO SAN CENTRAL HISPANO	0.50	0.50	0.49	0.49	0.51	0.51
SANOFI	0.50	0.51	0.51	0.51	0.50	0.50
SAP AG	0.51	0.51	0.50	0.50	0.51	0.51
SAINT GOBAIN	0.50	0.50	0.50	0.50	0.49	0.49

(continued)

Table 18 (continued)

Stock	Order book imbalance		Flow quantity		Past return	
	AUC	Accuracy	AUC	Accuracy	AUC	Accuracy
SIEMENS AG	0.52	0.53	0.48	0.48	0.50	0.50
SCHNEIDER ELECTRIC SA	0.50	0.50	0.49	0.49	0.50	0.50
TELEFONICA	0.50	0.50	0.50	0.50	0.51	0.51
UNICREDIT SPA	0.51	0.51	0.50	0.50	0.51	0.51
UNILEVER CERT	0.51	0.51	0.50	0.50	0.51	0.51
VIVENDI UNIVERSAL	0.50	0.50	0.50	0.50	0.51	0.51
VOLKSWAGEN	0.51	0.51	0.51	0.51	0.51	0.51

Table 19 The quality of the 4-class prediction: the daily gain average and standard deviation for the 30-min prediction (without trading costs)

Stock	Order book imbalance		Flow quantity		Past return	
	$\overline{Gain}$	$\sigma(Gain)$	$\overline{Gain}$	$\sigma(Gain)$	$\overline{Gain}$	$\sigma(Gain)$
INTERBREW	−11	887	−6	845	−41	823
AIR LIQUIDE	−57	669	−17	633	−21	624
ALLIANZ	−14	762	69	689	−41	729
ASML Holding NV	−87	862	43	1,075	−29	897
BASF AG	−20	807	−3	781	−67	722
BAYER AG	38	759	−93	774	−46	765
BBVARGENTARIA	−16	1,263	−63	1,138	16	1,084
BAY MOT WERKE	−25	783	−23	923	−13	901
DANONE	−61	744	19	726	−18	745
BNP PARIBAS	−28	998	−2	1,179	−9	1,151
CARREFOUR	4	1,108	−135	1,082	−52	972
CRH PLC IRLANDE	75	962	−105	1,161	−6	1,117
AXA	12	1,054	6	1,055	49	1,111
DAIMLER CHRYSLER	75	872	−51	825	9	961
DEUTSCHE BANK AG	54	1,054	−89	1,152	−35	996
VINCI	110	761	80	742	100	743
DEUTSCHE TELEKOM	27	722	81	700	−14	718
ESSILOR INTERNATIONAL	29	830	43	827	41	872
ENEL	27	991	−40	971	55	959
ENI	7	628	−18	651	−16	645
E.ON AG	−70	911	−4	963	65	826
TOTAL	49	660	108	689	73	669
GENERALI ASSIC	18	1,011	2	1,094	11	1,085
SOCIETE GENERALE	53	1,413	67	1,253	−5	1,335
GDF SUEZ	59	906	−24	847	25	823
IBERDROLA I	3	1,017	−73	960	51	949

(continued)

Table 19   (continued)

Stock	Order book imbalance		Flow quantity		Past return	
	$\overline{Gain}$	$\sigma(Gain)$	$\overline{Gain}$	$\sigma(Gain)$	$\overline{Gain}$	$\sigma(Gain)$
ING	−21	1,138	105	1,205	−80	1,142
INTESABCI	−128	1,359	−54	1,329	85	1,288
INDITEX	−8	894	−161	912	17	860
LVMH	−36	831	15	725	−26	675
MUNICH RE	29	641	−25	688	−7	727
LOREAL	−19	671	31	755	15	727
PHILIPS ELECTR	−24	844	24	789	−29	841
REPSOL	−87	878	−5	920	3	925
RWE ST	32	1,132	61	1,217	46	1,140
BANCO SAN CENTRAL HISPANO	2	1,150	−60	1,072	48	1,090
SANOFI	−29	810	25	856	7	794
SAP AG	4	683	−52	709	−15	682
SAINT GOBAIN	−66	996	22	994	−51	945
SIEMENS AG	127	771	−35	802	−59	725
SCHNEIDER ELECTRIC SA	−31	896	−79	837	8	838
TELEFONICA	−12	759	42	918	111	912
UNICREDIT SPA	130	1,529	58	1,498	81	1,357
UNILEVER CERT	5	543	31	546	−26	508
VIVENDI UNIVERSAL	21	874	−15	899	6	859
VOLKSWAGEN	71	929	120	994	75	1,055

Appendix 3: OLS Method

See Tables 20, 21, 22 and 23.

Table 20   The quality of the binary prediction: the daily gain average and standard deviation for the 30-min prediction (with 0.5 bp trading costs)

Stock	Order book imbalance		Flowquantity		Past return	
	$\overline{Gain}$	$\sigma(Gain)$	$\overline{Gain}$	$\sigma(Gain)$	$\overline{Gain}$	$\sigma(Gain)$
INTERBREW	−55	887	−51	845	−84	824
AIR LIQUIDE	−96	672	−61	635	−62	625
ALLIANZ	−57	764	23	687	−80	731
ASML Holding NV	−132	863	−6	1,072	−73	896
BASF AG	−61	809	−47	780	−108	724
BAYER AG	−7	758	−136	777	−84	767
BBVARGENTARIA	−58	1,265	−108	1,137	−25	1,082

(continued)

Table 20 (continued)

Stock	Order book imbalance		Flow quantity		Past return	
	$\overline{Gain}$	$\sigma(Gain)$	$\overline{Gain}$	$\sigma(Gain)$	$\overline{Gain}$	$\sigma(Gain)$
BAY MOT WERKE	−65	784	−69	923	−53	902
DANONE	−101	743	−25	726	−60	742
BNP PARIBAS	−71	997	−46	1,180	−51	1,149
CARREFOUR	−39	1,110	−182	1,085	−94	972
CRH PLC IRLANDE	31	960	−152	1,163	−48	1,116
AXA	−31	1,052	−37	1,054	4	1,109
DAIMLER CHRYSLER	36	874	−93	825	−31	961
DEUTSCHE BANK AG	9	1,053	−138	1,151	−77	997
VINCI	72	763	40	742	65	743
DEUTSCHE TELEKOM	−12	722	36	702	−53	720
ESSILOR INTERNATIONAL	−9	830	−1	828	−2	869
ENEL	−17	993	−81	974	17	959
ENI	−36	627	−58	652	−57	642
E.ON AG	−106	911	−45	965	22	824
TOTAL	10	661	66	690	34	666
GENERALI ASSIC	−26	1,011	−44	1,096	−32	1,087
SOCIETE GENERALE	10	1,415	19	1,252	−51	1,336
GDF SUEZ	14	905	−70	847	−16	818
IBERDROLA I	−40	1,016	−117	962	5	947
ING	−63	1,137	58	1,207	−122	1144
INTESABCI	−172	1,359	−97	1,327	47	1,290
INDITEX	−48	896	−204	913	−22	859
LVMH	−82	830	−30	725	−68	675
MUNICH RE	−13	641	−66	691	−49	728
LOREAL	−57	674	−9	754	−22	728
PHILIPS ELECTR	−65	845	−23	788	−71	839
REPSOL	−128	877	−52	920	−41	920
RWE ST	−7	1,130	15	1,218	5	1,140
BANCO SAN CENTRAL HISPANO	−37	1,149	−103	1,073	6	1,089
SANOFI	−67	810	−21	856	−34	797
SAP AG	−37	683	−100	709	−60	680
SAINT GOBAIN	−105	997	−23	995	−93	946
SIEMENS AG	84	772	−77	805	−98	725
SCHNEIDER ELECTRIC SA	−73	896	−123	836	−34	838
TELEFONICA	−49	760	−4	919	68	913
UNICREDIT SPA	84	1,529	15	1499	40	1,359
UNILEVER CERT	−39	543	−14	545	−67	509
VIVENDI UNIVERSAL	−24	874	−61	900	−37	856
VOLKSWAGEN	33	929	76	995	38	1,058

Table 21 The quality of the OLS prediction: the AUC and the accuracy per stock for the different horizons

Stock	1-min horizon		5-min horizon		30-min horizon	
	AUC	Accuracy	AUC	Accuracy	AUC	Accuracy
INTERBREW	0.54	0.54	0.50	0.50	0.50	0.50
AIR LIQUIDE	0.57	0.57	0.52	0.52	0.49	0.49
ALLIANZ	0.61	0.61	0.53	0.53	0.50	0.50
ASML Holding NV	0.55	0.55	0.51	0.51	0.51	0.51
BASF AG	0.54	0.54	0.52	0.52	0.50	0.50
BAYER AG	0.54	0.54	0.51	0.51	0.51	0.51
BBVARGENTARIA	0.54	0.54	0.51	0.51	0.49	0.49
BAY MOT WERKE	0.55	0.55	0.51	0.51	0.49	0.49
DANONE	0.56	0.56	0.51	0.51	0.49	0.49
BNP PARIBAS	0.53	0.53	0.51	0.51	0.50	0.50
CARREFOUR	0.55	0.55	0.51	0.51	0.52	0.52
CRH PLC IRLANDE	0.62	0.62	0.56	0.56	0.52	0.52
AXA	0.55	0.55	0.51	0.51	0.50	0.50
DAIMLER CHRYSLER	0.54	0.54	0.51	0.51	0.50	0.50
DEUTSCHE BANK AG	0.53	0.53	0.51	0.51	0.51	0.51
VINCI	0.55	0.55	0.52	0.52	0.51	0.51
DEUTSCHE TELEKOM	0.56	0.56	0.52	0.52	0.50	0.51
ESSILOR INTERNATIONAL	0.56	0.56	0.51	0.51	0.51	0.51
ENEL	0.62	0.62	0.53	0.53	0.48	0.48
ENI	0.64	0.64	0.54	0.54	0.50	0.50
E.ON AG	0.57	0.57	0.52	0.52	0.48	0.48
TOTAL	0.54	0.54	0.51	0.51	0.50	0.50
GENERALI ASSIC	0.61	0.61	0.54	0.54	0.50	0.50
SOCIETE GENERALE	0.53	0.53	0.50	0.50	0.52	0.52
GDF SUEZ	0.56	0.56	0.51	0.51	0.50	0.50
IBERDROLA I	0.57	0.57	0.52	0.52	0.51	0.51
ING	0.53	0.53	0.51	0.51	0.49	0.49
INTESABCI	0.59	0.59	0.51	0.51	0.50	0.50
INDITEX	0.59	0.59	0.53	0.53	0.52	0.52
LVMH	0.59	0.59	0.52	0.52	0.52	0.52
MUNICH RE	0.58	0.58	0.53	0.53	0.50	0.50
LOREAL	0.60	0.60	0.52	0.52	0.51	0.51
PHILIPS ELECTR	0.56	0.56	0.51	0.51	0.50	0.50
REPSOL	0.57	0.57	0.52	0.52	0.51	0.51
RWE ST	0.54	0.54	0.51	0.51	0.49	0.49
BANCO SAN CENTRAL HISPANO	0.54	0.54	0.51	0.51	0.49	0.49
SANOFI	0.54	0.54	0.51	0.51	0.49	0.49

(continued)

Table 21 (continued)

Stock	1-min horizon		5-min horizon		30-min horizon	
	AUC	Accuracy	AUC	Accuracy	AUC	Accuracy
SAP AG	0.54	0.54	0.51	0.51	0.51	0.51
SAINT GOBAIN	0.54	0.54	0.51	0.51	0.52	0.52
SIEMENS AG	0.54	0.54	0.51	0.51	0.50	0.50
SCHNEIDER ELECTRIC SA	0.54	0.54	0.52	0.52	0.51	0.51
TELEFONICA	0.59	0.59	0.52	0.52	0.50	0.50
UNICREDIT SPA	0.56	0.56	0.51	0.51	0.49	0.49
UNILEVER CERT	0.56	0.56	0.51	0.51	0.50	0.50
VIVENDI UNIVERSAL	0.57	0.57	0.51	0.51	0.51	0.51
VOLKSWAGEN	0.56	0.56	0.52	0.52	0.51	0.51

Table 22 The quality of the OLS prediction: the daily gain average and standard deviation for the different horizons (without trading costs)

Stock	1-min horizon		5-min horizon		30-min horizon	
	$\overline{Gain}$	$\sigma(Gain)$	$\overline{Gain}$	$\sigma(Gain)$	$\overline{Gain}$	$\sigma(Gain)$
INTERBREW	1,410	1,151	89	1,022	−54	1,022
AIR LIQUIDE	1,756	1,028	237	775	−22	707
ALLIANZ	2,832	1,332	355	907	−81	935
ASML Holding NV	1,693	1,237	55	1,208	156	1,080
BASF AG	1,220	1,109	143	883	5	877
BAYER AG	1,412	1,086	129	948	−32	853
BBVARGENTARIA	2,297	1,759	315	1,518	−27	1,178
BAY MOT WERKE	1,749	1,243	124	984	−22	904
DANONE	1,729	1,045	143	843	−121	791
BNP PARIBAS	1,362	1,580	263	1,386	−21	1,246
CARREFOUR	2,108	1,465	211	1,205	69	1,242
CRH PLC IRLANDE	4,302	1,924	1,121	1,352	94	1,239
AXA	2,139	1,450	239	1,334	−34	1,101
DAIMLER CHRYSLER	1,380	1,325	139	1,006	−6	1,139
DEUTSCHE BANK AG	1,431	1,493	118	1,302	105	1,106
VINCI	1,803	1,192	340	950	31	736
DEUTSCHE TELEKOM	1,780	1,380	218	858	25	784
ESSILOR INTERNATIONAL	1,934	1,244	299	1,041	−34	910
ENEL	3,632	1,526	298	1,115	−35	984
ENI	3,095	1,170	369	887	8	742
E.ON AG	2,119	1,412	182	1,247	−126	976
TOTAL	1,336	1,054	220	852	−20	780
GENERALI ASSIC	3,937	1,763	537	1,260	37	996

(continued)

Table 22 (continued)

Stock	1-min horizon		5-min horizon		30-min horizon	
	$\overline{Gain}$	$\sigma(Gain)$	$\overline{Gain}$	$\sigma(Gain)$	$\overline{Gain}$	$\sigma(Gain)$
SOCIETE GENERALE	1,499	1,787	98	1,627	155	1,448
GDF SUEZ	2,115	1,279	175	1,084	−72	964
IBERDROLA I	2,499	1,587	450	1,123	83	1,046
ING	1,358	1,477	135	1,351	38	1,159
INTESABCI	3,829	1,878	152	1,458	−41	1,482
INDITEX	2,729	1,515	486	1,043	77	957
LVMH	2,552	1,236	203	901	133	870
MUNICH RE	2,019	1,171	355	795	−21	812
LOREAL	2,447	1,107	196	966	55	828
PHILIPS ELECTR	2,152	1,174	264	952	−58	943
REPSOL	2,952	1,678	426	1,240	117	989
RWE ST	1,729	1,559	297	1,571	−10	1,158
BANCO SAN CENTRAL HISPANO	1,754	1,675	187	1,218	−46	1,091
SANOFI	1,258	1,045	93	978	−36	982
SAP AG	1,351	1,096	77	863	59	793
SAINT GOBAIN	1,800	1,414	154	1,053	73	1,021
SIEMENS AG	1,192	1,019	83	866	−45	810
SCHNEIDER ELECTRIC SA	1,668	1,297	310	1,118	92	937
TELEFONICA	2,768	1,317	269	934	27	970
UNICREDIT SPA	2,924	2,062	332	1,643	−256	1,459
UNILEVER CERT	1,385	878	89	654	−24	632
VIVENDI UNIVERSAL	2,259	1,363	129	1,073	65	1,046
VOLKSWAGEN	2,052	1,316	168	1,085	64	1,041

Table 23 The quality of the OLS prediction: the daily gain average and standard deviation for the different horizons (with 0.5 bp trading costs)

Stock	1-min horizon		5-min horizon		30-min horizon	
	$\overline{Gain}$	$\sigma(Gain)$	$\overline{Gain}$	$\sigma(Gain)$	$\overline{Gain}$	$\sigma(Gain)$
INTERBREW	−410	1,115	−247	1,027	−110	1,020
AIR LIQUIDE	−45	971	−118	766	−78	705
ALLIANZ	988	1,166	5	898	−137	933
ASML Holding NV	−165	1,177	−292	1,205	100	1,078
BASF AG	−581	1,102	−209	877	−48	877
BAYER AG	−414	1,042	−220	946	−88	852
BBVARGENTARIA	546	1,652	−34	1,508	−85	1,177
BAY MOT WERKE	−110	1,150	−229	978	−75	904

(continued)

Table 23 (continued)

Stock	1-min horizon		5-min horizon		30-min horizon	
	$\overline{Gain}$	$\sigma(Gain)$	$\overline{Gain}$	$\sigma(Gain)$	$\overline{Gain}$	$\sigma(Gain)$
DANONE	−163	965	−212	835	−179	790
BNP PARIBAS	−410	1,505	−81	1,383	−78	1,245
CARREFOUR	236	1,351	−139	1,206	14	1,240
CRH PLC IRLANDE	2,815	1,765	775	1,328	40	1,239
AXA	273	1,347	−109	1,334	−91	1,103
DAIMLER CHRYSLER	−399	1,279	−207	1,004	−63	1,138
DEUTSCHE BANK AG	−344	1,433	−224	1,302	48	1,106
VINCI	−14	1,113	−1	939	−24	736
DEUTSCHE TELEKOM	−7	1,287	−126	848	−30	782
ESSILOR INTERNATIONAL	80	1,126	−44	1,014	−90	909
ENEL	1,745	1,340	−56	1,099	−88	983
ENI	1,244	952	18	881	−45	740
E.ON AG	278	1,310	−169	1,243	−180	974
TOTAL	−474	998	−127	848	−78	778
GENERALI ASSIC	2,094	1,570	182	1,248	−12	995
SOCIETE GENERALE	−324	1,712	−257	1,615	99	1,448
GDF SUEZ	259	1,185	−173	1,071	−129	963
IBERDROLA I	714	1,462	118	1,107	27	1,046
ING	−414	1,425	−219	1,339	−20	1,158
INTESABCI	1,936	1,710	−210	1,456	−100	1,480
INDITEX	968	1,429	143	1,036	24	960
LVMH	692	1,069	−145	890	79	870
MUNICH RE	202	1,076	1	786	−78	812
LOREAL	581	966	−150	958	0	826
PHILIPS ELECTR	288	1,074	−91	946	−115	943
REPSOL	1,139	1,532	82	1,223	62	987
RWE ST	−86	1,509	−48	1,567	−68	1,158
BANCO SAN CENTRAL HISPANO	−40	1,585	−159	1,205	−101	1,090
SANOFI	−560	1,021	−253	978	−91	982
SAP AG	−456	1,008	−276	861	3	793
SAINT GOBAIN	−65	1,339	−194	1,051	16	1,019
SIEMENS AG	−595	1,003	−259	866	−98	810
SCHNEIDER ELECTRIC SA	−193	1,213	−42	1,108	38	934
TELEFONICA	952	1,172	−75	919	−31	969
UNICREDIT SPA	1,083	1,958	−32	1,639	−314	1,459
UNILEVER CERT	−491	795	−267	654	−81	634
VIVENDI UNIVERSAL	409	1,277	−220	1,073	9	1,043
VOLKSWAGEN	239	1,219	−186	1,076	7	1,039

Appendix 4: Ridge Method

Notice that:
For any $A \in \mathbb{R}^{p,p}$, $X \in \mathbb{R}^p$ such $||X||_2 \neq 0$:

$$\frac{||AX||_2}{||X||_2} \leq \max_{||Y||_2 \neq 0} \frac{||AY||_2}{||Y||_2} = ||A||_2$$
$$=> ||AX||_2 \leq ||A||_2 ||X||_2 \tag{1}$$

For any $A \in \mathbb{R}^{p,p}$, $B \in \mathbb{R}^{p,p}$:

$$||AB||_2 = \max_{||X||_2 \neq 0} \frac{||ABX||_2}{||X||_2} = \max_{||BX||_2 \neq 0} \frac{||ABX||_2}{||BX||_2} \frac{||BX||_2}{||X||_2}$$
$$\leq \max_{||Y||_2 \neq 0} \frac{||AY||_2}{||Y||_2} \max_{||X||_2 \neq 0} \frac{||BX||_2}{||X||_2} = ||A||_2 ||B||_2$$
$$=> ||AB||_2 \leq ||A||_2 ||B||_2 \tag{2}$$

Proof 1 Let A, B, X such that
$$AX = B \tag{3}$$

and
$$A(X + \delta X) = B + \delta B \tag{4}$$

From (3) and (4) $\delta X = A^{-1} \delta B$ and using (1)

$$||\delta X||_2 = ||A^{-1} \delta B||_2 \leq ||A^{-1}||_2 ||\delta B||_2 \tag{5}$$

From (3) $||B||_2 = ||AX||$ and using (1)

$$||B||_2 \leq ||A||_2 ||X||_2 \tag{6}$$

From (5) and (6), $||\delta X||_2 ||B||_2 \leq ||A^{-1}||_2 ||\delta B||_2 ||A||_2 ||X||_2$

Thus $\frac{||\delta X||_2}{||X||_2} \leq K(A) \frac{||\delta B||_2}{||B||_2}$

Proof 2 Let A, B, X such that $AX = B$ (3) and

$$(A + \delta A)(X + \delta X) = B \tag{7}$$

From (3) and (7), $\delta X = -A^{-1} \delta A (X + \delta X)$.

Using (1) and (2) follows $||\delta X||_2 \leq ||A^{-1}||_2 ||\delta A||_2 ||X + \delta X||_2$

Thus $\frac{||\delta X||_2}{||X+\delta X||_2} \leq K(A) \frac{||\delta A||_2}{||A||_2}$ (See Tables 24, 25, 26, 27, 28 and 29.)

Table 24 The quality of the Ridge HKB prediction: the AUC and the accuracy per stock for the different horizons

Stock	1-min horizon		5-min horizon		30-min horizon	
	AUC	Accuracy	AUC	Accuracy	AUC	Accuracy
INTERBREW	0.54	0.54	0.50	0.50	0.50	0.50
AIR LIQUIDE	0.57	0.57	0.52	0.52	0.50	0.50
ALLIANZ	0.61	0.61	0.53	0.53	0.49	0.49
ASML Holding NV	0.55	0.55	0.51	0.51	0.51	0.51
BASF AG	0.54	0.54	0.52	0.52	0.50	0.50
BAYER AG	0.54	0.54	0.51	0.51	0.50	0.50
BBVARGENTARIA	0.54	0.54	0.51	0.51	0.50	0.50
BAY MOT WERKE	0.55	0.55	0.51	0.51	0.50	0.50
DANONE	0.56	0.56	0.51	0.51	0.50	0.50
BNP PARIBAS	0.53	0.53	0.51	0.51	0.50	0.50
CARREFOUR	0.55	0.55	0.51	0.51	0.52	0.52
CRH PLC IRLANDE	0.62	0.62	0.56	0.56	0.52	0.52
AXA	0.56	0.55	0.51	0.51	0.50	0.50
DAIMLER CHRYSLER	0.54	0.54	0.51	0.51	0.50	0.50
DEUTSCHE BANK AG	0.53	0.53	0.51	0.51	0.51	0.51
VINCI	0.55	0.55	0.51	0.52	0.51	0.51
DEUTSCHE TELEKOM	0.56	0.56	0.52	0.52	0.51	0.52
ESSILOR INTERNATIONAL	0.56	0.56	0.51	0.51	0.51	0.51
ENEL	0.62	0.62	0.53	0.53	0.48	0.48
ENI	0.65	0.65	0.54	0.54	0.50	0.50
E.ON AG	0.57	0.57	0.52	0.52	0.48	0.48
TOTAL	0.54	0.54	0.51	0.51	0.50	0.50
GENERALI ASSIC	0.62	0.62	0.54	0.54	0.51	0.51
SOCIETE GENERALE	0.53	0.53	0.50	0.50	0.53	0.52
GDF SUEZ	0.57	0.57	0.52	0.52	0.50	0.50
IBERDROLA I	0.57	0.57	0.53	0.53	0.52	0.52
ING	0.53	0.53	0.51	0.50	0.50	0.50
INTESABCI	0.60	0.60	0.52	0.52	0.50	0.50
INDITEX	0.59	0.59	0.53	0.53	0.52	0.52
LVMH	0.59	0.59	0.52	0.52	0.50	0.50
MUNICH RE	0.59	0.59	0.53	0.53	0.50	0.50
LOREAL	0.60	0.60	0.52	0.52	0.51	0.51
PHILIPS ELECTR	0.56	0.56	0.51	0.51	0.49	0.49
REPSOL	0.58	0.58	0.52	0.52	0.52	0.52
RWE ST	0.54	0.54	0.51	0.51	0.50	0.50
BANCO SAN CENTRAL HISPANO	0.54	0.54	0.51	0.51	0.50	0.50
SANOFI	0.54	0.54	0.51	0.51	0.51	0.51

(continued)

Table 24 (continued)

Stock	1-min horizon		5-min horizon		30-min horizon	
	AUC	Accuracy	AUC	Accuracy	AUC	Accuracy
SAP AG	0.55	0.55	0.51	0.51	0.51	0.51
SAINT GOBAIN	0.54	0.54	0.51	0.51	0.52	0.52
SIEMENS AG	0.54	0.54	0.51	0.51	0.51	0.51
SCHNEIDER ELECTRIC SA	0.55	0.55	0.52	0.52	0.50	0.50
TELEFONICA	0.59	0.59	0.52	0.52	0.51	0.51
UNICREDIT SPA	0.57	0.57	0.51	0.51	0.49	0.49
UNILEVER CERT	0.56	0.56	0.51	0.51	0.49	0.49
VIVENDI UNIVERSAL	0.57	0.57	0.51	0.51	0.51	0.51
VOLKSWAGEN	0.57	0.57	0.52	0.52	0.51	0.51

Table 25 The quality of the Ridge HKB prediction: the daily gain average and standard deviation for the different horizons (without trading costs)

Stock	1-min horizon		5-min horizon		30-min horizon	
	$\overline{Gain}$	$\sigma(Gain)$	$\overline{Gain}$	$\sigma(Gain)$	$\overline{Gain}$	$\sigma(Gain)$
INTERBREW	1,476	1,155	108	973	−22	979
AIR LIQUIDE	1,793	976	264	805	6	703
ALLIANZ	2,884	1,343	399	958	−96	871
ASML Holding NV	1,817	1,233	141	1,155	189	1,037
BASF AG	1,244	1,138	218	942	−3	899
BAYER AG	1,475	1,096	125	921	−24	872
BBVARGENTARIA	2,429	1,763	288	1,321	−63	1,206
BAY MOT WERKE	1,784	1,208	143	1,006	29	978
DANONE	1,762	1,032	151	794	−36	840
BNP PARIBAS	1,548	1,554	258	1,403	18	1,201
CARREFOUR	2,159	1,485	239	1,285	62	1,277
CRH PLC IRLANDE	4,325	1,976	1,181	1,378	105	1,185
AXA	2,293	1,471	254	1,292	−48	1,131
DAIMLER CHRYSLER	1,439	1,354	94	1,052	20	1,083
DEUTSCHE BANK AG	1,469	1,497	127	1,305	43	1,055
VINCI	1,903	1,289	313	1,002	72	846
DEUTSCHE TELEKOM	1,826	1,420	203	846	70	767
ESSILOR INTERNATIONAL	2,002	1,268	238	1,010	50	915
ENEL	3,733	1,545	330	1,158	−60	996
ENI	3,158	1,196	413	852	−14	734
E.ON AG	2,253	1,392	249	1,222	−112	1,016
TOTAL	1,341	1,024	237	855	−25	767

(continued)

Table 25 (continued)

Stock	1-min horizon		5-min horizon		30-min horizon	
	$\overline{Gain}$	$\sigma(Gain)$	$\overline{Gain}$	$\sigma(Gain)$	$\overline{Gain}$	$\sigma(Gain)$
GENERALI ASSIC	4,025	1,839	576	1,257	1	1,020
SOCIETE GENERALE	1,521	1,793	131	1,617	202	1,504
GDF SUEZ	2,206	1,290	222	1,048	−59	921
IBERDROLA I	2,532	1,573	466	1,161	119	1,023
ING	1,487	1,473	174	1,298	77	1,177
INTESABCI	3,982	1,882	280	1,463	−82	1,498
INDITEX	2,816	1,492	566	1,042	114	958
LVMH	2,606	1,270	269	870	28	847
MUNICH RE	2,119	1,157	407	752	10	739
LOREAL	2,549	1,127	220	965	42	771
PHILIPS ELECTR	2,176	1,180	293	926	−70	939
REPSOL	3,016	1,648	502	1,227	207	946
RWE ST	1,812	1,551	336	1,486	91	1,146
BANCO SAN CENTRAL HISPANO	1,829	1,603	221	1,198	19	1,125
SANOFI	1,358	1,014	138	939	3	933
SAP AG	1,388	1,149	67	885	60	781
SAINT GOBAIN	1,899	1,433	242	954	87	979
SIEMENS AG	1,281	1,081	118	904	−55	951
SCHNEIDER ELECTRIC SA	1,744	1,315	387	1,166	−16	902
TELEFONICA	2,835	1,336	336	979	93	928
UNICREDIT SPA	3,060	2,056	317	1,605	−260	1,482
UNILEVER CERT	1,459	863	107	669	−63	637
VIVENDI UNIVERSAL	2,311	1,337	201	1,022	30	956
VOLKSWAGEN	2,171	1,341	227	1,205	21	1,085

Table 26 The quality of the Ridge HKB prediction: the daily gain average and standard deviation for the different horizons (with 0.5 bp trading costs)

Stock	1-min horizon		5-min horizon		30-min horizon	
	$\overline{Gain}$	$\sigma(Gain)$	$\overline{Gain}$	$\sigma(Gain)$	$\overline{Gain}$	$\sigma(Gain)$
INTERBREW	−334	1,110	−221	973	−76	976
AIR LIQUIDE	−12	911	−90	792	−47	701
ALLIANZ	1,042	1,167	60	945	−150	870
ASML Holding NV	−36	1,164	−191	1,148	135	1,036
BASF AG	−552	1,119	−126	935	−56	899
BAYER AG	−338	1,047	−214	916	−77	870
BBVARGENTARIA	683	1,653	−49	1,302	−119	1,206
BAY MOT WERKE	−71	1,104	−201	994	−21	976

(continued)

Table 26 (continued)

Stock	1-min horizon		5-min horizon		30-min horizon	
	$\overline{Gain}$	$\sigma(Gain)$	$\overline{Gain}$	$\sigma(Gain)$	$\overline{Gain}$	$\sigma(Gain)$
DANONE	−128	951	−194	787	−90	840
BNP PARIBAS	−217	1,463	−73	1,396	−39	1,199
CARREFOUR	290	1,373	−102	1,286	9	1274
CRH PLC IRLANDE	2,851	1,822	837	1,350	51	1,186
AXA	420	1,361	−82	1,288	−103	1,130
DAIMLER CHRYSLER	−345	1,302	−239	1,052	−35	1,080
DEUTSCHE BANK AG	−279	1,422	−205	1,304	−13	1,057
VINCI	80	1,213	−18	993	18	846
DEUTSCHE TELEKOM	45	1,335	−129	835	17	765
ESSILOR INTERNATIONAL	142	1,152	−92	986	−4	913
ENEL	1,841	1,355	−14	1,145	−110	996
ENI	1,306	964	71	846	−65	733
E.ON AG	406	1,284	−89	1,214	−164	1,014
TOTAL	−468	967	−101	850	−79	766
GENERALI ASSIC	2,174	1,648	230	1,245	−47	1,020
SOCIETE GENERALE	−286	1,711	−206	1,606	149	1,503
GDF SUEZ	337	1,199	−114	1,035	−112	919
IBERDROLA I	753	1,442	147	1,146	66	1,021
ING	−282	1,403	−169	1,285	21	1,175
INTESABCI	2,087	1,698	−78	1,450	−138	1,497
INDITEX	1,056	1,399	227	1,034	65	959
LVMH	736	1,110	−66	855	−24	847
MUNICH RE	299	1,049	60	743	−43	738
LOREAL	678	980	−117	953	−10	767
PHILIPS ELECTR	308	1,074	−54	920	−125	939
REPSOL	1,210	1,499	168	1,214	155	945
RWE ST	−6	1,492	−2	1,480	38	1,144
BANCO SAN CENTRAL HISPANO	48	1,500	−110	1,181	−36	1,124
SANOFI	−455	995	−194	936	−48	932
SAP AG	−421	1,055	−278	875	6	781
SAINT GOBAIN	27	1,339	−95	941	33	977
SIEMENS AG	−510	1,051	−210	906	−106	951
SCHNEIDER ELECTRIC SA	−112	1,214	46	1,149	−69	901
TELEFONICA	1,015	1,176	−1	961	39	925
UNICREDIT SPA	1,222	1,939	−40	1,596	−317	1,482
UNILEVER CERT	−409	772	−237	662	−118	638
VIVENDI UNIVERSAL	459	1,231	−139	1,012	−23	952
VOLKSWAGEN	366	1,240	−120	1,198	−35	1,084

Table 27 The quality of the Ridge LW prediction: the AUC and the accuracy per stock for the different horizons

	1-min horizon		5-min horizon		30-min horizon	
Stock	AUC	Accuracy	AUC	Accuracy	AUC	Accuracy
INTERBREW	0.55	0.55	0.52	0.52	0.50	0.50
AIR LIQUIDE	0.57	0.57	0.53	0.53	0.49	0.49
ALLIANZ	0.61	0.61	0.54	0.54	0.50	0.50
ASML Holding NV	0.56	0.56	0.52	0.52	0.52	0.52
BASF AG	0.54	0.54	0.52	0.52	0.50	0.50
BAYER AG	0.55	0.55	0.51	0.51	0.50	0.50
BBVARGENTARIA	0.54	0.54	0.51	0.51	0.50	0.50
BAY MOT WERKE	0.55	0.55	0.51	0.51	0.50	0.50
DANONE	0.56	0.56	0.51	0.51	0.49	0.49
BNP PARIBAS	0.54	0.54	0.52	0.52	0.50	0.50
CARREFOUR	0.55	0.55	0.51	0.51	0.51	0.51
CRH PLC IRLANDE	0.62	0.62	0.57	0.57	0.51	0.51
AXA	0.56	0.56	0.51	0.51	0.51	0.51
DAIMLER CHRYSLER	0.54	0.54	0.52	0.52	0.51	0.51
DEUTSCHE BANK AG	0.53	0.53	0.51	0.51	0.52	0.52
VINCI	0.56	0.56	0.52	0.53	0.51	0.52
DEUTSCHE TELEKOM	0.57	0.57	0.52	0.52	0.52	0.52
ESSILOR INTERNATIONAL	0.56	0.56	0.51	0.51	0.50	0.50
ENEL	0.63	0.63	0.54	0.54	0.50	0.50
ENI	0.65	0.65	0.55	0.55	0.50	0.50
E.ON AG	0.58	0.58	0.52	0.52	0.50	0.51
TOTAL	0.54	0.54	0.52	0.52	0.51	0.51
GENERALI ASSIC	0.62	0.62	0.55	0.55	0.49	0.49
SOCIETE GENERALE	0.53	0.53	0.50	0.50	0.52	0.52
GDF SUEZ	0.57	0.57	0.52	0.52	0.50	0.50
IBERDROLA I	0.57	0.57	0.53	0.53	0.52	0.52
ING	0.53	0.53	0.51	0.51	0.50	0.50
INTESABCI	0.60	0.60	0.53	0.53	0.48	0.48
INDITEX	0.60	0.60	0.54	0.54	0.51	0.51
LVMH	0.59	0.59	0.52	0.52	0.50	0.50
MUNICH RE	0.59	0.59	0.54	0.54	0.50	0.50
LOREAL	0.60	0.60	0.53	0.53	0.51	0.51
PHILIPS ELECTR	0.57	0.57	0.52	0.52	0.50	0.50
REPSOL	0.58	0.58	0.53	0.53	0.51	0.51
RWE ST	0.55	0.55	0.51	0.51	0.49	0.49
BANCO SAN CENTRAL HISPANO	0.54	0.54	0.52	0.52	0.51	0.51

(continued)

Table 27 (continued)

Stock	1-min horizon		5-min horizon		30-min horizon	
	AUC	Accuracy	AUC	Accuracy	AUC	Accuracy
SANOFI	0.55	0.55	0.51	0.51	0.50	0.50
SAP AG	0.55	0.55	0.51	0.51	0.51	0.51
SAINT GOBAIN	0.55	0.55	0.51	0.51	0.52	0.52
SIEMENS AG	0.55	0.55	0.52	0.52	0.51	0.52
SCHNEIDER ELECTRIC SA	0.55	0.55	0.52	0.52	0.50	0.50
TELEFONICA	0.60	0.60	0.53	0.53	0.51	0.51
UNICREDIT SPA	0.57	0.57	0.52	0.52	0.49	0.49
UNILEVER CERT	0.57	0.57	0.51	0.51	0.51	0.51
VIVENDI UNIVERSAL	0.58	0.58	0.52	0.52	0.51	0.51
VOLKSWAGEN	0.57	0.57	0.52	0.52	0.50	0.50

Table 28 The quality of the Ridge LW prediction: the daily gain average and standard deviation for the different horizons (without trading costs)

Stock	1-min horizon		5-min horizon		30-min horizon	
	$\overline{Gain}$	$\sigma(Gain)$	$\overline{Gain}$	$\sigma(Gain)$	$\overline{Gain}$	$\sigma(Gain)$
INTERBREW	1,651	1,145	288	882	−49	898
AIR LIQUIDE	1,848	1,048	337	809	−21	713
ALLIANZ	2,925	1,362	382	987	21	802
ASML Holding NV	1,963	1,221	253	1,023	194	1,096
BASF AG	1,401	1,177	177	1,022	−58	847
BAYER AG	1,621	1,109	157	869	13	804
BBVARGENTARIA	2,488	1,790	324	1,511	−87	1,149
BAY MOT WERKE	1,853	1,283	151	939	−46	912
DANONE	1,753	1,035	152	852	−123	790
BNP PARIBAS	1,544	1,683	242	1,417	11	1,311
CARREFOUR	2,334	1,468	244	1,331	19	1,254
CRH PLC IRLANDE	4,428	1,976	1,387	1,378	39	1,074
AXA	2,356	1,448	181	1,245	21	1,238
DAIMLER CHRYSLER	1,614	1,495	198	1,013	13	1,113
DEUTSCHE BANK AG	1,482	1,556	247	1,373	139	1,170
VINCI	1,958	1,301	386	1,143	108	851
DEUTSCHE TELEKOM	1,916	1,380	272	894	60	836
ESSILOR INTERNATIONAL	2,118	1,243	306	1,075	−22	912
ENEL	3,826	1,584	520	1,219	4	1,009
ENI	3,230	1,264	479	840	−66	770
E.ON AG	2,277	1,255	265	1,142	−55	1,023
TOTAL	1,428	1,079	224	798	75	787

(continued)

Table 28 (continued)

Stock	1-min horizon		5-min horizon		30-min horizon	
	$\overline{Gain}$	$\sigma(Gain)$	$\overline{Gain}$	$\sigma(Gain)$	$\overline{Gain}$	$\sigma(Gain)$
GENERALI ASSIC	4,044	1,813	686	1,323	−94	1,054
SOCIETE GENERALE	1,487	1,932	93	1,550	118	1,412
GDF SUEZ	2,307	1,296	282	976	−32	1,004
IBERDROLA I	2,721	1,542	522	1,194	107	969
ING	1,565	1,569	127	1,383	11	1,251
INTESABCI	4,060	1,876	504	1,504	−218	1,541
INDITEX	2,928	1,509	640	1,119	52	937
LVMH	2,700	1,264	363	894	−39	842
MUNICH RE	2,225	1,220	407	857	7	829
LOREAL	2,606	1,100	354	919	−7	827
PHILIPS ELECTR	2,284	1,254	320	962	4	889
REPSOL	3,053	1,694	537	1173	67	968
RWE ST	1,988	1,641	258	1,637	27	1,180
BANCO SAN CENTRAL HISPANO	1,981	1,535	415	1,326	83	1,282
SANOFI	1,466	1,047	101	967	15	926
SAP AG	1,522	1,207	89	841	−13	809
SAINT GOBAIN	2,060	1,473	194	1,088	64	979
SIEMENS AG	1,425	1,132	192	790	13	993
SCHNEIDER ELECTRIC SA	1,841	1,359	281	1,093	−26	907
TELEFONICA	2,859	1,269	458	1,000	111	932
UNICREDIT SPA	3,178	2,159	411	1,603	−98	1,563
UNILEVER CERT	1,539	841	158	681	−0	630
VIVENDI UNIVERSAL	2,486	1,370	295	999	−4	1,021
VOLKSWAGEN	2,280	1,432	288	1,047	−90	992

Table 29 The quality of the Ridge LW prediction: the daily gain average and standard deviation for the different horizons (with 0.5 bp trading costs)

Stock	1-min horizon		5-min horizon		30-min horizon	
	$\overline{Gain}$	$\sigma(Gain)$	$\overline{Gain}$	$\sigma(Gain)$	$\overline{Gain}$	$\sigma(Gain)$
INTERBREW	−169	1,072	−30	866	−98	895
AIR LIQUIDE	59	961	8	790	−69	709
ALLIANZ	1,091	1,178	49	978	−28	800
ASML Holding NV	100	1,133	−66	1,010	145	1,093
BASF AG	−358	1,098	−140	1,011	−108	844
BAYER AG	−182	1,069	−154	853	−33	802
BBVARGENTARIA	706	1,682	9	1,493	−132	1,148
BAY MOT WERKE	−1	1,172	−167	923	−92	911

(continued)

Table 29 (continued)

Stock	1-min horizon		5-min horizon		30-min horizon	
	$\overline{Gain}$	$\sigma(Gain)$	$\overline{Gain}$	$\sigma(Gain)$	$\overline{Gain}$	$\sigma(Gain)$
DANONE	−153	956	−167	840	−170	789
BNP PARIBAS	−205	1,577	−69	1,400	−37	1,309
CARREFOUR	459	1,334	−70	1,325	−27	1,252
CRH PLC IRLANDE	2,967	1,810	1,053	1,341	−5	1,073
AXA	451	1,325	−126	1,239	−27	1,235
DAIMLER CHRYSLER	−136	1,410	−113	1,014	−34	1,112
DEUTSCHE BANK AG	−267	1,454	−57	1,362	89	1,169
VINCI	96	1,218	73	1,128	63	848
DEUTSCHE TELEKOM	140	1,294	−34	872	16	833
ESSILOR INTERNATIONAL	223	1,106	−10	1,059	−71	909
ENEL	1,925	1,384	194	1,200	−41	1,003
ENI	1,380	1,036	152	824	−109	768
E.ON AG	430	1,137	−46	1,127	−99	1,022
TOTAL	−375	1,027	−82	787	28	785
GENERALI ASSIC	2,174	1,611	349	1,292	−140	1,052
SOCIETE GENERALE	−298	1,845	−213	1,537	70	1,408
GDF SUEZ	421	1,175	−31	957	−78	1,000
IBERDROLA I	925	1,393	224	1,178	61	968
ING	−201	1,468	−193	1,369	−38	1,248
INTESABCI	2,152	1,690	157	1,483	−265	1,541
INDITEX	1,138	1,403	312	1,105	6	938
LVMH	811	1,108	49	871	−86	840
MUNICH RE	399	1,097	75	849	−37	829
LOREAL	717	926	28	902	−52	825
PHILIPS ELECTR	391	1,145	−10	951	−45	885
REPSOL	1,245	1,557	218	1,152	19	968
RWE ST	188	1,554	−45	1,634	−19	1,177
BANCO SAN CENTRAL HISPANO	198	1,429	95	1,300	36	1,279
SANOFI	−336	1,005	−197	954	−31	923
SAP AG	−281	1,103	−234	836	−62	810
SAINT GOBAIN	137	1,373	−111	1,077	14	975
SIEMENS AG	−382	1,087	−112	789	−33	992
SCHNEIDER ELECTRIC SA	−9	1,250	−24	1,076	−74	903
TELEFONICA	1,021	1,123	128	983	64	928
UNICREDIT SPA	1,361	2,027	82	1,587	−150	1,561
UNILEVER CERT	−340	751	−166	666	−49	628
VIVENDI UNIVERSAL	593	1,245	−18	978	−54	1,021
VOLKSWAGEN	462	1,328	−36	1,039	−136	991

Appendix 5: LASSO Method

See Tables 30, 31 and 32.

Table 30 The quality of the LASSO prediction: the AUC and the accuracy per stock for the different horizons

	1-min horizon		5-min horizon		30-min horizon	
Stock	AUC	Accuracy	AUC	Accuracy	AUC	Accuracy
INTERBREW	0.54	0.54	0.51	0.51	0.50	0.50
AIR LIQUIDE	0.58	0.58	0.52	0.52	0.49	0.49
ALLIANZ	0.61	0.61	0.54	0.54	0.52	0.52
ASML Holding NV	0.56	0.56	0.52	0.52	0.51	0.51
BASF AG	0.53	0.53	0.51	0.51	0.51	0.51
BAYER AG	0.54	0.54	0.51	0.51	0.50	0.50
BBVARGENTARIA	0.54	0.54	0.51	0.51	0.49	0.49
BAY MOT WERKE	0.55	0.55	0.51	0.51	0.49	0.49
DANONE	0.56	0.56	0.51	0.51	0.50	0.50
BNP PARIBAS	0.54	0.54	0.51	0.51	0.49	0.49
CARREFOUR	0.55	0.55	0.51	0.51	0.50	0.50
CRH PLC IRLANDE	0.62	0.62	0.56	0.56	0.52	0.52
AXA	0.55	0.55	0.51	0.51	0.49	0.49
DAIMLER CHRYSLER	0.53	0.53	0.52	0.52	0.50	0.50
DEUTSCHE BANK AG	0.53	0.53	0.51	0.51	0.51	0.51
VINCI	0.55	0.55	0.52	0.53	0.52	0.52
DEUTSCHE TELEKOM	0.58	0.58	0.52	0.52	0.52	0.52
ESSILOR INTERNATIONAL	0.56	0.56	0.51	0.51	0.50	0.50
ENEL	0.62	0.62	0.53	0.53	0.50	0.50
ENI	0.64	0.64	0.55	0.55	0.49	0.49
E.ON AG	0.57	0.57	0.52	0.52	0.49	0.50
TOTAL	0.54	0.54	0.52	0.52	0.51	0.51
GENERALI ASSIC	0.62	0.62	0.54	0.54	0.51	0.51
SOCIETE GENERALE	0.53	0.53	0.50	0.50	0.52	0.52
GDF SUEZ	0.56	0.56	0.52	0.52	0.51	0.51
IBERDROLA I	0.56	0.56	0.53	0.53	0.53	0.53
ING	0.52	0.52	0.51	0.51	0.50	0.50
INTESABCI	0.60	0.60	0.53	0.53	0.50	0.50

(continued)

Table 30 (continued)

Stock	1-min horizon		5-min horizon		30-min horizon	
	AUC	Accuracy	AUC	Accuracy	AUC	Accuracy
INDITEX	0.59	0.59	0.53	0.53	0.52	0.52
LVMH	0.59	0.59	0.52	0.52	0.51	0.51
MUNICH RE	0.58	0.58	0.54	0.54	0.50	0.50
LOREAL	0.60	0.60	0.53	0.53	0.50	0.50
PHILIPS ELECTR	0.56	0.56	0.52	0.52	0.50	0.50
REPSOL	0.57	0.57	0.52	0.52	0.51	0.51
RWE ST	0.54	0.54	0.51	0.51	0.50	0.50
BANCO SAN CENTRAL HISPANO	0.54	0.54	0.52	0.52	0.50	0.50
SANOFI	0.54	0.54	0.51	0.51	0.50	0.50
SAP AG	0.53	0.53	0.52	0.52	0.50	0.50
SAINT GOBAIN	0.54	0.54	0.51	0.51	0.52	0.52
SIEMENS AG	0.54	0.54	0.51	0.51	0.50	0.50
SCHNEIDER ELECTRIC SA	0.54	0.54	0.51	0.51	0.49	0.49
TELEFONICA	0.59	0.59	0.53	0.53	0.51	0.51
UNICREDIT SPA	0.57	0.57	0.52	0.52	0.48	0.48
UNILEVER CERT	0.57	0.57	0.51	0.51	0.51	0.51
VIVENDI UNIVERSAL	0.57	0.57	0.52	0.52	0.52	0.52
VOLKSWAGEN	0.56	0.56	0.52	0.52	0.49	0.49

Table 31 The quality of the LASSO prediction: the daily gain average and standard deviation for the different horizons (without trading costs)

Stock	1-min horizon		5-min horizon		30-min horizon	
	$\overline{Gain}$	$\sigma(Gain)$	$\overline{Gain}$	$\sigma(Gain)$	$\overline{Gain}$	$\sigma(Gain)$
INTERBREW	621	1,094	246	916	−9	991
AIR LIQUIDE	952	1,241	294	822	21	665
ALLIANZ	2,758	1,232	368	959	13	846
ASML Holding NV	1,509	1,514	232	1,007	100	1,003
BASF AG	294	771	74	979	−1	870
BAYER AG	563	1,116	157	864	−61	829
BBVARGENTARIA	1,780	1,847	397	1,307	−74	1,107
BAY MOT WERKE	1,158	1,429	127	989	−67	921
DANONE	926	1,167	132	942	−51	763
BNP PARIBAS	867	1,664	199	1,462	−21	1,171

(continued)

Table 31 (continued)

Stock	1-min horizon		5-min horizon		30-min horizon	
	$\overline{Gain}$	$\sigma(Gain)$	$\overline{Gain}$	$\sigma(Gain)$	$\overline{Gain}$	$\sigma(Gain)$
CARREFOUR	1,738	1,663	110	1,299	63	1,189
CRH PLC IRLANDE	4,301	1,951	1,293	1,421	85	1,142
AXA	1,861	1,610	98	1,261	−37	1,217
DAIMLER CHRYSLER	610	1,399	199	1,045	36	999
DEUTSCHE BANK AG	657	1,311	190	1,288	112	1,125
VINCI	693	1,172	377	1,088	114	860
DEUTSCHE TELEKOM	943	1,448	176	922	72	790
ESSILOR INTERNATIONAL	1,419	1,477	223	1,043	35	969
ENEL	3,631	1,654	365	1,198	14	1,105
ENI	3,010	1,190	491	867	−78	787
E.ON AG	2,009	1,316	327	1,129	−91	1,017
TOTAL	304	778	234	767	72	727
GENERALI ASSIC	3,923	1,787	642	1,283	65	1,201
SOCIETE GENERALE	783	1,859	92	1,515	78	1,478
GDF SUEZ	1,821	1,340	280	1,046	34	1,001
IBERDROLA I	2,340	1,640	495	1,170	139	1,002
ING	819	1,545	129	1,427	23	1,189
INTESABCI	3,966	1,850	488	1,435	−151	1,525
INDITEX	2,359	1,670	559	1,160	122	935
LVMH	2,490	1,255	321	925	71	836
MUNICH RE	1,657	1,341	421	853	37	792
LOREAL	2,320	1,089	326	925	6	857
PHILIPS ELECTR	1,640	1,298	304	1,008	−8	941
REPSOL	2,770	1,671	489	1,225	83	1,000
RWE ST	989	1,515	161	1,335	77	1,158
BANCO SAN CENTRAL HISPANO	1,229	1,658	368	1,269	−46	1,099
SANOFI	513	960	206	942	−40	900
SAP AG	313	801	130	831	25	809
SAINT GOBAIN	1,059	1,544	195	1,153	37	973
SIEMENS AG	334	941	99	733	−101	1,010
SCHNEIDER ELECTRIC SA	674	1,201	222	1,051	21	930
TELEFONICA	2,647	1,293	386	1,031	109	893
UNICREDIT SPA	2,859	2,055	331	1,540	−243	1,499
UNILEVER CERT	344	713	126	724	2	677
VIVENDI UNIVERSAL	1,991	1,341	284	1,015	75	1,048
VOLKSWAGEN	1,709	1,480	232	1,158	−32	953

Table 32 The quality of the LASSO prediction: the daily gain average and standard deviation for the different horizons (with 0.5 bp trading costs)

Stock	1-min horizon		5-min horizon		30-min horizon	
	$\overline{Gain}$	$\sigma(Gain)$	$\overline{Gain}$	$\sigma(Gain)$	$\overline{Gain}$	$\sigma(Gain)$
INTERBREW	−99	839	−48	905	−60	988
AIR LIQUIDE	181	857	3	792	−30	662
ALLIANZ	1,136	1,028	55	942	−38	843
ASML Holding NV	354	1,100	−57	992	48	1,002
BASF AG	−114	614	−215	979	−55	869
BAYER AG	−65	806	−135	849	−110	829
BBVARGENTARIA	368	1,553	90	1,282	−128	1,105
BAY MOT WERKE	73	1,012	−158	973	−117	920
DANONE	27	809	−153	933	−102	762
BNP PARIBAS	−32	1,412	−100	1,443	−74	1,169
CARREFOUR	322	1,377	−180	1,297	13	1,188
CRH PLC IRLANDE	2,965	1,760	967	1,381	34	1,142
AXA	371	1,367	−201	1,256	−91	1,217
DAIMLER CHRYSLER	−291	1,277	−95	1,038	−17	997
DEUTSCHE BANK AG	−240	1,166	v83	1,276	58	1,125
VINCI	25	792	88	1,066	62	857
DEUTSCHE TELEKOM	292	1,011	−108	891	22	788
ESSILOR INTERNATIONAL	94	1,257	−69	1,029	−16	966
ENEL	2,051	1,450	52	1,182	−37	1,102
ENI	1,507	965	169	851	−127	786
E.ON AG	429	1,172	12	1,112	−144	1,015
TOTAL	−77	548	−32	745	20	724
GENERALI ASSIC	2278	1,562	316	1,250	15	1,200
SOCIETE GENERALE	−197	1,650	−198	1,503	25	1,476
GDF SUEZ	336	1,124	−18	1,033	−17	997
IBERDROLA I	856	1,425	216	1,147	88	1,001
ING	−165	1,346	−175	1,408	−31	1,186
INTESABCI	2,267	1,662	153	1,407	−206	1,523
INDITEX	1,030	1,435	246	1,148	72	935
LVMH	845	1,120	36	900	22	833
MUNICH RE	312	1,119	107	838	−13	790
LOREAL	751	977	20	917	−47	855
PHILIPS ELECTR	186	1,079	−7	999	−60	938
REPSOL	1,198	1,502	192	1,195	31	1,001
RWE ST	−96	1,313	−133	1,344	26	1,156
BANCO SAN CENTRAL HISPANO	87	1,351	62	1,244	−95	1,098
SANOFI	−63	620	−77	932	−88	898

(continued)

Table 32 (continued)

	1-min horizon		5-min horizon		30-min horizon	
Stock	$\overline{Gain}$	$\sigma(Gain)$	$\overline{Gain}$	$\sigma(Gain)$	$\overline{Gain}$	$\sigma(Gain)$
SAP AG	−163	672	−178	822	−28	808
SAINT GOBAIN	27	1,234	−90	1,135	−17	969
SIEMENS AG	−135	839	−175	735	−150	1,011
SCHNEIDER ELECTRIC SA	−122	957	−71	1,029	−30	926
TELEFONICA	1,073	1,163	71	1,023	57	889
UNICREDIT SPA	1,306	1,911	13	1,527	−298	1,499
UNILEVER CERT	−29	328	−148	714	−50	675
VIVENDI UNIVERSAL	459	1,143	6	997	20	1046
VOLKSWAGEN	294	1,265	−83	1,147	−84	952

References

1. L. Bachelier, Theorie de la speculation (1900)
2. E.F. Fama, Efficient capital markets: a review of theory and empirical work. J. Financ. (1969)
3. B.G. Malkiel, *A Random Walk Down Wall Street* (1973)
4. N. Taleb, *Fooled by Randomness the Hidden Role of Chance in the Markets and in Life* (2001)
5. W. Buffett, The superinvestors of Graham-and-Doddsville (1984)
6. A. Chakraborti, I.M. Toke, M. Patriarca, F. Abergel, Econophysics: empirical facts and agent-based models (2010)
7. F. Lillo, J.D. Farmer, The long memory of the efficient market (2008)
8. B. Zheng, F. Abergel, Price jump prediction in limit order book (2012)
9. J. Murphy, *Technical Analysis of the Financial Markets: A Comprehensive Guide to Trading Methods and Applications*
10. G. Vidyamurthy, *Pairs Trading: Quantitative Methods and Analysis* (2004)
11. T. Hastie, R. Tibshirani, J. Friedman, *The Elements of Statistical Learning* (Springer, 2011)
12. G.A.F. Seber, A.J. Lee, *Linear Regression Analysis* (Wiley, 2003)
13. J.D. Riley, Solving systems of linear equations with a positive definite, symmetric, but possibly ill-conditioned matrix (1955)
14. W. Cheney, D. Kincaid, *Numerical Mathematics and Computing* (2008)
15. A.E. Hoerl, R.W. Kennard, Ridge regression: biased estimation for nonorthogonal problems (1970)
16. A.E. Hoerl, R.W. Kennard, K.F. Baldwin, Ridge regression: some simulations (1975)
17. J.F. Lawless, P. Wang, A simulation study of Ridge and other regression estimators (1976)
18. R. Tibshirani, *Regression Shrinkage and Selection via the Lasso* (1996)
19. H. Zou, T. Hastie, Regularization and variable selection via the elastic net (2004)

Calibration of a Stock's Beta Using Option Prices

Sofiene El Aoud and Frédéric Abergel

Abstract In this paper, we present a continuous time Capital Asset Pricing Model where the volatilities of the market index and the stock are both stochastic. Using a singular perturbation technique, we provide approximations for the prices of European options on both the stock and the index. We derive then an estimator of the parameter beta under the risk-neutral pricing measure $\mathbb{P}^*$ using option and underlying prices. Following that, we study empirically the discrepancy between the implied value of the parameter β under $\mathbb{P}^*$ and its realized value under the real-world probability measure $\mathbb{P}$. Finally, we show that the parameter β is crucial for the hedging of stock options using instruments on the index, and we study numerically the performance of the proposed hedging strategies.

1 Introduction

The capital asset pricing model is an econometric model that provides an estimation of an asset's return in function of the systematic risk or the market risk. This model is an expansion of an earlier work of Markowitz on portfolio construction (see [1]). The CAPM model was considered to be an original and innovative model because it introduced he concept of systematic and specific risk. The parameter β, which is a key parameter in this model, enables to separate the stock risk into two parts: the first part represents the systematic risk contained in the market index, while the second part is the idiosyncratic risk that reflects the specific performance of the stock. The parameter beta in the capital asset pricing model is very useful for portfolio construction purposes (see [2–4]), thus its estimation is a matter of interest and has been a subject of study for several authors in the last decades. This parameter

S.E. Aoud (✉) · F. Abergel (✉)
Laboratoire de Mathématiques Appliquées aux Systèmes, Ecole Centrale Paris,
Grande Voie des Vignes, 92290 Châtenay Malabry, France
e-mail: sofiene.elaoud@ecp.fr

F. Abergel
e-mail: frederic.abergel@ecp.fr

© Springer International Publishing Switzerland 2015   67
F. Abergel et al. (eds.), *Econophysics and Data Driven Modelling of Market Dynamics*,
New Economic Windows, DOI 10.1007/978-3-319-08473-2_2

was traditionally estimated using historical data of daily returns of the stock and the market index where it is obtained as the slope of the linear regression of stock returns on market index returns (see [5]). This approach is backward-looking as it estimates the realized value of the parameter in the past using historical data. The authors in [6, 7] showed that the parameter β is not constant but rather time-varying. Thus, the value of the realized beta in the future can be remarkably different from its value in the past, and the backward-looking estimation may be inefficient.

In the recent literature, different authors have focused on the estimation of the beta coefficient using option data, which provides a different estimation method for this parameter. Indeed, whereas classical methods allow an historical estimation of this parameter, the option based estimation method enables to obtain a "forward looking" measure of this parameter. Thus, the obtained estimator represents the information contained in derivatives prices and then summarizes the expectation of market participants for the forward realization of this parameter.

In [8], Christoffersen, Jacobs and Vainberg provided an estimation of this parameter using the risk-neutral variance and skewness of the stock and the index. More recently, Fouque and Kollman proposed in [9] a continuous-time CAPM model in which the market index has a stochastic volatility driven by a fast mean-reverting process. Using a singular perturbation method, they managed to obtain an approximation of the beta parameter depending on the skews of implied volatilities of both the stock and the index. Fouque and Tashman introduced in [10] a "Stressed-Beta model" in which the parameter β can take two values depending on the market regime. Using this model, Fouque et al. provided a method to price options on the index and the stock. This method enables also to estimate the parameter β based on options data. In [11], Carr and Madan used the CAPM model to price options on the stock when options on the index are liquid. Their approach didn't aim to estimate the parameter beta using option prices, but to price options on the stock given the parameter beta and options prices on the market index.

This work deals with the estimation of the coefficient beta under the risk-neutral measure using options prices and highlights the utility of the obtained estimator for diverse applications. The paper is organized as follows. In the second section, we present the capital asset pricing model with constant idiosyncratic volatility for the stock. We recall briefly the method presented in [9], which allows to estimate the parameter beta using implied volatility data. In the third section, we consider a new model in which the stock's idiosyncratic volatility is stochastic. This choice was motivated by several empirical studies which confirm the random character of the idiosyncratic volatility process, thus the new model is more adapted to describe the joint dynamics of the stock and the market index. In the setting of the new model, we provide approximations for European option prices on both the stock and the index through the use of a singular perturbation technique. Afterward, we deduce an estimator of the parameter beta using option and underlying prices, and we call it the implied beta. In the fourth section, we present some possible applications which emphasize the utility of the beta estimator. First, we investigate the capacity of the implied beta to predict the forward realized beta under the historic probability measure. Then, in a second application, we show that the parameter beta can be used for the purpose of hedging stock options using instruments on the index. We run Monte-Carlo simulations to support the theoretical study and test numerically the hedging methods.

2 Model with Constant Idiosyncratic Volatility

2.1 Presentation of the Model

Consider a financial market living on a stochastic basis $(\Omega, \mathscr{F}, \mathbb{F}, \mathbb{P})$, where the filtration $\mathbb{F} = \{\mathscr{F}_t\}$ satisfies the usual conditions, and where $\mathbb{P}$ is the objective probability measure. The authors in [9] proposed a continuous time capital asset pricing model where the market index has a stochastic volatility. Under the probability measure P, the dynamics of the stock S and the index I are described as follows:

$$\frac{dI_t}{I_t} = \mu_I dt + f(Y_t) dW_t^{(1)},$$

$$\frac{dS_t}{S_t} = \mu_S dt + \beta \frac{dI_t}{I_t} + \sigma dW_t^{(2)},$$

$$dY_t = \frac{1}{\varepsilon}(m - Y_t)dt + \frac{\nu\sqrt{2}}{\sqrt{\varepsilon}}dW_t^{(3)},$$

where $W_t^{(3)} = \rho W_t^{(1)} + \sqrt{1-\rho^2} W_t^{(4)}$ and $W = \begin{pmatrix} W^{(1)} \\ W^{(2)} \\ W^{(4)} \end{pmatrix}$ is a Wiener process under $\mathbb{P}$

The authors made the assumption that $0 < \varepsilon \ll 1$ which implies that the process (Y), which drives the index volatility, is a fast mean-reverting Ornstein-Uhlenbeck process.

Let $\lambda_t = \begin{pmatrix} \frac{\mu_I - r}{f(Y_t)} \\ \frac{\mu_S + r(\beta - 1)}{\sigma} \\ \gamma(Y_t) \end{pmatrix}$ where the function γ denotes the volatility risk-premium.

The probability measure $\mathbb{P}^*$, equivalent to $\mathbb{P}$, is defined in the following way:

$$\frac{dP^*}{dP}\bigg|_{\mathscr{F}_t} = exp(-\int_0^t \lambda_u' dW_u - \frac{1}{2}\int_0^t |\lambda_u|^2 du),$$

Let $W^* = \begin{pmatrix} W^{*,(1)} \\ W^{*,(2)} \\ W^{*,(4)} \end{pmatrix}$ such that: $W_t^* = W_t + \int_0^t \lambda_u du$. Using Girsanov's theorem, it follows that W^* is a $(\mathbb{P}^*, \{\mathscr{F}_t\})$ Brownian motion.

Under continuity and boundedness conditions on the function γ, $\mathbb{P}^*$ is a risk-neutral probability measure under which the index and the stock have the following dynamics:

$$\frac{dI_t}{I_t} = rdt + f(Y_t)dW_t^{*,(1)},$$

$$\frac{dS_t}{S_t} = rdt + \beta f(Y_t)dW_t^{*,(1)} + \sigma dW_t^{*,(2)},$$

$$dY_t = (\frac{1}{\varepsilon}(m - Y_t) - \frac{v\sqrt{2}}{\sqrt{\varepsilon}}\chi(Y_t))dt + \frac{v\sqrt{2}}{\sqrt{\varepsilon}}dW_t^{*,(3)},$$

where $\chi(Y_t) = \rho\frac{\mu_I - r}{f(Y_t)} + \sqrt{1 - \rho^2}\gamma(Y_t)$ and $W_t^{*,(3)} = \rho W_t^{*,(1)} + \sqrt{1 - \rho^2}W_t^{*,(4)}$.

2.2 Calibration of Implied Beta

Using a singular perturbation method with respect to the small parameter ε, the authors in [9] obtained an approximation $\tilde{P}^{I,\varepsilon}(K_I, T)$ for the price of an European call on the index with strike K_I and maturity T, and an approximation $\tilde{P}^{S,\varepsilon}(K_S, T)$ for the price of an European call on the stock with strike K_S and maturity T. Afterward, through the use of a Taylor expansion in $\sqrt{\varepsilon}$ for the implied volatility of the stock and the index, they provided an approximation for the implied volatilities $\Sigma_I(K_I, T)$ and $\Sigma_S(K_S, T)$ of the index and the stock respectively:

$$\Sigma_I(K_I, T) = b_I + a_I\frac{\log(\frac{F_I}{K_I})}{T},$$

$$\Sigma_S(K_S, T) = b_S + a_S\frac{\log(\frac{F_S}{K_S})}{T}.$$

It should be precised that he quantities F_I and F_S denote the forward prices for maturity T of the index and the stock respectively, while the quantities b_I, a_I, b_S, a_S are functions of the model parameters. Thus, the parameter β can be approximated by $\hat{\beta}$ which is defined as :

$$\hat{\beta} = (\frac{a_S}{a_I})^{\frac{1}{3}}\frac{b_S}{b_I}. \tag{1}$$

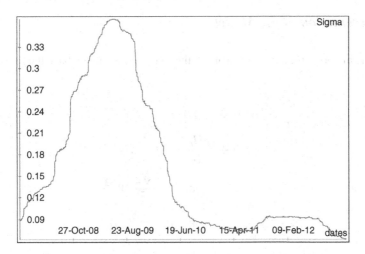

Fig. 1 Evolution of the idiosyncratic volatility of XLF with respect to the SPX index

2.3 Limits of the Model

In the model described so far, the stock's idiosyncratic volatility is supposed to be constant and equal to σ. This hypothesis can be considered too strong, indeed the authors in [12–15] conducted empirical studies on the idiosyncratic volatility and gave empirical evidence of the randomness of this process.

In order to have an idea about the magnitude of fluctuations of the idiosyncratic volatility, the graph of the parameter σ of the Financial Select Sector (named XLF) when projected on the SPX index is given below. The considered period ranges from 01/01/2008 to 31/12/2012. The parameter σ is obtained through the computation of the standard deviation of errors in the linear regression of the daily returns of the Financial Select Sector (XLF) on the daily returns of the SPX index using a sliding window of one month (Fig. 1).

The inspection of the graph above shows that the assumption of constant idiosyncratic volatility can be strong, and thus can induce a misleading understanding of the joint dynamics of the stock and the index. Consequently, a new model will be introduced in the next section in order to account for this characteristic.

3 Model with Stochastic Idiosyncratic Volatility

A new continuous-time capital asset pricing model is presented here. In this model setting, the stock's idiosyncratic volatility is driven by a fast mean-reverting Ornstein-Uhlenbeck process. The aim of this section is to derive approximations for European option prices on both the stock and the index, and to provide an estimator of the parameter β using option prices.

3.1 Presentation of the Model

Under the historic probability measure $\mathbb{P}$, the stock and the index have the following dynamics:

$$\frac{dI_t}{I_t} = \mu_I dt + f_1(Y_t)dW_t^{(1)},$$

$$\frac{dS_t}{S_t} = \mu_S dt + \beta\frac{dI_t}{I_t} + f_2(Z_t)dW_t^{(2)},$$

$$dY_t = \frac{1}{\varepsilon}(m_Y - Y_t)dt + \frac{v_Y\sqrt{2}}{\sqrt{\varepsilon}}dW_t^{(3)},$$

$$dZ_t = \frac{\alpha}{\varepsilon}(m_Z - Z_t)dt + \frac{v_Z\sqrt{2\alpha}}{\sqrt{\varepsilon}}dW_t^{(4)},$$

where $W_t^{(3)} = \rho_Y W_t^{(1)} + \sqrt{1 - \rho_Y^2}W_t^{(5)}$, $W_t^{(4)} = \rho_Z W_t^{(2)} + \sqrt{1 - \rho_Z^2}W_t^{(6)}$ and

$W = \begin{pmatrix} W^{(1)} \\ W^{(2)} \\ W^{(5)} \\ W^{(6)} \end{pmatrix}$ is a $(\mathbb{P}, \{\mathscr{F}_t\})$ Wiener process. It is supposed that $0 < \varepsilon \ll 1$

and $0 < \frac{\varepsilon}{\alpha} \ll 1$. This hypothesis implies that the processes (Y) and (Z) are fast mean-reverting Ornstein-Uhlenbeck processes.

Let the process λ be defined as $\lambda_t = \begin{pmatrix} \frac{\mu_I - r}{f_1(Y_t)} \\ \frac{\mu_S + r(\beta - 1)}{f_2(Z_t)} \\ \gamma_1(Y_t) \\ \gamma_2(Z_t) \end{pmatrix}$ where the functions γ_1 and

γ_2 denote the volatility risk premiums related to the processes Y and Z respectively. The probability measure P^*, equivalent to P, can then be defined as follows:

$$\frac{dP^*}{dP} = exp(-\int_0^t \lambda_u' dW_u - \frac{1}{2}\int_0^t |\lambda_u|^2 du).$$

Let $W_t^* = W_t + \int_0^t \lambda_u du$. It can be deduced, through the use of Girsanov's theorem, that W^* is a $(\mathbb{P}^*, \{\mathscr{F}_t\})$ Brownian motion. Thus, the dynamics of the stock and the index, and under $\mathbb{P}^*$, can be described in the following way:

$$\frac{dI_t}{I_t} = rdt + f_1(Y_t)dW_t^{*,(1)},$$

$$\frac{dS_t}{S_t} = rdt + \beta f_1(Y_t)dW_t^{*,(1)} + f_2(Z_t)dW_t^{*,(2)},$$

$$dY_t = \frac{1}{\varepsilon}(m_Y - Y_t)dt - \frac{v_Y\sqrt{2}}{\sqrt{\varepsilon}}\chi_1(Y_t)dt + \frac{v_Y\sqrt{2}}{\sqrt{\varepsilon}}dW_t^{*,(3)},$$

$$dZ_t = \frac{\alpha}{\varepsilon}(m_Z - Z_t)dt - \frac{v_Z\sqrt{2\alpha}}{\sqrt{\varepsilon}}\chi_2(Z_t)dt + \frac{v_Z\sqrt{2\alpha}}{\sqrt{\varepsilon}}dW_t^{*,(4)},$$

where: $\chi_1(Y_t) = \rho_Y\frac{\mu_I-r}{f_1(Y_t)} + \sqrt{1 - \rho_Y^2}\gamma_1(Y_t)$ and $\chi_2(Z_t) = \rho_Z\frac{\mu_S+r(\beta-1)}{f_2(Z_t)} + \sqrt{1 - \rho_Z^2}\gamma_2(Z_t)$.

The processes $W^{*,(3)}$ and $W^{*,(4)}$ are Brownian motions under $\mathbb{P}^*$ such that:

$$W^{*,(3)} = \rho_Y W^{*,(1)} + \sqrt{1 - \rho_Y^2}W^{*,(5)},$$

$$W^{*,(4)} = \rho_Z W^{*,(2)} + \sqrt{1 - \rho_Z^2}W^{*,(6)}.$$

3.2 Pricing Options on the Index and the Stock

3.2.1 Approximation Formula for Index Option Price

Let $P^{I,\varepsilon}(K_I, T) = E^{P^*}(e^{-r(T-t)}(I_T - K_I)^+|\mathscr{F}_t)$ be the price of an European call on the index with strike K_I and maturity T. The processes (I) and (Y) have the same dynamics as in the model with constant idiosyncratic volatility, which implies that the pricing of index options remains the same.

By means of a singular perturbation method with respect to the parameter ε, the authors obtained in [9] an approximation $\tilde{P}^{I,\varepsilon}(K_I, T)$ for the price $P^{I,\varepsilon}(K_I, T)$ at order 1 in ε. The proof of the approximation result is given in [9] and also in (Appendix 1) for completeness. The results are recalled here:

$$\tilde{P}^{I,\varepsilon} = \tilde{P}_0^{I,\varepsilon} - (T - t)V_3^{I,\varepsilon}I_t\frac{\partial}{\partial I_t}(I_t^2\frac{\partial^2\tilde{P}_0^{I,\varepsilon}}{\partial I_t^2}), \qquad (2)$$

where the quantities $\tilde{P}_0^{I,\varepsilon}$ and $V_3^{I,\varepsilon}$ are defined as:

$$\tilde{P}_0^{I,\varepsilon} = P_{BS}^I(\bar{\sigma}_I^*), \qquad (3)$$

$$V_2^{I,\varepsilon} = -\frac{\sqrt{\varepsilon}}{\sqrt{2}}v_Y\langle\phi_I'\chi_1\rangle_1, \qquad (4)$$

$$V_3^{I,\varepsilon} = \frac{\sqrt{\varepsilon}}{\sqrt{2}}\rho_Y v_Y\langle\phi_I'f_1\rangle_1, \qquad (5)$$

$$(\bar{\sigma}_I^*)^2 = \langle f_1^2\rangle_1 - 2V_2^{I,\varepsilon}. \qquad (6)$$

The operator $\langle . \rangle_1$ denotes the average with respect to the invariant distribution of the Ornstein-Uhlenbeck process (Y_1) whose dynamics are described as follows:

$$dY_{1,t} = (m_Y - Y_{1,t})dt + v_Y\sqrt{2}dW_t^{(3)}.$$

Besides, the function ϕ_I is defined as the solution of the following Poisson equation:

$$\mathcal{L}_0^I \phi_I(y) = f_1^2(y) - \langle f_1^2 \rangle_1, \tag{7}$$

where $\mathcal{L}_0^I$ is the infinitesimal generator of the process (Y_1) and can be written:

$$\mathcal{L}_0^I = \frac{\partial}{\partial y}(m_Y - y) + v_Y^2\frac{\partial^2}{\partial y^2}.$$

3.2.2 Approximation Formula for Stock Option Price

Let $P_t^{S,\varepsilon}(K_S, T)$ be the price, at time t, of an European call on the stock with strike K_S and maturity T:

$$P_t^{S,\varepsilon}(K_S, T) = E^{P^*}(e^{-r(T-t)}(S_T - K_S)^+|\mathcal{F}_t).$$

The notation $P_t^{S,\varepsilon}$ is used instead of $P_t^{S,\varepsilon}(K_S, T)$ for simplification purposes.

Using a singular perturbation technique on the parameter ε, an approximation $\tilde{P}^{S,\varepsilon}$ for the option's price $P_t^{S,\varepsilon}$ is obtained. The approximation error is at order 1 in ε. The results can be detailed as follows:

Proposition 3.1

$$\tilde{P}^{S,\varepsilon} = \tilde{P}_0^{S,\varepsilon} - (T-t)V_3^{S,\varepsilon}S_t\frac{\partial}{\partial S_t}(S_t^2\frac{\partial^2 \tilde{P}_0^{S,\varepsilon}}{\partial S_t^2}), \tag{8}$$

where the quantities $\tilde{P}_0^{S,\varepsilon}$ and $V_3^{S,\varepsilon}$ are defined as:

$$\tilde{P}_0^{S,\varepsilon} = P_{BS}^S(t, S_t, \bar{\sigma}_S^*), \tag{9}$$

$$(\bar{\sigma}_S^*)^2 = \bar{\sigma}_S^2 - 2V_2^{S,\varepsilon}, \tag{10}$$

$$V_2^{S,\varepsilon} = -\frac{\sqrt{\varepsilon}}{\sqrt{2}}(\beta^2 v_Y \langle \phi_I' \chi_1 \rangle_{1,2} + v_Z\sqrt{\alpha}\langle \phi_{Idios}' \chi_2 \rangle_{1,2}), \tag{11}$$

$$V_3^{S,\varepsilon} = \frac{\sqrt{\varepsilon}}{\sqrt{2}}(\beta^3 \rho_Y v_Y \langle \phi_I' f_1 \rangle_{1,2} + \rho_Z v_Z\sqrt{\alpha}\langle \phi_{Idios}' f_2 \rangle_{1,2}). \tag{12}$$

The operator $\langle . \rangle_{1,2}$ denotes the averaging with respect to the invariant distribution of $\begin{pmatrix} Y_1 \\ Y_2 \end{pmatrix}_t$ which has the following dynamics:

$$d\begin{pmatrix} Y_{1,t} \\ Y_{2,t} \end{pmatrix} = \begin{pmatrix} 1 & 0 \\ 0 & \alpha \end{pmatrix}\left(\begin{pmatrix} m_Y \\ m_Z \end{pmatrix} - \begin{pmatrix} Y_{1,t} \\ Y_{2,t} \end{pmatrix}\right)dt + \sqrt{2}\begin{pmatrix} v_Y & 0 \\ 0 & \sqrt{\alpha}v_Z \end{pmatrix}d\begin{pmatrix} W_t^{(3)} \\ W_t^{(4)} \end{pmatrix}.$$

The function ϕ_{Idios} is the solution of the equation:

$$\mathscr{L}_0^S \phi_{Idios}(z) = f_2^2(z) - \langle f_2^2 \rangle_{1,2}.$$

where $\mathscr{L}_0^S$ is the infinitesimal generator of the two-dimensional Ornstein-Uhlenbeck process $\begin{pmatrix} Y_1 \\ Y_2 \end{pmatrix}.$

Proof The price of an European call on the stock writes:

$$P_t^{S,\varepsilon} = E^{P^*}(e^{-r(T-t)}(S_T - K_S)^+ | S_t = x, Y_t = y, Z_t = z).$$

Since the process (S, Y, Z) is markovian, the Feynman-Kac theorem can be used to prove that $\mathscr{L}^S P_t^{S,\varepsilon} = 0$ where $\mathscr{L}^S$ is a differential operator expanded in powers of $\sqrt{\varepsilon}$:

$$\mathscr{L}^S = \mathscr{L}_2^S + \frac{1}{\sqrt{\varepsilon}}\mathscr{L}_1^S + \frac{1}{\varepsilon}\mathscr{L}_0^S,$$

and $\mathscr{L}_0^S, \mathscr{L}_1^S, \mathscr{L}_2^S$ are defined as follows:

$$\mathscr{L}_0^S = (m_Y - y)\frac{\partial}{\partial y} + v_Y^2\frac{\partial^2}{\partial y^2} + \alpha(m_Z - z)\frac{\partial}{\partial z} + \alpha v_Z^2\frac{\partial^2}{\partial z^2},$$

$$\mathscr{L}_1^S = -v_Y\sqrt{2}\chi_1(y)\frac{\partial}{\partial y} + \beta S_t f_1(y)\sqrt{2}\rho_Y v_Y\frac{\partial^2}{\partial S\partial y}$$

$$- v_Z\sqrt{2\alpha}\chi_2(z)\frac{\partial}{\partial z} + S_t f_2(z)\sqrt{2\alpha}\rho_Z v_Z\frac{\partial^2}{\partial S\partial z},$$

$$\mathscr{L}_2^S = \frac{\partial}{\partial t} + r(\frac{\partial}{\partial S}S_t - .) + \frac{1}{2}\frac{\partial^2}{\partial S_t^2}S_t^2(\beta^2 f_1(y)^2 + f_2(z)^2).$$

The differential operator $\mathscr{L}_0^S$ is the infinitesimal generator of the two-dimensional Ornstein-Uhlenbeck process $\begin{pmatrix} Y_1 \\ Y_2 \end{pmatrix}$ which has the following dynamics:

$$d\begin{pmatrix} Y_{1,t} \\ Y_{2,t} \end{pmatrix} = \begin{pmatrix} 1 & 0 \\ 0 & \alpha \end{pmatrix}\left(\begin{pmatrix} m_Y \\ m_Z \end{pmatrix} - \begin{pmatrix} Y_{1,t} \\ Y_{2,t} \end{pmatrix}\right) dt + \sqrt{2}\begin{pmatrix} v_Y & 0 \\ 0 & \sqrt{\alpha}v_Z \end{pmatrix} d\begin{pmatrix} W_t^{(3)} \\ W_t^{(4)} \end{pmatrix}.$$

The following notations are used:

- The operator $\langle.\rangle_1$ denotes the averaging with respect to the invariant distribution of the process $(Y_{1,t})_t$.
- The operator $\langle.\rangle_2$ denotes the averaging with respect to the invariant distribution of the process $(Y_{2,t})_t$.
- The operator $\langle.\rangle_{1,2}$ denotes the averaging with respect to the invariant distribution of $\begin{pmatrix} Y_{1,t} \\ Y_{2,t} \end{pmatrix}_t$.

The option price $P^{S,\varepsilon}$ is expanded in powers of $\sqrt{\varepsilon}$:

$$P^{S,\varepsilon} = \sum_{i=0}^{\infty}(\sqrt{\varepsilon})^i P_i^{S,\varepsilon},$$

then, the expression $\mathscr{L}^S P_t^{S,\varepsilon}$ can be written as follows:

$$(\mathscr{L}_2^S + \frac{1}{\sqrt{\varepsilon}}\mathscr{L}_1^S + \frac{1}{\varepsilon}\mathscr{L}_0^S)(\sum_{i=0}^{\infty}(\sqrt{\varepsilon})^i P_i^{S,\varepsilon}) = 0.$$

By classifying the terms of the last equation by powers of $\sqrt{\varepsilon}$, the terms of orders $(-2), (-1), 0, 1, 2$ in $\sqrt{\varepsilon}$ are obtained:

$$(-2): \mathscr{L}_0^S P_0^{S,\varepsilon} = 0, \tag{13}$$

$$(-1): \mathscr{L}_1^S P_0^{S,\varepsilon} + \mathscr{L}_0^S P_1^{S,\varepsilon} = 0, \tag{14}$$

$$(0): \mathscr{L}_2^S P_0^{S,\varepsilon} + \mathscr{L}_1^S P_1^{S,\varepsilon} + \mathscr{L}_0^S P_2^{S,\varepsilon} = 0, \tag{15}$$

$$(1): \mathscr{L}_2^S P_1^{S,\varepsilon} + \mathscr{L}_1^S P_2^{S,\varepsilon} + \mathscr{L}_0^S P_3^{S,\varepsilon} = 0, \tag{16}$$

$$(2): \mathscr{L}_2^S P_2^{S,\varepsilon} + \mathscr{L}_1^S P_3^{S,\varepsilon} + \mathscr{L}_0^S P_4^{S,\varepsilon} = 0. \tag{17}$$

The term of order (-2) in $\sqrt{\varepsilon}$ states that $\mathscr{L}_0^S P_0^{S,\varepsilon} = 0$. Since the operator $\mathscr{L}_0^S$ contains only derivatives with respect to y and z, this equation can be solved by choosing $P_0^{S,\varepsilon} = P_0^{S,\varepsilon}(t, S_t)$ independent of Y_t and Z_t.

The term of order (-1) in $\sqrt{\varepsilon}$ states that $\mathscr{L}_1^S P_0^{S,\varepsilon} + \mathscr{L}_0^S P_1^{S,\varepsilon} = 0$. The differential operator $\mathscr{L}_1^S$ contains first and second order derivatives with respect to y and z, thus $\mathscr{L}_1^S P_0^{S,\varepsilon} = 0$. The equation becomes then $\mathscr{L}_0^S P_1^{S,\varepsilon} = 0$. The equation is satisfied for $P_1^{S,\varepsilon} = P_1^{S,\varepsilon}(t, S_t)$ independent of Y_t and Z_t.

The quantities $P_0^{S,\varepsilon}$ and $P_1^{S,\varepsilon}$ being independent of Y_t and Z_t, it can be stated that:

$$\mathcal{L}_0^S P_0^{S,\varepsilon} = \mathcal{L}_1^S P_0^{S,\varepsilon} = \mathcal{L}_0^S P_1^{S,\varepsilon} = \mathcal{L}_1^S P_1^{S,\varepsilon} = 0.$$

Since $\mathcal{L}_1^S P_1^{S,\varepsilon} = 0$, the term of order 0 in $\sqrt{\varepsilon}$ becomes:

$$\mathcal{L}_2^S P_0^{S,\varepsilon} + \mathcal{L}_0^S P_2^{S,\varepsilon} = 0,$$

which is a Poisson equation for $P_2^{S,\varepsilon}$ with respect to $\mathcal{L}_0^S$. The solvability condition for this equation is:

$$\left\langle \mathcal{L}_2^S P_0^{S,\varepsilon} \right\rangle_{1,2} = \left\langle \mathcal{L}_2^S \right\rangle_{1,2} P_0^{S,\varepsilon} = 0.$$

The average $\left\langle \mathcal{L}_2^S \right\rangle_{1,2}$ of the generator $\mathcal{L}_2^S$ verifies:

$$\left\langle \mathcal{L}_2^S \right\rangle_{1,2} = \frac{\partial}{\partial t} + r\left(\frac{\partial}{\partial S_t} S_t - . \right) + \frac{1}{2} \frac{\partial^2}{\partial S_t^2} S_t^2 \left\langle \beta^2 f_1^2(y) + f_2^2(z) \right\rangle_{1,2}.$$

It can be deduced that $\left\langle \mathcal{L}_2^S \right\rangle_{1,2} = \mathcal{L}_{BS}^S(\bar{\sigma}_S)$ where $\bar{\sigma}_S^2 = \beta^2 \left\langle f_1^2 \right\rangle_{1,2} + \left\langle f_2^2 \right\rangle_{1,2}$.

Consequently, $P_0^{S,\varepsilon}$ is the solution of the following problem:

$$\mathcal{L}_{BS}(\bar{\sigma}_S) P_0^{S,\varepsilon} = 0,$$
$$P_0^{S,\varepsilon}(T, S_T) = h(S_T).$$

It can be easily seen that the quantity $P_0^{S,\varepsilon}$ represents the Black-Scholes price of the option with implied volatility equal to $\bar{\sigma}_S$:

$$P_0^{S,\varepsilon} = P_{BS}^S(t, S_t, \bar{\sigma}_S).$$

The term of order 1 in $\sqrt{\varepsilon}$ is a Poisson equation for $P_3^{S,\varepsilon}$ with respect to $\mathcal{L}_0^S$, whose solvability condition is:

$$\left\langle \mathcal{L}_2^S \right\rangle_{1,2} P_1^{S,\varepsilon} = -\left\langle \mathcal{L}_1^S P_2^{S,\varepsilon} \right\rangle_{1,2} = \left\langle \mathcal{L}_1^S (\mathcal{L}_0^S)^{-1} (\mathcal{L}_2^S - \left\langle \mathcal{L}_2^S \right\rangle_{1,2}) \right\rangle_{1,2} P_0^{S,\varepsilon}. \quad (18)$$

The quantity $P_1^{S,\varepsilon}$ is the solution of the last equation with terminal condition $P_1^{S,\varepsilon}(T, S_T) = 0$.

The function f_1 is independent of z and the function f_2 is independent of y, it follows that: $\left\langle f_1^2 \right\rangle_{1,2} = \left\langle f_1^2 \right\rangle_1$ and $\left\langle f_2^2 \right\rangle_{1,2} = \left\langle f_2^2 \right\rangle_2$. In addition, the function ϕ_I which is the solution of (7), doesn't depend on z. This implies:

$$\mathcal{L}_0^S \phi_I(y) = \mathcal{L}_0^I \phi_I(y) = f_1^2(y) - \left\langle f_1^2 \right\rangle_{1,2}.$$

Let the function ϕ_{Idios} be the solution of the following equation:

$$\mathscr{L}_0^S \phi_{Idios}(z) = f_2^2(z) - \langle f_2^2 \rangle_{1,2}. \tag{19}$$

The function ϕ_{Idios} doesn't depend on y, thus it can be deduced:

$$\mathscr{L}_0^S(\beta^2 \phi_I(y) + \phi_{Idios}(z)) = \beta^2(f_1^2(y) - \langle f_1^2 \rangle_{1,2}) + (f_2^2(z) - \langle f_2^2 \rangle_{1,2}).$$

Building on this, it can be obtained:

$$\mathscr{L}_1^S(\mathscr{L}_0^S)^{-1}(\mathscr{L}_2^S - \langle \mathscr{L}_2^S \rangle_{1,2}) = (\beta^2 \mathscr{L}_1^S \phi_I(y) + \mathscr{L}_1^S \phi_{Idios}(z)) \frac{1}{2} S_t^2 \frac{\partial^2}{\partial S_t^2}.$$

The development of the right-hand side of the previous equation yields:

$$\frac{1}{2}(\beta^2 \langle \mathscr{L}_1^S \phi_I(y) \rangle_{1,2} + \langle \mathscr{L}_1^S \phi_{Idios}(z) \rangle_{1,2})$$

$$= (\frac{\beta^3 v_Y \rho_Y}{\sqrt{2}} \langle \phi_I' f_1 \rangle_{1,2} + \frac{\rho_Z v_Z \sqrt{\alpha}}{\sqrt{2}} \langle \phi_{Idios}' f_2 \rangle_{1,2}) S_t \frac{\partial}{\partial S_t}$$

$$- (\frac{\beta^2 v_Y}{\sqrt{2}} \langle \phi_I' \chi_1 \rangle_{1,2} + \frac{v_Z \sqrt{\alpha}}{\sqrt{2}} \langle \phi_{Idios}' \chi_2 \rangle_{1,2}).$$

Let the quantities $V_2^{S,\varepsilon}$ and $V_3^{S,\varepsilon}$ be defined as follows:

$$V_3^{S,\varepsilon} = \frac{\sqrt{\varepsilon}}{\sqrt{2}}(\beta^3 v_Y \rho_Y \langle \phi_I' f_1 \rangle_{1,2} + \rho_Z v_Z \sqrt{\alpha} \langle \phi_{Idios}' f_2 \rangle_{1,2}),$$

$$V_2^{S,\varepsilon} = -\frac{\sqrt{\varepsilon}}{\sqrt{2}}(\beta^2 v_Y \langle \phi_I' \chi_1 \rangle_{1,2} + v_Z \sqrt{\alpha} \langle \phi_{Idios}' \chi_2 \rangle_{1,2}).$$

The Eq. (18) becomes:

$$\langle \mathscr{L}_2^S \rangle_{1,2} \sqrt{\varepsilon} P_1^{S,\varepsilon} = V_2^{S,\varepsilon} S_t^2 \frac{\partial^2 P_0^{S,\varepsilon}}{\partial S_t^2} + V_3^{S,\varepsilon} S_t \frac{\partial}{\partial S_t}(S_t^2 \frac{\partial^2 P_0^{S,\varepsilon}}{\partial S_t^2}). \tag{20}$$

Consequently, $P_1^{S,\varepsilon}$ is the solution of (20) with the final condition $P_1^{S,\varepsilon}(T, S_T) = 0$. Since the differential operator $\langle \mathscr{L}_2^S \rangle_{1,2} = \mathscr{L}_{BS}^S(\bar{\sigma}_S)$ commits with the operators $\mathscr{D}_{1,S} = S_t \frac{\partial}{\partial S_t}$ and $\mathscr{D}_{2,S} = S_t^2 \frac{\partial^2}{\partial S_t^2}$, and that $\langle \mathscr{L}_2^S \rangle_{1,2} P_0^{S,\varepsilon} = 0$, the solution to the last problem can be given explicitly as:

$$\sqrt{\varepsilon} P_1^{S,\varepsilon} = -(T - t)(V_2^{S,\varepsilon} S_t^2 \frac{\partial^2 P_0^{S,\varepsilon}}{\partial S_t^2} + V_3^{S,\varepsilon} S_t \frac{\partial}{\partial S_t}(S_t^2 \frac{\partial^2 P_0^{S,\varepsilon}}{\partial S_t^2})). \tag{21}$$

In order to check the validity of the solution, the following verification can be made:

$$
\begin{aligned}
\left\langle \mathcal{L}_2^S \right\rangle_{1,2} \sqrt{\varepsilon} P_1^{S,\varepsilon} &= (V_2^{S,\varepsilon} \mathcal{D}_{2,S} P_0^{S,\varepsilon} + V_3^{S,\varepsilon} \mathcal{D}_{1,S} \mathcal{D}_{2,S} P_0^{S,\varepsilon}) \left\langle \mathcal{L}_2^S \right\rangle_{1,2} ((t-T)) \\
&\quad - (T-t)(V_2^{S,\varepsilon} \mathcal{D}_{2,S} \left\langle \mathcal{L}_2^S \right\rangle_{1,2} (P_0^{S,\varepsilon}) \\
&\quad + V_3^{S,\varepsilon} \mathcal{D}_{1,S} \mathcal{D}_{2,S} \left\langle \mathcal{L}_2^S \right\rangle_{1,2} (P_0^{S,\varepsilon})) \\
&= V_2^{S,\varepsilon} \mathcal{D}_{2,S} P_0^{S,\varepsilon} + V_3^{S,\varepsilon} \mathcal{D}_{1,S} \mathcal{D}_{2,S} P_0^{S,\varepsilon}.
\end{aligned}
$$

By neglecting terms of order higher to 1 in $\sqrt{\varepsilon}$, the stock option price can be approximated by $(P_0^{S,\varepsilon} + \sqrt{\varepsilon} P_1^{S,\varepsilon})$. In order to reduce the number of the parameters in the approximation, a second approximation is derived here. Let $\mathcal{L}_{BS}(\bar{\sigma}_S^*)$ be the Black-Scholes differential operator with volatility $\bar{\sigma}_S^*$:

$$(\bar{\sigma}_S^*)^2 = \bar{\sigma}_S^2 - 2V_2^{S,\varepsilon}.$$

Let the quantity $\tilde{P}_0^{S,\varepsilon}$ be introduced as the solution of the following problem:

$$
\begin{aligned}
\mathcal{L}_{BS}(\bar{\sigma}_S^*)\tilde{P}_0 &= 0, \\
\tilde{P}_0^{S,\varepsilon}(T, S_T) &= (S_T - K_S)^+.
\end{aligned}
$$

It follows that $\tilde{P}_0^{S,\varepsilon} = P_{BS}(t, S_t, \bar{\sigma}_S^*)$. Following that, the quantity $\tilde{P}_1^{S,\varepsilon}$ is defined as the solution of the following problem:

$$
\begin{aligned}
\mathcal{L}_{BS}(\bar{\sigma}_S^*)\sqrt{\varepsilon}\tilde{P}_1^{S,\varepsilon} &= V_3^{S,\varepsilon} S_t \frac{\partial}{\partial S_t}(S_t^2 \frac{\partial^2}{\partial S_t^2} \tilde{P}_0^{S,\varepsilon}), \\
\tilde{P}_1^{S,\varepsilon}(T, S_T) &= 0.
\end{aligned}
$$

Using the same arguments as before, it can be deduced that $\tilde{P}_1^{S,\varepsilon} = -(T-t) V_3^{S,\varepsilon} S_t \frac{\partial}{\partial S_t}(S_t^2 \frac{\partial^2}{\partial S_t^2} \tilde{P}_0^{S,\varepsilon})$.

It can be proved that the option price $P^{S,\varepsilon}$ can be approximated up to order 1 in $\sqrt{\varepsilon}$ by $\tilde{P}^{S,\varepsilon}$ which is defined as:

$$\tilde{P}^{S,\varepsilon} = \tilde{P}_0^{S,\varepsilon} + \sqrt{\varepsilon}\tilde{P}_1^{S,\varepsilon}.$$

The proof of this result is detailed in (Appendix 2).

3.3 Calibration of Implied Beta Using Options Prices

In the last section, approximations for the prices of European options on the index and the stock are given respectively in (2) and (8). Using these results, an estimator of the parameter β is provided in this model setting.

3.3.1 Approximation Formula of the Implied Volatility Smile

For the purpose of the beta estimation, a Taylor expansion is carried out in (Appendix 3) in order to provide approximations of the implied volatilities of the stock and the index. The following results are obtained:

Proposition 3.2 *The implied volatility of an European call on the index with strike K_I and maturity T can be approximated, at order 1 in $\sqrt{\varepsilon}$, by $\Sigma_I(K_I, T)$:*

$$\Sigma_I(K_I, T) = b_I + a_I \frac{\log(\frac{F_I}{K_I})}{T}, \tag{22}$$

where $b_I = \bar{\sigma}_I^* - \frac{V_3^{I,\varepsilon}}{2\bar{\sigma}_I^*}$, $a_I = \frac{V_3^{I,\varepsilon}}{(\bar{\sigma}_I^*)^3}$ and $F_I = I_t e^{r(T-t)}$.

Likewise, the implied volatility of an European call on the stock with strike K_S and maturity T can be approximated, at order 1 in $\sqrt{\varepsilon}$, by $\Sigma_S(K_S, T)$:

$$\Sigma_S(K, T) = b_S + a_S \frac{\log(\frac{F_S}{K_S})}{T}, \tag{23}$$

where $b_S = \bar{\sigma}_S^* - \frac{V_3^{S,\varepsilon}}{2\bar{\sigma}_S^*}$, $a_S = \frac{V_3^{S,\varepsilon}}{(\bar{\sigma}_S^*)^3}$ and $F_S = S_t e^{r(T-t)}$.

3.3.2 Comparison with the Model with Constant Idiosyncratic Volatility

The approximations of the smiles of the stock and the index, given in (22) and (23), are used here to estimate the parameter β. Indeed, based on the definitions of $V_3^{S,\varepsilon}$ and $V_3^{I,\varepsilon}$, it can be written that:

$$V_3^{S,\varepsilon} = \beta^3 V_3^{I,\varepsilon} + \frac{\sqrt{\varepsilon}}{\sqrt{2}} \rho_Z \nu_Z \sqrt{\alpha} \langle \phi'_{Idios} f_2 \rangle_{1,2},$$

$$\frac{V_3^{S,\varepsilon}}{V_3^{I,\varepsilon}} = \beta^3 + \frac{\rho_Z \nu_Z \sqrt{\alpha} \langle \phi'_{Idio} f_2 \rangle_{1,2}}{\rho_Y \nu_Y} \frac{\langle \phi'_{Idio} f_2 \rangle_{1,2}}{\langle \phi'_I f_1 \rangle_{1,2}}.$$

The estimator $\hat{\beta}$ proposed in [9] and introduced in (1) verifies:

$$\hat{\beta}^3 = \frac{V_3^{S,\varepsilon}}{V_3^{I,\varepsilon}} = \beta^3 + \frac{\rho_Z v_Z \sqrt{\alpha}}{\rho_Y v_Y} \frac{\langle \phi'_{Idio} f_2 \rangle_{1,2}}{\langle \phi'_I f_1 \rangle_{1,2}}.$$

it can be deduced that, in the case of stochastic idiosyncratic volatility, the quantity $\hat{\beta}$ is a biased estimator of the parameter β. Thus, it would be useful to provide an unbiased estimator of β in the new model setting.

3.3.3 Alternative Method for the Estimation

It can be recalled that:

$$\bar{\sigma}_S^2 = \beta^2 \bar{\sigma}_I^2 + \langle f_2^2 \rangle_{1,2}.$$

Using the relations between $\bar{\sigma}_I^2$ and $(\bar{\sigma}_I^*)^2$ as well as between $\bar{\sigma}_S^2$ and $(\bar{\sigma}_S^*)^2$, the following result is deduced:

$$\beta^2 = \frac{(\bar{\sigma}_S^*)^2 - \langle f_2^2 \rangle_{1,2}}{(\bar{\sigma}_I^*)^2 + 2V_2^{I,\varepsilon}} + \frac{2V_2^{S,\varepsilon}}{(\bar{\sigma}_I^*)^2 + 2V_2^{I,\varepsilon}}.$$

Based on the smile approximation formula, given in (15), for an asset A denoting either the stock S or the index I, it can be stated:

$$b_A = \bar{\sigma}_A^* - \frac{1}{2} a_A (\bar{\sigma}_A^*)^2.$$

The latter second order equation in $\bar{\sigma}_A^*$ has two admissible solutions:

$$x_1 = \frac{1 - \sqrt{1 - 2a_A b_A}}{a_A},$$

$$x_2 = \frac{1 + \sqrt{1 - 2a_A b_A}}{a_A}.$$

Since $V_2^{A,\varepsilon}$ and $V_3^{A,\varepsilon}$ are of order 1 in $\sqrt{\varepsilon}$, then a_A is of order 1 in $\sqrt{\varepsilon}$ and $b_A = \bar{\sigma}_A^* + o(\sqrt{\varepsilon})$. Thus, it can be deduced that the appropriate solution of the second-order equation is x_1, and that the quantities $\bar{\sigma}_S^*$ and $\bar{\sigma}_I^*$ can be written as below:

$$\bar{\sigma}_S^* = \frac{1 - \sqrt{1 - 2a_S b_S}}{a_S}, \quad \bar{\sigma}_I^* = \frac{1 - \sqrt{1 - 2a_I b_I}}{a_I}.$$

Using previous results, the parameter β can be approximated using $\tilde{\beta}$:

$$\tilde{\beta} = \sqrt{\frac{(\frac{1-\sqrt{1-2a_S b_S}}{a_S})^2 - \langle f_2^2 \rangle_{1,2}}{(\frac{1-\sqrt{1-2a_I b_I}}{a_I})^2 + 2V_2^{I,\varepsilon}} + \frac{2V_2^{S,\varepsilon}}{(\frac{1-\sqrt{1-2a_I b_I}}{a_I})^2 + 2V_2^{I,\varepsilon}}}. \qquad (24)$$

Through the use of a Taylor expansion, it can be shown that $\tilde{\beta}$ writes:

$$\tilde{\beta} = \sqrt{\frac{(\bar{\sigma}_S^*)^2 - \langle f_2^2 \rangle_{1,2}}{(\bar{\sigma}_I^*)^2}} \left(1 + \frac{V_2^{S,\varepsilon}}{(\bar{\sigma}_S^*)^2 - \langle f_2^2 \rangle_{1,2}} - \frac{V_2^{I,\varepsilon}}{(\bar{\sigma}_I^*)^2}\right) + o(\varepsilon).$$

In order to compute the value of $\tilde{\beta}$, the quantities $\langle f_2^2 \rangle_{1,2}$ and $V_2^{S,\varepsilon}$ can be estimated statistically using historical data from both option and underlying prices. The estimation procedure will be detailed in the next section.

3.3.4 Numerical Simulations

The accuracy of the estimators $\tilde{\beta}$ and $\hat{\beta}$ are tested in this subsection on simulated data. Indeed, Monte Carlo simulations of the CAPM model with stochastic idiosyncratic volatility are performed using the following parameters:

$$\begin{cases} \varepsilon = 0.1, & \alpha = 1, \\ \rho_Y = -0.8, & \rho_Z = -0.5, \\ \nu_Y = 0.15, & \beta = 1.5, \\ r = 0, & \mu_S = \mu_I = 0, \\ \gamma_1 = 0, & \gamma_2 = 0, \end{cases}$$

The choice of the parameters $r = \mu_S = \mu_I = 0$ and $\gamma_1 = \gamma_2 = 0$ enables to have $V_2^{S,\varepsilon} = V_2^{I,\varepsilon} = 0$. Consequently, the parameter $\tilde{\beta}$ in (24) becomes:

$$\tilde{\beta} = \sqrt{\frac{(\frac{1-\sqrt{1-2a_S b_S}}{a_S})^2 - \langle f_2^2 \rangle_{1,2}}{(\frac{1-\sqrt{1-2a_I b_I}}{a_I})^2}}.$$

Several experiments are carried out here according to the value of the parameter ν_Z. For each value of $\nu_Z \in \{0.05, 0.15, 0.25, 0.4\}$, we launch 100 Monte Carlo pricing algorithms whose random number generators have different seeds. In each of the 100 pricing algorithms, we generate 20,000 paths of the Brownian motions $\left(W_t^{*,(1)}, W_t^{*,(2)}, W_t^{*,(3)}, W_t^{*,(4)}\right)_{\{0 \le t \le T\}}$ as well as their antithetic. We deduce afterward the paths of the processes $(I_t, S_t, y_t, z_t)_{\{0 \le t \le T\}}$ and we price, using the Monte Carlo method, options on the stock and the index with different strikes and with

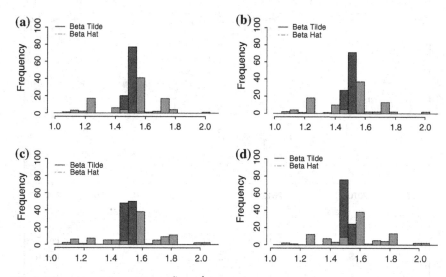

Fig. 2 Histogram of the estimators $\tilde{\beta}$ and $\hat{\beta}$ for $\beta = 1.5$. **a** $v_Z = 0.05$. **b** $v_Z = 0.15$. **c** $v_Z = 0.25$. **d** $v_Z = 0.4$

maturity $T = 0.5$. Finally, we compute the estimators $\tilde{\beta}$ and $\hat{\beta}$ using (1) and (24) respectively, and we obtain the following histograms:

The histograms show that the estimator $\tilde{\beta}$ insures a faster convergence towards the parameter β. Indeed, by using only 40,000 antithetic paths in each Monte Carlo pricing algorithm, the bias of the estimator $\tilde{\beta}$ is lower than the one of $\hat{\beta}$ (Figs. 2, 3).

4 Applications for the Estimation of the Parameter β

In this section, we highlight the importance of the estimation of the parameter β under the risk-neutral measure $\mathbb{P}^*$. We present two main applications which emphasize the utility of this estimation for econometric studies or also for derivatives hedging. In the first subsection, we investigate whether the implied beta $\tilde{\beta}$ can predict the future value of the parameter β under the real-world probability measure $\mathbb{P}$. This question is legitimate to be posed since option-implied information under the risk-neutral pricing measure $\mathbb{P}^*$ may help to predict information under the physical measure $\mathbb{P}$. In the second subsection, we show that the estimator $\tilde{\beta}$ may be crucial for hedging purposes. Indeed, we point out that the quantity $\tilde{\beta}$ can be used in order to hedge the volatility or the delta risk of stock options using instruments on the index. Thus, we conclude that having a good estimation of the value of the parameter β under the risk-neutral measure $\mathbb{P}^*$ can be very useful from a risk-management perspective (Figs. 4, 5).

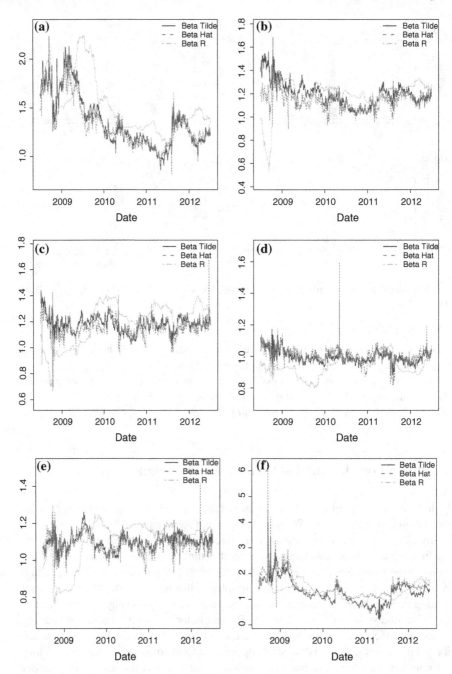

Fig. 3 Comparison between $\tilde{\beta}$ (*blue* line) and $\hat{\beta}$ (*red* line). **a** XLF vs SPX. **b** XLE vs SPX. **c** XLB vs SPX. **d** XLK vs SPX. **e** XLI vs SPX. **f** GS vs SPX

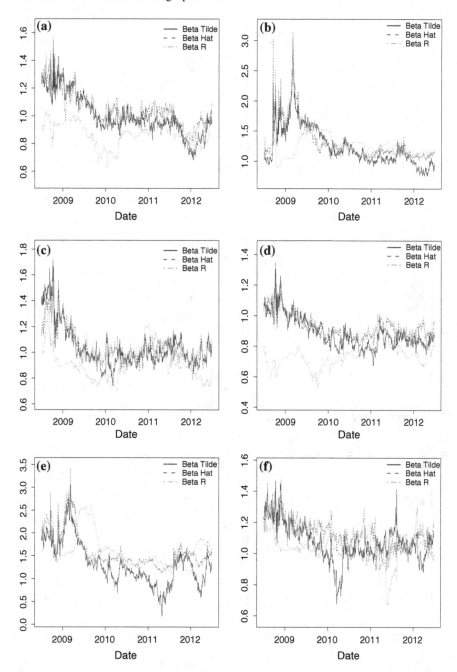

Fig. 4 Comparison between $\tilde{\beta}$ (*blue line*) and $\hat{\beta}$ (*red line*). **a** MSFT versus SPX. **b** GE versus SPX. **c** GOOG versus SPX. **d** IBM versus SPX. **e** JPM versus SPX. **f** CSCO versus SPX

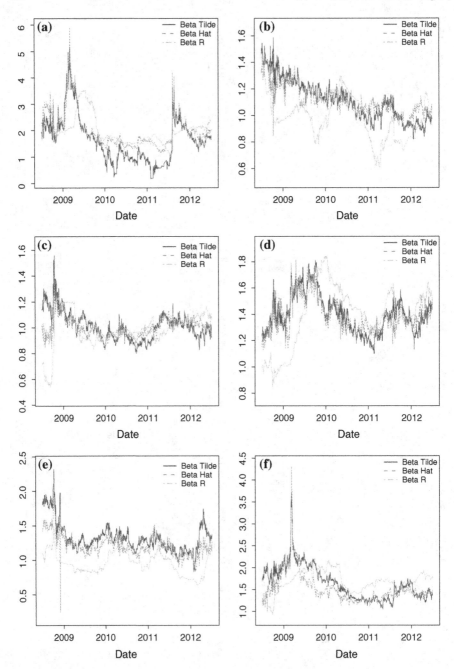

Fig. 5 Comparison between $\tilde{\beta}$ (*blue line*) and $\hat{\beta}$ (*red line*). **a** BAC versus SPX. **b** INTC versus SPX. **c** CVX versus SPX. **d** CAT versus SPX. **e** AAPL versus SPX. **f** AA versus SPX

4.1 Prediction of Forward Beta

We use spot and option data of several US stocks and ETF between 01/01/2008 and 31/12/2012. The option prices have a maturity equal to $T = 0.5$ and a moneyness ranging between 80 to 120 %. The data sample used in this empirical study includes the following instruments: Financial Select Sector (XLF), Energy Select Sector (XLE), Materials Select Sector (XLB), Technology Select Sector (XLK), Industrial Select Sector (XLI), Goldman Sachs (GS), Microsoft (MSFT), General Electric (GE), Google (GOOG), IBM (IBM), JP Morgan (JPM), Cisco Systems (CSCO), Bank Of America (BAC), Intel Corporation (INTC), Chevron Corporation (CVX), Caterpillar Inc(CAT), Apple Inc (AAPL), Alcoa Inc (AA).

For every date t of the sample, the estimators $\hat{\beta}(t)$ and $\tilde{\beta}(t)$ are computed according to (1) and (24) respectively. In order to compute $\tilde{\beta}$, the methodology given below is followed:

1. The quantities $\bar{\sigma}_I$ and $\bar{\sigma}_S$ are approximated by the historical volatilities of the market index and the stock respectively using underlying log-returns.
2. The implied volatilities $\Sigma_I(K_I, T)$ and $\Sigma_S(K_S, T)$ are regressed on the variables $\frac{\log(\frac{F_I(T)}{K_I})}{T-t}$ and $\frac{\log(\frac{F_S(T)}{K_S})}{T-t}$ respectively, thus the slopes a_S, a_I and the intercepts b_S, b_I are deduced.
3. From the estimated slope a_S, the intercept b_S and the effective volatility $\bar{\sigma}_S$, the following quantities can be calculated:
   - $\bar{\sigma}_S^* = \frac{1-\sqrt{1-2a_S b_S}}{a_S}$
   - $V_2^{S,\varepsilon} = \frac{\bar{\sigma}_S^2 - (\bar{\sigma}_S^*)^2}{2}$
4. Likewise, from the estimated slope a_I, the intercept b_I and the effective volatility $\bar{\sigma}_I$, the following quantities are computed:
   - $\bar{\sigma}_I^* = \frac{1-\sqrt{1-2a_I b_I}}{a_I}$
   - $V_2^{I,\varepsilon} = \frac{\bar{\sigma}_I^2 - (\bar{\sigma}_I^*)^2}{2}$
5. The time series of the idiosyncratic volatility $(f(Z_t))_t$ is obtained through the regression of the log-returns of the stock (S) on those of the index (I). Using these data, the parameters m_Z, v_Z and ρ_Z are calibrated using the maximum likelihood method as suggested in [16]. The quantity $\langle f_2^2 \rangle_{1,2}$ is then evaluated. For example, if f_2 denotes the exponential function, then:

$$\langle f_2^2 \rangle_{1,2} = e^{2m_Z + 2v_Z^2}$$

6. The estimator $\tilde{\beta}$ can finally be computed using (24).

Let the quantity β_H denote the historical measure of the parameter beta. Thus, at a given date t, $\beta_H(t)$ is defined as the slope of the linear regression of the stock log-returns on the index log-returns between $t - T$ and t. The estimator β_H is computed

Table 1 Bias of the
estimators $\hat{\beta}$, $\tilde{\beta}$ and β_H

Stock	$E(\hat{\beta} - \beta_F)$	$E(\tilde{\beta} - \beta_F)$	$E(\beta_H - \beta_F)$
XLF.US	−0.148	−0.121	0.054
XLE.US	−0.05	0.0094	−0.0313
XLB.US	−0.0231	0.0002	−0.0345
XLK.US	0.0679	0.0542	−0.0158
XLI.US	−0.0258	−0.0183	−0.0215
AAPL.US	0.2649	0.4073	0.0005
AA.US	−0.1814	−0.0738	−0.0333
BAC.US	−0.1414	−0.3557	0.0222
CAT.US	−0.0453	−0.0574	−0.0512
CSCO.US	0.0854	0.0216	−0.0073
CVX.US	0.0252	0.0366	−0.0187
GE.US	0.0375	−0.0534	0.0016
GOOG.US	0.1693	0.1742	0.0193
GS.US	0.1306	−0.0962	0.0113
IBM.US	0.2063	0.1575	−0.0233
INTC.US	0.1462	0.1451	−0.0069
JPM.US	−0.0415	−0.347	0.0383
MSFT.US	0.1717	0.1102	−0.0234

on a backward window of length T, and so can be compared to the estimators $\hat{\beta}(t)$ and $\tilde{\beta}(t)$ which are estimated using option prices with maturity T.

The graphs below represent the time series $(\hat{\beta}(t))_{t_0 \le t \le t_N}$, $(\tilde{\beta}(t))_{t_0 \le t \le t_N}$ and $(\beta_H(t))_{t_0 \le t \le t_N}$ for $t_0 = 01/01/2008$ and $t_N = 31/12/2012$.

Let the parameter β_F denote the forward realized beta, which is obtained as the slope of the linear regression of stock returns on index returns performed on a forward window with length T. In other words, $\beta_F(t) = \beta_H(t + T)$. It is then interesting to see how well the estimators $\hat{\beta}$, $\tilde{\beta}$ and β_H predict the quantity β_F. The predictive power of these estimators is tested using several statistical measures, the bias and the root-mean-square error are retained here (Tables 1, 2):

It can be seen through the tables above that the historical beta estimator β_H is unbiased for all the considered stocks and has in average the lowest RMSE compared to the estimators $\hat{\beta}$ and $\tilde{\beta}$. It can also be noticed that the estimator $\tilde{\beta}$ has a significant positive bias for the stocks AAPL, GOOG, IBM, INTC, MSFT and a negative bias for XLF and BAC. Thus, the RMSE of $\tilde{\beta}$ for these stocks is higher than the one of β_H. It can be deduced that option market participants may have an expectation of the parameter β which is temporarily different from its value under the physical probability measure $\mathbb{P}$, this finding proves the existence of discrepancies between the risk-neutral measure and the physical measure.

Table 2 RMSE of the estimators $\hat{\beta}$, $\tilde{\beta}$ and β_H

Stock	$\sqrt{E((\frac{\hat{\beta}-\beta_F}{\beta_F})^2)}$	$\sqrt{E((\frac{\tilde{\beta}-\beta_F}{\beta_F})^2)}$	$\sqrt{E((\frac{\beta_H-\beta_F}{\beta_F})^2)}$
XLF.US	0.143	0.138	0.195
XLE.US	0.183	0.207	0.128
XLB.US	0.196	0.209	0.116
XLK.US	0.132	0.124	0.095
XLI.US	0.128	0.112	0.108
AAPL.US	0.41	0.547	0.252
AA.US	0.187	0.188	0.181
BAC.US	0.234	0.407	0.337
CAT.US	0.118	0.123	0.149
CSCO.US	0.159	0.162	0.147
CVX.US	0.198	0.238	0.165
GE.US	0.154	0.189	0.158
GOOG.US	0.277	0.31	0.165
GS.US	0.341	0.292	0.303
IBM.US	0.369	0.348	0.134
INTC.US	0.279	0.294	0.244
JPM.US	0.158	0.292	0.265
MSFT.US	0.297	0.27	0.163

4.2 Hedging of Options on the Stock by Instruments on the Index

The capital asset pricing model offers a practical framework where the stock dynamics are linked to those of the market index. This section aims to show that the parameter beta may be very useful to hedge the delta risk or volatility risk of a stock option using instruments on the index.

4.2.1 Hedging Volatility Risk

The natural way to hedge the volatility risk of a stock option is to use other stock options with a longer maturity. Nevertheless, if options on this stock are not sufficiently liquid, it won't be possible to carry out this hedging strategy. In this case, it may be judicious to use options on the index in order to perform the hedging.

Let X be the value of the portfolio containing an option on the stock, a quantity $-\vartheta$ of an option on the index, a quantity $-\Delta_S$ of the stock and a quantity Δ_I of the index. The value of this portfolio at time t is $X_t = P_t^{S,\varepsilon} - \vartheta_t P_t^{I,\varepsilon} + \Delta_{I,t} I_t - \Delta_{S,t} S_t$ and X_t satisfies the following SDE:

$$dX_t = \left(\mathscr{L}^S P_t^{S,\varepsilon} + r P_t^{S,\varepsilon}\right) dt + \frac{\partial P_t^{S,\varepsilon}}{\partial S}\left(dS_t - rS_t dt\right) + \frac{\partial P_t^{S,\varepsilon}}{\partial y}\frac{v_Y\sqrt{2}}{\sqrt{\varepsilon}} dW_t^{*,(3)}$$

$$+ \frac{\partial P_t^{S,\varepsilon}}{\partial z}\frac{v_Z\sqrt{2\alpha}}{\sqrt{\varepsilon}} dW_t^{*,(4)} - \vartheta_t\left(\mathscr{L}^I P_t^{I,\varepsilon} + r P_t^{I,\varepsilon}\right) dt - \vartheta_t \frac{\partial P_t^{I,\varepsilon}}{\partial I}\left(dI_t - rI_t dt\right)$$

$$- \vartheta_t \frac{\partial P_t^{I,\varepsilon}}{\partial y}\frac{v_Y\sqrt{2}}{\sqrt{\varepsilon}} dW_t^{*,(3)} - \Delta_{S,t} dS_t + \Delta_{I,t} dI_t.$$

By rearranging the terms, the last equation becomes:

$$dX_t = r\left(P_t^{S,\varepsilon} - \vartheta_t P_t^{I,\varepsilon} + \vartheta_t \frac{\partial P_t^{I,\varepsilon}}{\partial I} I_t - \frac{\partial P_t^{S,\varepsilon}}{\partial S} S_t\right) dt + \left(\frac{\partial P_t^{S,\varepsilon}}{\partial S} - \Delta_{t,S}\right) dS_t$$

$$+ \left(\Delta_{I,t} - \vartheta_t \frac{\partial P_t^{I,\varepsilon}}{\partial I}\right) dI_t + \left(\frac{\partial P_t^{S,\varepsilon}}{\partial y} - \vartheta_t \frac{\partial P_t^{I,\varepsilon}}{\partial y}\right)\frac{v_Y\sqrt{2}}{\sqrt{\varepsilon}} dW_t^{*,(3)}$$

$$+ \frac{\partial P_t^{S,\varepsilon}}{\partial z}\frac{v_Z\sqrt{2\alpha}}{\sqrt{\varepsilon}} dW_t^{*,(4)},$$

If the parameters $(\vartheta_t, \Delta_{S,t}, \Delta_{I,t})$ are chosen in the following way:

$$\vartheta_t = \frac{\frac{\partial P^{S,\varepsilon}}{\partial y}}{\frac{\partial P^{I,\varepsilon}}{\partial y}}, \quad \Delta_{S,t} = \frac{\partial P^{S,\varepsilon}}{\partial S_t}, \quad \Delta_{I,t} = \vartheta_t \frac{\partial P^{I,\varepsilon}}{\partial I_t},$$

then the portfolio X is delta-hedged continuously in the stock and the index, and is made insensitive to the variations of the process y. It follows that:

$$dX_t = r\left(P_t^{S,\varepsilon} - \vartheta_t P_t^{I,\varepsilon} + \Delta_{I,t} I_t - \Delta_{S,t} S_t\right) dt + \frac{\partial P_t^{S,\varepsilon}}{\partial z}\frac{v_Z\sqrt{2\alpha}}{\sqrt{\varepsilon}} dW_t^{*,(4)},$$

$$= rX_t dt + \frac{\partial P_t^{S,\varepsilon}}{\partial z}\frac{v_Z\sqrt{2\alpha}}{\sqrt{\varepsilon}} dW_t^{*,(4)},$$

which means that the risk of the volatility of the index is canceled, and the only remaining risk comes from the idiosyncratic volatility.

The hedging parameter ϑ_t can be obtained through the computation of the terms $\frac{\partial P^{S,\varepsilon}}{\partial y}$ and $\frac{\partial P^{I,\varepsilon}}{\partial y}$ using Monte-Carlo simulations. This computation method is time-consuming, so it may be interesting to have a closed-form approximation for this hedging ratio. For this purpose, the quantities $\frac{\partial P^{S,\varepsilon}}{\partial y}$ and $\frac{\partial P^{I,\varepsilon}}{\partial y}$ should be approximated at order 1 in ε as explained below:

$$P(t, I_t, Y_t) = P_0^{I,\varepsilon}(t, I_t) + \sqrt{\varepsilon} P_1^{I,\varepsilon}(t, I_t) + \varepsilon P_2^{I,\varepsilon}(t, I_t, Y_t) + o(\varepsilon)$$

$$P(t, S_t, Y_t, Z_t) = P_0^{S,\varepsilon}(t, S_t) + \sqrt{\varepsilon} P_1^{S,\varepsilon}(t, S_t) + \varepsilon P_2^{S,\varepsilon}(t, S_t, Y_t, Z_t) + o(\varepsilon)$$

where:

$$P_2^{I,\varepsilon}(t, I_t, Y_t) = -\frac{1}{2} I_t^2 \frac{\partial^2 P_0^{I,\varepsilon}}{\partial I_t^2} \phi_I(Y_t),$$

$$P_2^{S,\varepsilon}(t, S_t, Y_t, Z_t) = -\frac{1}{2} S_t^2 \frac{\partial^2 P_0^{S,\varepsilon}}{\partial S_t^2} \left(\beta^2 \phi_I(Y_t) + \phi_{Idios}(Z_t) \right).$$

Consequently, the quantity ϑ_t writes:

$$\vartheta_t = \beta^2 \frac{S_t^2 \frac{\partial^2 P_0^{S,\varepsilon}}{\partial S_t^2}}{I_t^2 \frac{\partial^2 P_0^{I,\varepsilon}}{\partial I_t^2}} + o(\varepsilon).$$

Since $P_0^{S,\varepsilon}$ and $P_0^{I,\varepsilon}$ have known closed-form expressions, the quantity ϑ_t can be approximated analytically by $\hat{\vartheta}_t = \beta^2 \frac{S_t^2 \frac{\partial^2 P_0^{S,\varepsilon}}{\partial S_t^2}}{I_t^2 \frac{\partial^2 P_0^{I,\varepsilon}}{\partial I_t^2}}$, the approximation error is at order 1 in ε.

4.2.2 Delta and Vega Hedging

The lack of liquidity or the presence of transaction costs on the stock can make the delta-hedging of the stock option costly. Therefore, it may be useful to have an alternative hedging strategy in circumstances where there are trading constraints on the stock. The setting of the CAPM model enables the decomposition of the stock risk into two parts: the index risk and the idiosyncratic risk. It is then reasonable in this framework to hedge the delta of a stock option using the index.

Let L be the portfolio containing an option on the stock, a quantity $-\vartheta$ of an option on the index and a quantity $-\varphi_I$ of the index. We have then $L_t = P_t^{S,\varepsilon} - \vartheta_t P_t^{I,\varepsilon} - \varphi_{I,t} I_t$. It follows that:

$$dL_t = r \left(P_t^{S,\varepsilon} - \vartheta_t P_t^{I,\varepsilon} - \varphi_{I,t} I_t \right) dt$$

$$+ \left(\frac{\partial P_t^{S,\varepsilon}}{\partial S} \beta S_t f_1(y_t) - \vartheta_t \frac{\partial P_t^{I,\varepsilon}}{\partial I} I_t f_1(y_t) - \varphi_{I,t} I_t f_1(y_t) \right) dW_t^{*,(1)}$$

$$+ \left(\frac{\partial P_t^{S,\varepsilon}}{\partial y} - \vartheta_I \frac{\partial P_t^{I,\varepsilon}}{\partial y} \right) \frac{v_Y \sqrt{2}}{\sqrt{\varepsilon}} dW_t^{*,(3)} + \frac{\partial P_t^{S,\varepsilon}}{\partial S} S_t f_2(z_t) dW_t^{*,(2)}$$

$$+ \frac{\partial P_t^{S,\varepsilon}}{\partial z} \frac{v_Z \sqrt{2}}{\sqrt{\varepsilon}} dW_t^{*,(4)}.$$

By taking $(\vartheta_t, \varphi_{I,t})$ such that:

$$\vartheta_t = \frac{\frac{\partial P^{S,\varepsilon}}{\partial y}}{\frac{\partial P^{I,\varepsilon}}{\partial y}}, \quad \varphi_{I,t} = \frac{\beta S_t}{I_t}\frac{\partial P^{S,\varepsilon}}{\partial S} - \vartheta_t \frac{\partial P^{I,\varepsilon}}{\partial I},$$

the risks related to the index and its volatility are hedged. It follows that:

$$dL_t = rL_t dt + \frac{\partial P^{S,\varepsilon}}{\partial S} S_t f_2(z_t) dW_t^{*,(2)} + \frac{\partial P^{S,\varepsilon}}{\partial z}\frac{v_z\sqrt{2}}{\sqrt{\varepsilon}} dW_t^{*,(4)}.$$

Here again, the computation of the quantities ϑ_t and $\varphi_{I,t}$ can be made easier through the estimation of the parameter β. Indeed, $\varphi_{I,t}$ and $\varphi_{I,t}$ can be approximated respectively by $\hat{\vartheta}_t$ and $\hat{\varphi}_{I,t} = \frac{\beta S_t}{I_t}\frac{\partial P^{S,\varepsilon}}{\partial S} - \hat{\vartheta}_t \frac{\partial P^{I,\varepsilon}}{\partial I}$.

4.2.3 Numerical Simulations

We perform in this subsection numerical simulations in order to test the hedging strategies proposed so far. First, we generate 200 paths of the processes $(I_t), (S_t), (y_t), (z_t)$ for $t \in [0, T]$ with a time step equal to $\delta = \frac{1}{256}$. On each of these simulated paths indexed by $i \in \{1, \ldots, 200\}$, and for $k \in [0, \frac{T}{\delta}]$, we generate 50,000 paths of $(I_t^{(i)}, S_t^{(i)}, y_t^{(i)}, z_t^{(i)})_{\{t\in[k\delta,T]\}}$ with a time step equal to δ and we compute the quantities $\Delta_{S,k\delta}^{(i)}, \Delta_{I,k\delta}^{(i)}, \left(P_{k\delta}^{S,\varepsilon}\right)^{(i)}, \left(P_{k\delta}^{I,\varepsilon}\right)^{(i)}$ using a Monte Carlo method. The quantities $\vartheta_{k\delta}$ and $\varphi_{I,k\delta}$ are approximated using $\hat{\vartheta}_{k\delta}$ and $\hat{\varphi}_{I,k\delta}$ respectively. We can then obtain $X_{k\delta}$ denoting the portfolio value at time $k\delta$, and we deduce the final value X_T at the maturity date T. It should be precised here that $\tilde{\beta}$ is used instead of β for the computation of $\hat{\vartheta}$ and $\hat{\varphi}_I$, so that the error of estimation is accounted in the hedging error (Figs. 6, 7 and Tables 3, 4).

For these numerical simulations, the following parameters are used:

$$
\begin{cases}
\varepsilon = 0.1, & \alpha = 1, \\
\rho_Y = -0.8, & \rho_Z = -0.5, \\
v_Y = 0.15, & v_Z = 0.05, \\
r = 0, & \mu_I = 0, \\
\beta = 1.5, & \mu_S = 0, \\
\gamma_1 = 0, & \gamma_2 = 0, \\
y_0 = \log(0.4), & z_0 = \log(0.2), \\
I_0 = 1000, & S_0 = 50,
\end{cases}
$$

It can be precised here, that since the hedging strategies are tested on 200 independent paths, the simulations are done in parallel using the computing cluster of Ecole

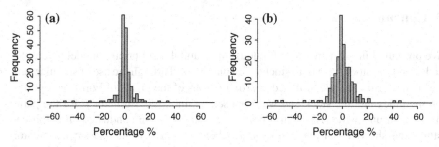

Fig. 6 Histograms of hedging errors for $T = 1$ Month. **a** Histogram of $\frac{X_T-X_0}{X_0} \times 100$. **b** Histogram of $\frac{L_T-L_0}{L_0} \times 100$

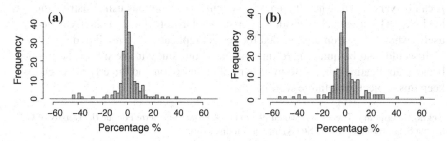

Fig. 7 Histograms of hedging errors for $T = 3$ Months. **a** Histogram of $\frac{X_T-X_0}{X_0} \times 100$. **b** Histogram of $\frac{L_T-L_0}{L_0} \times 100$

Table 3 Statistics of hedging errors for $T = 1$ Month

Statistics of hedging 1M stock options	Median	Mean	Std.	Skewness	Kurtosis
$100 \times \frac{X_T-X_0}{X_0}$	−0.04	0.00	7.51	−1.61	15.87
$100 \times \frac{L_T-L_0}{L_0}$	−0.43	0.07	9.64	−0.63	10.26

Table 4 Statistics of hedging errors for $T = 3$ Months

Statistics of hedging 3M stock options	Median	Mean	Std.	Skewness	Kurtosis
$100 \times \frac{X_T-X_0}{X_0}$	0.07	0.29	10.11	−0.30	8.08
$100 \times \frac{L_T-L_0}{L_0}$	−1.34	−0.15	12.36	0.07	6.33

Centrale Paris. The histograms of the hedging errors $\left(\frac{X_T-X_0}{X_0} \times 100\right)$ in the case of volatility risk hedging and $\left(\frac{L_T-L_0}{L_0} \times 100\right)$ in the case of Delta and volatility risk hedging, are given below:

The same procedure is repeated here in order to perform the numerical simulations for $T = 3$ Months.

5 Conclusion

We presented in this paper a continuous time capital asset pricing model where the index and the stock have both stochastic volatilities. Through the use of singular perturbation technique, we provided approximations of the prices of European options on the index and the stock and we proposed an unbiased estimator $\tilde{\beta}$ for the parameter beta under the risk-neutral pricing measure $\mathbb{P}^*$. We conducted an empirical study and showed that the estimator $\tilde{\beta}$, when compared to the classical estimator β_H, doesn't insure a better prediction of the future realized β under the physical measure $\mathbb{P}$. This proves again that there are some discrepancies between the pricing measure $\mathbb{P}^*$ and the real-world measure $\mathbb{P}$. Besides, we showed that the estimator $\tilde{\beta}$ can be very useful when it comes to hedging stock options using instruments on the index. This approach, considered here as a relative-hedging method, can be very useful when the instruments needed for perfect replication are not liquid.

It would also be quite interesting to extend our study to the use of the implied beta for applications on portfolio construction on option and underlying assets. We keep this subject for a future work.

Acknowledgments The authors would like to thank Prof. Jean-Pierre Fouque (University of California Santa Barbara, United States) for fruitful discussions.

Appendices

Appendix 1: Pricing Options on the Index

Let $P_t^{I,\varepsilon} = E^P(h(I_T)|I_t = x, Y_t = y)$ be the price of a european option on the index with payoff $h(I_T)$ and maturity T.

$$P_t^{I,\varepsilon} = P^{I,\varepsilon}(t, I_t, Y_t).$$

Since the process (I, Y) is markovian, applying the Feynman-Kac theorem yields:

$$\mathscr{L}^I P_t^{I,\varepsilon} = 0,$$

where $\mathscr{L}^I$ is a differential operator whose elements can be classified by powers of $\sqrt{\varepsilon}$:

$$\mathscr{L}^I = \mathscr{L}_2^I + \frac{1}{\sqrt{\varepsilon}}\mathscr{L}_1^I + \frac{1}{\varepsilon}\mathscr{L}_0^I,$$

$$\mathscr{L}_0^I = \frac{\partial}{\partial y}(m_Y - y) + v_Y^2 \frac{\partial^2}{\partial y^2},$$

$$\mathscr{L}_1^I = -v_Y\sqrt{2}\chi_1(y)\frac{\partial}{\partial y} + \sqrt{2}\rho_Y v_Y I_t f_1(y)\frac{\partial^2}{\partial I \partial y},$$

$$\mathscr{L}_2^I = \frac{\partial}{\partial t} + r(\frac{\partial}{\partial I_t}I_t - .) + \frac{1}{2}\frac{\partial^2}{\partial I_t^2}I_t^2 f_1(y)^2.$$

The differential operator $\mathscr{L}_0^I$ represents the infinitesimal generator of the Ornstein-Uhlenbeck process $(Y_{1,t})_t$ which has the following dynamics:

$$dY_{1,t} = (m_Y - Y_{1,t})dt + v_Y\sqrt{2}dW_t.$$

The price $P^{I,\varepsilon}$ can be expanded in powers of $\sqrt{\varepsilon}$:

$$P^{I,\varepsilon} = \sum_{i=0}^{\infty}(\sqrt{\varepsilon})^i P_i^{I,\varepsilon}.$$

Next to that, the term $\mathscr{L}^I P_t^{I,\varepsilon} = 0$ can be expanded and its elements can be classified by powers of $\sqrt{\varepsilon}$. The terms of orders $-2, -1, 0$ and 1 in $\sqrt{\varepsilon}$ are written below:

$$(-2) : \mathscr{L}_0^I P_0^{I,\varepsilon} = 0,$$
$$(-1) : \mathscr{L}_1^I P_0^{I,\varepsilon} + \mathscr{L}_0^I P_1^{I,\varepsilon} = 0,$$
$$(0) : \mathscr{L}_2^I P_0^{I,\varepsilon} + \mathscr{L}_1^I P_1^{I,\varepsilon} + \mathscr{L}_0^I P_2^{I,\varepsilon} = 0,$$
$$(1) : \mathscr{L}_2^I P_1^{I,\varepsilon} + \mathscr{L}_1^I P_2^{I,\varepsilon} + \mathscr{L}_0^I P_3^{I,\varepsilon} = 0,$$

The term $P_0^{I,\varepsilon}$ is a solution of $\mathscr{L}_0^I P_0^{I,\varepsilon} = 0$ with final condition $P_0^{I,\varepsilon}(T, I_T, y_T) = h(I_T)$ (h is independent of y_T). By solving this equation, it can be found that:

$$P_0(t, I_t, y_t) = C_1(t, I_t)\int_0^y e^{\frac{u^2}{2\mu_Y^2} - \frac{um_Y}{\mu^2}} du + C_2(t, I_t).$$

If C_1 is not the null function then the solution diverges when $y \to +\infty$. Nevertheless, in the case of a call option, the option price is bounded ($0 \le P(t, I_t) \le I_t$). Then the quantity C_1 has to be null, and then P_0 is independent of y.

The term of order (-1) in $\sqrt{\varepsilon}$ yields $\mathscr{L}_1^I P_0^{I,\varepsilon} + \mathscr{L}_0^I P_1^{I,\varepsilon} = 0$ which reduces to $\mathscr{L}_0^I P_1^{I,\varepsilon} = 0$ (since P_0 doesn't depend on y). Using the same reasoning as before, it can proved that $P_1^{I,\varepsilon} = P_1^{I,\varepsilon}(t, I_t)$ which is independent of y. Consequently:

$$\mathscr{L}_0^I P_0^{I,\varepsilon} = \mathscr{L}_1^I P_0^{I,\varepsilon} = 0,$$
$$\mathscr{L}_0^I P_1^{I,\varepsilon} = \mathscr{L}_1^I P_1^{I,\varepsilon} = 0.$$

Using $\mathcal{L}_1^I P_1^{I,\varepsilon} = 0$, the term of order 0 in $\sqrt{\varepsilon}$ becomes:

$$\mathcal{L}_0^I P_2^{I,\varepsilon} + \mathcal{L}_2^I P_0^{I,\varepsilon} = 0.$$

This is a Poisson equation for $P_2^{I,\varepsilon}$. Its solvability condition is:

$$\left\langle \mathcal{L}_0^I P_2^{I,\varepsilon} \right\rangle_1 + \left\langle \mathcal{L}_2^I P_0^{I,\varepsilon} \right\rangle_1 = 0,$$

where the operator $\langle . \rangle_1$ is the average with respect to the invariant distribution $N(m_Y, v_Y^2)$ of the Ornstein-Uhlenbeck process $(Y_{1,t})_t$.

Since $\mathcal{L}_0^I$ is the infinitesimal generator of the process (Y_1), $\left\langle \mathcal{L}_0^I P_2^{I,\varepsilon} \right\rangle_1 = 0$. Then, the solvability condition reduces to:

$$\left\langle \mathcal{L}_2^I \right\rangle_1 P_0^{I,\varepsilon} = 0. \tag{25}$$

The operator $\left\langle \mathcal{L}_2^I \right\rangle_1$ has the following form:

$$\left\langle \mathcal{L}_2^I \right\rangle_1 = \frac{\partial}{\partial t} + r(\frac{\partial}{\partial I} I - .) + \frac{1}{2} \frac{\partial^2}{\partial I^2} I^2 \left\langle f_1^2 \right\rangle_1.$$

Consequently $\left\langle \mathcal{L}_2^I \right\rangle_1 = \mathcal{L}_{BS}(\bar{\sigma}_I)$ where $\bar{\sigma}_I^2 = \left\langle f_1^2 \right\rangle_1$. The term $P_0^{I,\varepsilon}$ is the solution of the following problem:

$$\mathcal{L}_{BS}(\bar{\sigma}_I) P_0^{I,\varepsilon} = 0,$$
$$P_0^\varepsilon(T, I_T) = h(I_T).$$

Therefore, $P_0^{I,\varepsilon} = P_{BS}(t, I_t, \bar{\sigma}_I)$ meaning that $P_0^{I,\varepsilon}$ is the Black-Scholes price of the index option with implied volatility equal to $\bar{\sigma}_I$.

As a result, the term $P_2^{I,\varepsilon}$ can be written as $P_2^{I,\varepsilon} = -(\mathcal{L}_0^I)^{-1}(\mathcal{L}_2^I - \left\langle \mathcal{L}_2^I \right\rangle_1) P_0^{I,\varepsilon}$. The term of order 1 in $\sqrt{\varepsilon}$ is a poisson equation for $P_3^{I,\varepsilon}$. Its solvability condition is :

$$\left\langle \mathcal{L}_2^I P_1^{I,\varepsilon} \right\rangle_1 = -\left\langle \mathcal{L}_1^I P_2^{I,\varepsilon} \right\rangle_1, \tag{26}$$

$$\left\langle \mathcal{L}_2^I \right\rangle_1 P_1^{I,\varepsilon} = \left\langle \mathcal{L}_1^I (\mathcal{L}_0^I)^{-1}(\mathcal{L}_2^I - \left\langle \mathcal{L}_2^I \right\rangle_1) \right\rangle_1 P_0^{I,\varepsilon}. \tag{27}$$

Let ϕ_I the solution of the following Poisson equation:

$$\mathcal{L}_0^I \phi_I(y) = f_1^2(y) - \left\langle f_1^2 \right\rangle_1. \tag{28}$$

Since the difference term between the differential operator $\mathcal{L}_2^I$ and its average is:

$$\mathcal{L}_2^I - \langle\mathcal{L}_2^I\rangle_1 = \frac{1}{2}(f_1^2(y) - \langle f_1^2\rangle_1)I_t^2\frac{\partial^2}{\partial I_t^2},$$

it can be written that:

$$(\mathcal{L}_0^I)^{-1}(\mathcal{L}_2 - \langle\mathcal{L}_2\rangle_1) = \frac{1}{2}\phi_I(y)I_t^2\frac{\partial^2}{\partial I_t^2}.$$

By applying the operator $\mathcal{L}_1$ to the last equation, it follows that:

$$\mathcal{L}_1^I(\mathcal{L}_0^I)^{-1}(\mathcal{L}_2 - \langle\mathcal{L}_2\rangle_1) = (-v_Y\sqrt{2}\langle\chi_1\phi_I'\rangle_1 + \sqrt{2}\rho_Y v_Y\langle f_1\phi_I'\rangle_1 I_t\frac{\partial}{\partial I_t})\frac{1}{2}I_t^2\frac{\partial^2}{\partial I_t^2}.$$

Let the quantities $V_2^{I,\varepsilon}$ and $V_3^{I,\varepsilon}$ be defined as below:

$$V_2^{I,\varepsilon} = -\frac{\sqrt{\varepsilon}}{\sqrt{2}}v_Y\langle\phi_I'\chi_1\rangle_1,$$

$$V_3^{I,\varepsilon} = \frac{\sqrt{\varepsilon}}{\sqrt{2}}\rho_Y v_Y\langle\phi_I' f\rangle_1.$$

Using $V_2^{I,\varepsilon}$ and $V_3^{I,\varepsilon}$, the Eq. (27) becomes:

$$\langle\mathcal{L}_2^I\rangle_1\sqrt{\varepsilon}P_1^{I,\varepsilon} = V_2^{I,\varepsilon}I_t^2\frac{\partial^2 P_0}{\partial I_t^2} + V_3^{I,\varepsilon}I_t\frac{\partial}{\partial I_t}(I_t^2\frac{\partial^2 P_0}{\partial I_t^2}).$$

Therefore, $P_1^{I,\varepsilon}$ is the solution of the following problem :

$$\langle\mathcal{L}_2^I\rangle_1\sqrt{\varepsilon}P_1^{I,\varepsilon} = V_2^{I,\varepsilon}I_t^2\frac{\partial^2 P_0}{\partial I_t^2} + V_3^{I,\varepsilon}I_t\frac{\partial}{\partial I_t}(I_t^2\frac{\partial^2 P_0}{\partial I_t^2}), \qquad (29)$$

$$P_1^{I,\varepsilon}(T, I_T) = 0. \qquad (30)$$

In order to simplify the notations, the following differential operators are defined:

$$\mathcal{D}_{1,I} = I_t\frac{\partial}{\partial I_t},$$

$$\mathcal{D}_{2,I} = I_t^2\frac{\partial^2}{\partial I_t^2}.$$

Using the fact that $\langle\mathcal{L}_2^I\rangle_1 = \mathcal{L}_{BS}(\bar{\sigma}_I)$ commits with $\mathcal{D}_{1,I}$ and $\mathcal{D}_{2,I}$, and that $\langle\mathcal{L}_2^I\rangle_1 P_0 = 0$, the solution to the last problem can be given explicitly by:

$$\sqrt{\varepsilon} P_1^{I,\varepsilon} = -(T-t)(V_2^{I,\varepsilon} I_t^2 \frac{\partial^2 P_0}{\partial I_t^2} + V_3^{I,\varepsilon} I_t \frac{\partial}{\partial I_t}(I_t^2 \frac{\partial^2 P_0}{\partial I_t^2})).$$

By neglecting terms of order higher or equal to 2 in $\sqrt{\varepsilon}$, the option's price can be approximated by $(P_0^{I,\varepsilon} + \sqrt{\varepsilon} P_1^{I,\varepsilon})$. As it was proven by the authors in [9], it is possible to carry out a parameter reduction method and approximate $P^{I,\varepsilon}$ by the following formula:

$$P^{I,\varepsilon} \sim \tilde{P}_0^{I,\varepsilon} + \sqrt{\varepsilon} \tilde{P}_1^{I,\varepsilon},$$

such as:

$$\tilde{P}_0^{I,\varepsilon} = P_{BS}(\bar{\sigma}_I^*), \tag{31}$$

$$(\bar{\sigma}_I^*)^2 = \bar{\sigma}_I^2 - 2V_2^{I,\varepsilon}, \tag{32}$$

$$\sqrt{\varepsilon} \tilde{P}_1^{I,\varepsilon} = -(T-t)V_3^{I,\varepsilon} I_t \frac{\partial}{\partial I_t}(I_t^2 \frac{\partial^2 \tilde{P}_0}{\partial I_t^2}). \tag{33}$$

Appendix 2: Accuracy of the Approximation

It can be seen here that:

$$\mathscr{L}_{BS}(\bar{\sigma}_S^*) = \mathscr{L}_{BS}(\bar{\sigma}_S) - V_2^{S,\varepsilon} S_t^2 \frac{\partial^2}{\partial S_t^2}. \tag{34}$$

Using (34), it can be proved that:

$$\mathscr{L}_{BS}(\bar{\sigma}_S)(P_0^{S,\varepsilon} - \tilde{P}_0^{S,\varepsilon}) = -V_2^{S,\varepsilon} S_t^2 \frac{\partial^2 \tilde{P}_0^{S,\varepsilon}}{\partial S_t^2},$$

$$(P_0^{S,\varepsilon} - \tilde{P}_0^{S,\varepsilon})(T, S_T) = 0.$$

The source term is $O(\sqrt{\varepsilon})$ because of $V_2^{S,\varepsilon}$, then the difference term $(P_0^{S,\varepsilon} - \tilde{P}_0^{S,\varepsilon})$ is also $O(\sqrt{\varepsilon})$. Consequently, it follows that:

$$|P^{S,\varepsilon} - (\tilde{P}_0^{S,\varepsilon} + \sqrt{\varepsilon} \tilde{P}_1^{S,\varepsilon})| \le |P^{S,\varepsilon} - (P_0^{S,\varepsilon} + \sqrt{\varepsilon} P_1^{S,\varepsilon})| + |(P_0^{S,\varepsilon} + \sqrt{\varepsilon} P_1^{S,\varepsilon})$$
$$- (\tilde{P}_0^{S,\varepsilon} + \sqrt{\varepsilon} \tilde{P}_1^{S,\varepsilon})|,$$

The first term $|P^{S,\varepsilon,\delta} - (P_0^{S,\varepsilon} + \sqrt{\varepsilon} P_1^{S,\varepsilon})|$ is already $o(\varepsilon)$. Therefore, the second term should be studied. To simplify the notations, the error term $\mathscr{R}$ is introduced as following:

$$\mathcal{R} = (P_0^{S,\varepsilon} + \sqrt{\varepsilon}P_1^{S,\varepsilon}) - (\tilde{P}_0^{S,\varepsilon} + \sqrt{\varepsilon}\tilde{P}_1^{S,\varepsilon}).$$

Besides the differential operators $\mathcal{H}_\varepsilon$ and $\mathcal{H}_\varepsilon^*$ are defined:

$$\mathcal{H}_\varepsilon = V_2^{S,\varepsilon}\mathcal{D}_{2,S} + V_3^{S,\varepsilon}\mathcal{D}_{1,S}\mathcal{D}_{2,S},$$
$$\mathcal{H}_\varepsilon^* = V_3^{S,\varepsilon}\mathcal{D}_{1,S}\mathcal{D}_{2,S}.$$

Using the previous notations, the quantity $\mathcal{L}_{BS}(\bar{\sigma}_S)\mathcal{R}$ can be computed:

$$\begin{aligned}
\mathcal{L}_{BS}(\bar{\sigma}_S)\mathcal{R} &= \mathcal{L}_{BS}(\bar{\sigma}_S)((P_0^{S,\varepsilon} + \sqrt{\varepsilon}P_1^{S,\varepsilon}) - (\tilde{P}_0^{S,\varepsilon} + \sqrt{\varepsilon}\tilde{P}_1^{S,\varepsilon})), \\
&= \mathcal{H}_\varepsilon P_0^{S,\varepsilon} - (\mathcal{L}_{BS}(\bar{\sigma}_S^*) + V_2^{S,\varepsilon}\mathcal{D}_{2,S})(\tilde{P}_0^{S,\varepsilon} + \sqrt{\varepsilon}\tilde{P}_1^{S,\varepsilon}), \\
&= \mathcal{H}_\varepsilon P_0^{S,\varepsilon} - \mathcal{H}_\varepsilon^* \tilde{P}_0^{S,\varepsilon} - V_2^{S,\varepsilon}\mathcal{D}_{2,S}(\tilde{P}_0^{S,\varepsilon} + \sqrt{\varepsilon}\tilde{P}_1^{S,\varepsilon}), \\
&= \mathcal{H}_\varepsilon^*(P_0^{S,\varepsilon} - \tilde{P}_0^{S,\varepsilon}) - V_2^{S,\varepsilon}\mathcal{D}_{2,S}(\tilde{P}_0^{S,\varepsilon} - P_0^{S,\varepsilon} + \sqrt{\varepsilon}\tilde{P}_1^{S,\varepsilon}).
\end{aligned}$$

Knowing that:

- $(P_0^{S,\varepsilon} - \tilde{P}_0^{S,\varepsilon})$ is $O(\sqrt{\varepsilon})$.
- $\mathcal{H}_\varepsilon^*$ is $O(\sqrt{\varepsilon})$.
- $V_2^{S,\varepsilon}\mathcal{D}_{2,S}$ is $O(\sqrt{\varepsilon})$.
- $\sqrt{\varepsilon}\tilde{P}_1^{S,\varepsilon}$ is $O(\sqrt{\varepsilon})$.

and additionally $\mathcal{R}(T) = 0$, then it follows that $\mathcal{R} = O(\varepsilon)$. This concludes the derivation of the following result:

$$P_0^{S,\varepsilon} + \sqrt{\varepsilon}P_1^{S,\varepsilon} = \tilde{P}_0^{S,\varepsilon} + \sqrt{\varepsilon}\tilde{P}_1^{S,\varepsilon} + O(\varepsilon).$$

So up to order 1 in $\sqrt{\varepsilon}$, the option price $P^{S,\varepsilon}$ can be approximated by $\tilde{P}^{S,\varepsilon}$ which is defined as:

$$\tilde{P}^{S,\varepsilon} = \tilde{P}_0^{S,\varepsilon} + \sqrt{\varepsilon}\tilde{P}_1^{S,\varepsilon}.$$

The estimation error obtained, when approximating $P^{S,\varepsilon}$ by $\tilde{P}^{S,\varepsilon}$, is at order 1 in ε. Indeed, by neglecting terms of order higher to 1 in $\sqrt{\varepsilon}$, the term $(P_0^{S,\varepsilon} + \sqrt{\varepsilon}P_1^{S,\varepsilon})$ is obtained as an approximation of the price $P(t, S_t, Y_t)$. It is then important to show that this approximation is of order 1 in ε meaning that:

$$|P(t, S_t, Y_t) - (P_0^{S,\varepsilon} + \sqrt{\varepsilon}P_1^{S,\varepsilon})| \leq C\varepsilon.$$

The proof of this property is given in [17] in the case where the payoff h is smooth. A summary of this proof is given here in order to make this paper self contained.

Let us introduce the quantity $Z^{S,\varepsilon}$ which verifies that:

$$P(t, S_t, Y_t) = P_0^{S,\varepsilon} + \sqrt{\varepsilon}P_1^{S,\varepsilon} + \varepsilon P_2^{S,\varepsilon} + \varepsilon^{\frac{3}{2}}P_3^{S,\varepsilon} - Z^{S,\varepsilon}.$$

Since $\mathscr{L}^S P(t, S_t, Y_t) = 0$, it follows that:

$$\mathscr{L}^S Z^{S,\varepsilon} = \mathscr{L}_S(P_0^{S,\varepsilon} + \sqrt{\varepsilon} P_1^{S,\varepsilon} + \varepsilon P_2^{S,\varepsilon} + \varepsilon^{\frac{3}{2}} P_3^{S,\varepsilon}).$$

The differential operator $\mathscr{L}^S$ can be written as $\mathscr{L}^S = \mathscr{L}_2^S + \frac{1}{\sqrt{\varepsilon}} \mathscr{L}_1^S + \frac{1}{\varepsilon} \mathscr{L}_0^S$. By developing $\mathscr{L}^S Z^{S,\varepsilon}$ and regrouping the terms by orders of $\sqrt{\varepsilon}$, it follows that:

$$\mathscr{L}^S Z^{S,\varepsilon} = \frac{1}{\varepsilon} \mathscr{L}_0^S P_0^{S,\varepsilon} + \frac{1}{\sqrt{\varepsilon}} (\mathscr{L}_0^S P_1^{S,\varepsilon} + \mathscr{L}_1^S P_0^{S,\varepsilon}) + (\mathscr{L}_0^S P_2^{S,\varepsilon} + \mathscr{L}_1^S P_1^{S,\varepsilon} + \mathscr{L}_2^S P_0^{S,\varepsilon})$$
$$+ \sqrt{\varepsilon}(\mathscr{L}_0^S P_3^{S,\varepsilon} + \mathscr{L}_1^S P_2^{S,\varepsilon} + \mathscr{L}_2^S P_1^{S,\varepsilon}) + \varepsilon(\mathscr{L}_1^S P_3^{S,\varepsilon} + \mathscr{L}_2^S P_2^{S,\varepsilon})$$
$$+ \varepsilon^{\frac{3}{2}} \mathscr{L}_2^S P_3^{S,\varepsilon}.$$

The terms $P_0^{S,\varepsilon}$, $P_1^{S,\varepsilon}$ and $P_2^{S,\varepsilon}$ are chosen to nullify the first four terms in the previous equation, therefore:

$$\mathscr{L}^S Z^{S,\varepsilon} = \varepsilon(\mathscr{L}_1^S P_3^{S,\varepsilon} + \mathscr{L}_2^S P_2^{S,\varepsilon}) + \varepsilon^{\frac{3}{2}} \mathscr{L}_2^S P_3^{S,\varepsilon},$$

and $Z^{S,\varepsilon}$ satisfies the final condition:

$$Z^{S,\varepsilon}(T, S_T, Y_T, Z_T) = \varepsilon P_2(T, S_T, Y_T, Z_T) + \varepsilon^{\frac{3}{2}} P_3(T, S_T, Y_T, Z_T).$$

Using the Feynman-Kac theorem, it follows that:

$$Z^{S,\varepsilon}(t, x, y, z) = \varepsilon E(e^{-r(T-t)}(P_2(T, S_T, Y_T, Z_T) + \varepsilon^{\frac{1}{2}} P_3(T, S_T, Y_T, Z_T))$$
$$- \int_t^T e^{-r(u-t)}((\mathscr{L}_1^S P_3^{S,\varepsilon} + \mathscr{L}_2^S P_2^{S,\varepsilon}) + \varepsilon^{\frac{1}{2}} \mathscr{L}_2^S P_3^{S,\varepsilon})$$
$$\times (u, S_u, Y_u, Z_u) du | S_t = x, Y_t = y, Z_t = z).$$

Under assumptions on the smoothness of the payoff function h and boundedness of the functions χ_1 and χ_2, the term $Z^{S,\varepsilon}$ is at most linearly growing in $|y|$ and $|z|$ and then $Z^{S,\varepsilon}(t, x, y, z) = O(\varepsilon)$

The demonstration of the accuracy of the approximation for a non smooth payoff h (as in the case of a call option) is derived in [18].

Appendix 3: Approximation of the Implied Volatility

The method developed here was suggested in [9] for the model with constant idio-syncratic volatility.

The symbol A is used to denote either the stock S or the index I. The price of an European option $P^{A,\varepsilon}$ on the asset A can be written as:

$$P^{A,\varepsilon} = \tilde{P}_0^{A,\varepsilon} - (T-t)V_3^{A,\varepsilon}A_t\frac{\partial}{\partial A_t}(A_t^2\frac{\partial^2\tilde{P}_0^{A,\varepsilon}}{\partial A_t^2}) + o(\varepsilon),\qquad (35)$$

where $\tilde{P}_0^{A,\varepsilon}$ is defined as:

$$\tilde{P}_0^{A,\varepsilon} = P_{BS}(t, A_t, \bar{\sigma}_A^*).$$

The term $P^{A,\varepsilon}$ could represent the price of the option on the index $P^{I,\varepsilon}$ if $A = I$ or the price of the option on the stock $P^{S,\varepsilon}$ in the case where $A = S$.

Let I_A be the implied volatility associated to the asset's option price $P^{A,\varepsilon}$ meaning that $P^{A,\varepsilon} = P_{BS}(t, A_t, I_A(K_A, T))$. An expansion of $I_A(K, T)$ could be made around $\bar{\sigma}_A^*$ in powers of $\sqrt{\varepsilon}$:

$$I_A(K_A, T) = \bar{\sigma}_A^* + \sqrt{\varepsilon}I_1(K_A, T) + O(\varepsilon).$$

Using Taylor's formula, it follows that:

$$P^{A,\varepsilon} = P_{BS}(t, A_t, \bar{\sigma}_A^*) + \frac{\partial P_{BS}}{\partial\sigma}\Big|_{\sigma=\bar{\sigma}_A^*}\sqrt{\varepsilon}I_1 + o(\varepsilon).\qquad (36)$$

Combining the Eqs. (35) and (36) gives:

$$\frac{\partial P_{BS}(t, A_t, \bar{\sigma}_A^*)}{\partial\sigma}\Big|_{\sigma=\bar{\sigma}_A^*}\sqrt{\varepsilon}I_1(K, T) = -(T-t)V_3^{A,\varepsilon}A_t\frac{\partial}{\partial A_t}(A_t^2\frac{\partial^2\tilde{P}_0}{\partial A_t^2}).\qquad (37)$$

Performing simple computations on the derivatives of the Black-Scholes price yields that:

$$\mathscr{D}_{2,A}\tilde{P}_0 = \frac{1}{\bar{\sigma}_A^*(T-t)}\frac{\partial P_{BS}}{\partial\bar{\sigma}_A^*}(t, A_t, \bar{\sigma}_A^*).$$

By applying then the operator $\mathscr{D}_{1,A}$ to the last equation, it can be obtained that:

$$\mathscr{D}_{1,A}\mathscr{D}_{2,A}\tilde{P}_0 = \frac{A_t}{\bar{\sigma}_A^*(T-t)}\frac{\partial^2 P_{BS}}{\partial A_t\partial\bar{\sigma}_A^*}(t, A_t, \bar{\sigma}_A^*).$$

Using closed-form formulas of Black-Scholes greeks, it can be written that:

$$A_t \frac{\partial^2 P_{BS}}{\partial A_t \partial \bar{\sigma}_A^*}(t, A_t, \bar{\sigma}_A^*) = -\frac{d_2}{\bar{\sigma}_A^* \sqrt{T-t}} \frac{\partial P_{BS}}{\partial \bar{\sigma}_A^*}(t, A_t, \bar{\sigma}_A^*).$$

The Eq. (37) can be written as:

$$\sqrt{\varepsilon} I_1(K_A, T) = -\frac{V_3^{A,\varepsilon}}{\bar{\sigma}_A^*} \frac{A_t \frac{\partial^2 \tilde{P}_0}{\partial A_t \partial \bar{\sigma}_A^*}}{\frac{\partial \tilde{P}_0}{\partial \bar{\sigma}_A^*}}.$$

Then, it is straightforward that:

$$\sqrt{\varepsilon} I_1(K_A, T) = \frac{V_3^{A,\varepsilon} d_2(K_A, T)}{(\bar{\sigma}_A^*)^2 \sqrt{T-t}},$$

where:

$$d_2(K_A, T) = \frac{\log(\frac{A_t e^{r(T-t)}}{K_A}) - \frac{(\bar{\sigma}_A^*)^2}{2}(T-t)}{\bar{\sigma}_A^* \sqrt{T-t}}.$$

The implied volatility can then be approximated using the following formula:

$$I_A(K_A, T) = \bar{\sigma}_A^* - \frac{V_3^{A,\varepsilon}}{2\bar{\sigma}_A^*} + \frac{V_3^{A,\varepsilon}}{(\bar{\sigma}_A^*)^3} \frac{\log(\frac{F_A(T)}{K_A})}{T-t}. \qquad (38)$$

Then, the following smile approximation can be obtained:

$$I_A(K_A, T) = b_A + a_A \frac{\log(\frac{F_A(T)}{K_A})}{T-t},$$

with:

$$b_A = \bar{\sigma}_A^* - \frac{V_3^{A,\varepsilon}}{2\bar{\sigma}_A^*},$$

$$a_A = \frac{V_3^{A,\varepsilon}}{(\bar{\sigma}_A^*)^3}.$$

References

1. H. Markowitz, Portfolio selection. J. Financ. **7**(1), 77–91 (1952)
2. M.E. Blume, I. Friend, A new look at the capital asset pricing model. J. Financ. **28**(1), 19–34 (1973)
3. F.K. Reilly, K.C. Brown, *Investment Analysis and Portfolio Management* (Cengage Learning, 2011)
4. J.F. Weston, Investment decisions using the capital asset pricing model. Financ. Manag. **2**(1), 25–33 (1973)
5. D.U.A. Galagedera, An alternative perspective on the relationship between downside beta and CAPM beta. Emerg. Mark. Rev. **8**(1), 4–19 (2007)
6. A. Abdymomunov, J. Morley, Time variation of CAPM betas across market volatility regimes. Appl. Financ. Econ. **21**(19), 1463–1478 (2011)
7. T.G. Andersen, T. Bollerslev, F.X. Diebold, G. Wu, Realized beta: persistence and predictability. Adv. Econ. **20**, 1–39 (2006)
8. P. Christoffersen, K. Jacobs, G. Vainberg, *Forward-Looking Betas, Manuscript* (McGill University, Québec, 2008)
9. J.-P. Fouque, E. Kollman, Calibration of stock betas from skews of implied volatilities. Appl. Math. Financ. **18**(2), 119–137 (2011)
10. J.-P. Fouque, A.P. Tashman, Option pricing under a stressed-beta model. Ann. Financ. **8**(2–3), 183–203 (2012)
11. P. Carr, D. Madan, Factor models for option pricing. Forthcom. Asia Pac. Financ. Mark. **19**(4), 319–329 (2000)
12. J.Y. Campbell, M. Lettau, B.G. Malkiel, Y. Xu, Have individual stocks become more volatile? An empirical exploration of idiosyncratic risk. J. Financ. **56**(1), 1–43 (2001)
13. C. Kearney, V. Potì, Have european stocks become more volatile? an empirical investigation of idiosyncratic and market risk in the euro area. Eur. Financ. Manag. **14**(3), 419–444 (2008)
14. S.X. Wei, C. Zhang, Why did individual stocks become more volatile? J. Bus. **79**(1), 259–292 (2006)
15. Y. Xu, B.G. Malkiel, Investigating the behavior of idiosyncratic volatility. J. Bus. **76**(4), 613–645 (2003)
16. J.C.G. Franco, Maximum likelihood estimation of mean reverting processes. Real Options Pract. (2003)
17. F. Jean-Pierre, G. Papanicolaou, K. Ronnie Sircar, *Derivatives in Financial Markets with Stochastic Volatility* (Cambridge University Press, Cambridge, 2000)
18. G. Papanicolaou, J.-P. Fouque, K. Solna, R. Sircar, Singular perturbations in option pricing. SIAM J. App. Math. **63**(5), 1648–1665 (2003)

Long-Term Evolution of the Topological Structure of Interactions Among Stocks in the New York Stock Exchange 1925–2012

Chandrashekar Kuyyamudi, Anindya S. Chakrabarti and Sitabhra Sinha

Abstract Financial markets are complex systems that comprise of many agents interacting with each other as well as responding to external information. Earlier studies on the cross-correlations of price movements of different stocks have revealed the interaction structure of various financial markets—which has resulted in the intriguing speculation that the evolution of a market from emerging or developing to developed status is accompanied by systematic changes in its interaction structure. Using a very large data-base of daily price changes of equities listed in the New York Stock Exchange we have investigated the long-term changes that this financial market has undergone over a period of nearly nine decades (1925–2012). We have used spectral analysis of the daily log-return cross-correlations in order to reveal the network of significant interactions between equities. We find that the distribution of interaction strengths varies with the state of the economy. In particular, the skewness of the distribution shows a remarkable increase in recent years. We have investigated the strength distribution over the network in different periods by treating the network as resulting from a percolation process where the threshold value of interaction strength for deciding whether to connect a pair of nodes is varied. We find that the formation of the giant component can occur very differently in different periods— which reflects the micro-structure of the interactions between the equities.

C. Kuyyamudi (✉) · S. Sinha
The Institute of Mathematical Sciences, CIT Campus, Taramani, Chennai 600113, India
e-mail: kachandra@imsc.res.in

S. Sinha
e-mail: sitabhra@imsc.res.in

A.S. Chakrabarti
Department of Economics, Boston University, 270 Bay State Road, Boston,
MA 02134, USA
e-mail: anindya@bu.edu

© Springer International Publishing Switzerland 2015   105
F. Abergel et al. (eds.), *Econophysics and Data Driven Modelling of Market Dynamics*,
New Economic Windows, DOI 10.1007/978-3-319-08473-2_3

1 Introduction

Economic phenomena provide a rich source for observing and analyzing complex collective phenomena emerging from interactions between a large number of agents [1]. Financial markets in particular has attracted the attention of many physicists trying to understand of the behavior of large socio-economic systems because of the availability of large volumes of data available for processing through automated algorithms. One of the goals of the 'econophysics' program is to develop a general description of the broad features of economic systems, especially the various dynamical regimes that they exhibit and the transitions that occur between them. Such a physics-based description necessarily focuses on features that are not sensitively dependent on idiosyncratic aspects of specific systems—in analogy with the theory of critical phenomena that describes universal features of phase transitions occurring in a large variety of physical systems (ranging from fluids to magnets to superconductors) which differ greatly in terms of their microscopic details. However, before we can arrive at such a theoretical description for economic systems, one first needs to identify empirical evidence of universality, i.e., features that are quantitatively invariant across many systems (sometimes referred to as 'stylized facts' in the economics literature). The most prominent candidate for such an universal property of markets is the so-called "inverse cubic law", the occurrence of heavy tails in the cumulative distribution of fluctuations—in stock price or market index—with a characteristic exponent of -3 [2]. First observed for stocks in the S&P 100 list [3] and the German exchange [4], it has later been established in a wide range of markets at various stages of development—and has been reported to occur even at the earliest stages of an emerging financial market [5, 6]. Other candidates for universality, such as the distributions of transaction size and trading volume, have also been proposed. However, their validity across different markets is yet to be well-established [7].

The universal features discussed above are related to steady-state distributions that are essentially static. However, physicists are also interested in the dynamical aspects of markets, for instance, the process by which a financial market evolves from an emerging to a developed one. While it is obvious that the historical route through which a particular market develops are affected significantly by its unique local environment, as well as, chance events, one might ask if there are also certain universal aspects that most, if not all, markets would exhibit during their evolution. For instance, does the topological structure of the interactions between market components, e.g., obtained by analyzing the correlated movements of stocks of different companies, change as the market evolves? Does the network of stocks evolve from being essentially disconnected (i.e., the stocks are relatively independent of each other) to a situation where stocks belonging to the same industrial sector cluster together into strongly interacting modules? This can result from the stocks being more affected by general, market-wide effects in the earlier, emerging period, to being more sensitive to sector-specific signals in the developed stage. This scenario was proposed in Ref. [8], where the Indian market was shown to have only three prominent clusters (two corresponding to the Information Technology and

Pharmaceutical sectors respectively, while the third consisted of stocks belonging to several sectors) in contrast to the NYSE which was shown in Ref. [9] to have nine clusters of stocks that corresponded very well with specific industrial sectors (such as Energy, Utilities, Technology, Healthcare&Consumer, Basic Materials, etc.). A plausible hypothesis could be forwarded, based on these observations, that as a market evolves the network of interactions between stocks in a market develops from being homogeneous to one having a strong community (or modular) organization. While it is intriguing to explore this idea using empirical data, a major limitation to such an exercise is the lack of detailed transaction records of financial markets over a sufficiently long period that would allow one to relate their evolution with changes in topological structure of the network of interactions among their stocks. It is only fairly recently that a database comprising information about daily prices of a large number of stocks that spans several decades has become available. This data pertains to the New York Stock Exchange (NYSE), the largest exchange in the world at present. While the more recent trading records of this market (dating from the 1990s) has been analyzed by many groups, the analysis of long-term trends which this newly available data-set promises to reveal has so far not been reported. Here we describe the preliminary results of our analysis of the patterns of evolution over nine decades in the interaction structure of a major financial market. Our observations provide a brief glimpse into aspects of the dynamical evolution of a complex economic system and suggests intriguing connections between historical phenomena such as crashes and changes in the topological nature of the stock interaction network.

2 The NYSE

The New York Stock Exchange is the largest stock exchange in the world both in terms of the volume of shares traded (6.9 trillion US Dollars in 2013) and the market capitalization of listed companies (15.6 trillion US Dollars as of May 2013) [10]. Although not the oldest stock exchange in the world (usually considered to be the Amsterdam Exchange that began in the early 17th century), or indeed even in the Unites States of America (Philadelphia exchange being the oldest), the New York exchange has dominated the world of finance for over a century. It traces its origin to the Buttonwood Agreement of 1,792 among dealers and auctioneers of stocks and bonds in New York, in the wake of a crash caused by the failed speculative dealings of an influential New York merchant William Duer [11]. Following a period of uncertainty in the financial market resulting from the War of 1812 between USA and Britain, the New York dealers organized themselves further with the formal establishment of the New York Stock and Exchange Board in 1817, which changed its name to New York Stock Exchange in 1863. In the 19th century, the transportation sector dominated the financial market, initially with stocks of roadways and canal construction companies and later with railroad companies as the American West was opened up. With the expansion of the US economy in the late 19th century—although there were periods of economic downturns, especially the one following the large financial panic of 1873—the NYSE developed from trading mostly in stocks of local companies to becoming more of a national marketplace [11]. The last decades of the 19th and the first decades of the 20th century saw a boom in financial sector, that infamously ended with the Great Crash of 1929. This would eventually result in governmental regulation and oversight of market activities, although traders would continue to struggle against this for decades. The economic upturn of USA following the end of the second world war that saw unprecedented growth and record profits for various industries, also saw a massive surge in investment in the financial sector (often attributed to the optimistic mood of the public in this era). In addition, the Bretton Woods agreement between the major industrialized countries that established the US Dollar as the basis for determining foreign exchange rates, effectively formalized the dominant role of the USA in the world financial market. The "bull market" phase that started around 1952 continued for more than one and a half decades. However, the following decade saw rising volatility in the financial market with stock values showing large changes over short periods that were not necessarily connected to major changes in the real economy. Rising inflation through 1960s and 1970s, the 1973 "oil shock" when the Organization of Petroleum Exporting Countries (OPEC) suddenly increased the price of oil and the collapse of Bretton Woods, resulted in a "bear market" phase that ended with another boom starting in the 1980s. Stocks of high-tech companies emerged as market favorites while, trading in types of financial products other than stocks and bonds, such as futures and options, became popular. In 1987 the market saw one of its worst crashes, with market indices dropping about 20% of their value on what is referred to as "Black Monday" (Oct. 19). The 1990s saw the onset of what is probably the largest bull market in Wall Street history, with dramatic increase in trading volumes for NYSE as well as other exchanges such as NASDAQ. This period also saw NASDAQ emerging as a strong competitor of NYSE with major information technology companies such as Cisco, Intel and Microsoft being listed in the former and not in the latter. Despite occasional scares, the boom in the market continued till the 2008–2009 financial crisis, that has been compared to the 1929 crash for its severity, and from which the market can still be said to be recovering.

Description of the data set. We obtained time series of daily closing prices for all the stocks in NYSE between December 31, 1925 and February 1, 2012 from a database maintained by the Center for Research in Security Prices (CRSP) [12]. We split the data into 85 overlapping periods of 1000 days, with an overlap of 260 days. The choice of 260 days as the shift between consecutive periods is motivated by the fact that the number of working days in a typical year is around 260, but it does fluctuate around that value. We rejected all stock timeseries with more than 50 missing values during the 1000 days time period over which they are divided. With the exception of first two periods we have more than 500 stocks in all periods, we randomly sampled 500 stocks from these periods. In the case of first and second period we have 460 and 492 stocks respectively. We filled the missing value using backward fill method where we replace the missing value by the most recent value in the past. We use unique stock identifier assigned by CSRP to label to uniquely identify the stocks. We use the four digit Standard Industrial Classification (SIC) code to determine the sector to which a particular stock belongs. To obtain the state of the US economy over the period we have looked at the Dow Jones Industrial Average (DJIA). It reflects the performance of 30 large USA-based publicly owned companies during a standard trading session (which are listed either in NYSE or NASDAQ). The inflation adjusted DJIA with monthly resolution has been obtained from the *macrotrends* website [13].

3 Results

3.1 Return Cross-Correlation Matrix

In order to quantify the correlation between the price movements of various stocks, we calculate the cross-correlation between return time series of each pair of stock. To measure the price variations such that the result is independent of the scale of measurement, we have considered the logarithmic return of the price of a stock. This is defined for the ith stock and a time interval Δt (often taken as 1 day) as:

$$R_i(t, \Delta t) \equiv \ln P_i(t + \Delta t) - \ln P_i(t). \tag{1}$$

As the range of fluctuations for different stocks can be quite diverse, we define the normalized return,

$$r_i(t, \Delta t) \equiv \frac{R_i - \langle R_i \rangle}{\sigma_i}, \tag{2}$$

where the standard deviation of R_i, $\sigma_i = \sqrt{\langle R_i^2 \rangle - \langle R_i \rangle^2}$, is a measure of the volatility of the stock, and $\langle ... \rangle$ represents time average over the period of observation. Using the normalised returns for different stocks we can compute the cross-correlation matrix $\mathbf{C}$, whose elements $C_{ij} \equiv \langle r_i r_j \rangle$ represents the correlation between returns for stocks i and j, having a value lying in the domain $[-1, 1]$. By construction, $\mathbf{C}$ is symmetric with $C_{ii} = 1$.

We have computed the correlation matrix for all the 85 overlapping periods (each of 1000 days duration) between 1925–2012 using the above described procedure. Figure 1 shows how the probability distribution of elements of the cross-correlation matrix $\mathbf{C}$ varies with time over the period that we have considered. The distributions are highly assymetric, being skewed towards positive values. The degree of asymmetry varies with time, with the distributions in the earliest (i.e., the 1930s and 1940s) and the most recent (i.e., the 1990s onwards) eras of the period under study having the largest degree of skewness. As the preponderance of positive matrix elements during these periods appear to be indicative of increased correlation among the stocks, it is intriguing to note that these years happen to coincide with periods of marked upheaval in the economy of the United States of America (and by extension, the world). As high volatility and subsequent crashes in markets have been seen to

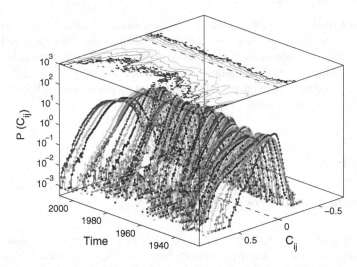

Fig. 1 Time evolution of probability distribution function of correlation matrix (**C**) elements over 85 different periods spanning the period between December 31, 1925 to February 1, 2012. Note that the probability density scale is logarithmic. The contour plot shown on *top* is obtained for equidistant intervals of the logarithmic probability denisty (viz., 10^{-3}, 10^{-2}, 10^{-1}, 1, 10 and 100) and highlights certain characteristic features, such as the highly skewed nature of the distributions towards positive values

be associated with strong correlations between the movements of different market components [14], the asymmetry of the **C** distribution appears to be acting as an indicator of the state of the market at different times.

3.2 Eigenvalue Analysis of the Correlation Matrix

In order to analyze in greater detail the temporal aspects of the correlation distributions, we have performed spectral analysis of the correlation matrices. If we consider N different time series, each of length T, which are mutually uncorrelated, then the resulting random correlation matrix (known as a Wishart matrix) has an eigenvalue distribution with well-defined statistical properties in the asymptotic limit (viz., $N \rightarrow \infty$, $T \rightarrow \infty$ such that $T/N \equiv Q \geq 1$) [15]:

$$P_{wm}(\lambda) = \frac{Q}{2\pi} \frac{\sqrt{(\lambda_{max} - \lambda)(\lambda - \lambda_{min})}}{\lambda}, \tag{3}$$

for $\lambda_{min} \leq \lambda \leq \lambda_{max}$ and 0 otherwise. The bounds of the distribution are given by $\lambda_{max,min} = \left[1 \pm (1/\sqrt{Q})\right]^2$. To see how the properties of empirical correlation matrices compare with the corresponding Wishart matrices, let us consider the first interval in the period we are considering, viz., Dec 31,1925–May 9,1929. As there

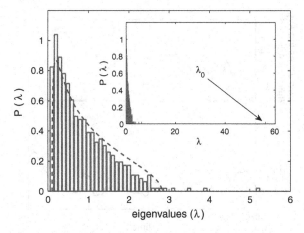

Fig. 2 Probability distribution of eigenvalues of Correlation matrix, **C** of stock return timeseries of period between Dec 31, 1925 and May 9, 1929. The *dotted line* indicates the probability distribution of eigenvalues of the corresponding Wishart matrix. The *dotted line* emphasises the existence of eigenvalues that deviate from the Wishart matrix *upper bound*. The inset shows the entire distribution of eigenvalues, highlighting the separation of the largest eigenvalue from rest of the eigenvalues

are $N = 460$ stocks in the data set for this interval, each of which has $T = 1000$ points, we obtain a value of $Q = 2.173$. Thus, if the different time-series were uncorrelated, the eigenvalues from the resulting cross-correlation matrix would lie within the bounds $\lambda_{max} = 2.816$ and $\lambda_{min} = 0.104$. As seen from Fig. 2, the bulk of the eigenvalues of the empirical correlation matrix are indeed in this range. However, about six of the largest eigenvalues deviate from this random bulk, with the largest eigenvalue $\lambda_0 = 57.337$ being much larger than the other (e.g., the second largest eigenvalue $\lambda_1 = 6.425$). It is through analysis of these spectral modes which deviate from the expected behavior of a random Wishart matrix that once can obtain an understanding of the structure of interactions between the components (viz., the stocks) of a market.

3.3 The Largest Eigenvalue

The largest eigenvalue is generally considered to represent the global (or market-wide) signal that affects all stocks equally. In order to verify how well the largest eigenvalue represents the overall behavior of the market, we compare it with the composite index of the stock market. We have considered the Dow Jones Index as it is a much older index compared to the NYSE Composite index. Although between 1926 to 2012 the value of the index has seen very large increases, when the inflation adjusted Dow Jones Index is considered the range of variation is significantly narrowed down. In Fig. 3 we see that the largest eigenvalue indeed appears to be

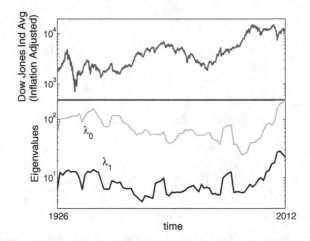

Fig. 3 Comparison between the time-series of inflation adjusted Dow Jones Industrial Average (DJIA) index and the first and second largest eigenvalues of the correlation matrices for the entire period between December 31, 1925 and February 1, 2012. Both eigenvalues show similar time variation and resemble the trend observed in the time-series of the DJIA index

representative of the overall trend in the market (as reflected in the Dow Jones Index). We speculate that if more intervals between 1926 and 2012 are considered by reducing the shift between consecutive periods and the corresponding largest eigenvalues are calculated, better agreement with the Dow Jones Index may be seen.

3.4 Filtering the Correlation Matrix

As seen from Fig. 1, the bulk of the eigenvalues of the empirical cross-correlation matrices occur within the bounds given by the corresponding Wishart matrices, while the largest eigenvalue lies far apart from the rest of the deviating eigenvalues. The eigenvector corresponding to the largest eigenvalue has alomst uniform contribution from each of the component stocks, which is indicative of the fact that the largest eigenvalue represents the global signal that has a similar effect on all stocks. In order to focus mainly on the inter-stock interactions which are neither purely stochastic in nature nor due to the all-encompassing global signal, we need to focus on the eigenvalues deviating from the random bulk with the exception of the largest eigenvalue, λ_0. The correlation matrix **C** can be decomposed as

$$\mathbf{C} = \sum_{i=0}^{N-1} \lambda_i \, u_i \, u_i^T , \tag{4}$$

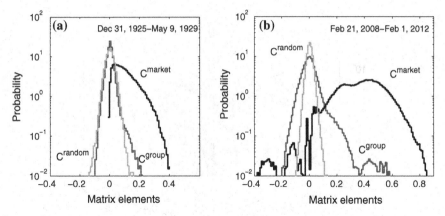

Fig. 4 Probability density function of elements of $\mathbf{C}^{market}$, $\mathbf{C}^{random}$, and $\mathbf{C}^{group}$ for two different periods which are separated by about 83 years. In both of these periods, $\mathbf{C}^{random}$ is distributed symmetrically about the origin. On the other hand, the distribution of elements of $\mathbf{C}^{group}$ is skewed to the *right* (to different extents in the two periods) while the bulk of the $\mathbf{C}^{market}$ distribution lies on the *right side* of the origin

where λ_i are the eigenvalues of $\mathbf{C}$ sorted in descending order and u_i are the corresponding eigenvectors. In order to separate the contributions arising from the random modes , those arising from interactions between stocks and the market mode (the global driving signal), we decompose $\mathbf{C}$ into three parts, corresponding to the *market*, *group*, and *random* components:

$$\mathbf{C} = \mathbf{C}^{market} + \mathbf{C}^{group} + \mathbf{C}^{random}, \tag{5}$$

$$\mathbf{C} = \lambda_0 u_0 u_0^T + \sum_{i=1}^{N_g} \lambda_i u_i u_i^T + \sum_{i=N_g+1}^{N-1} \lambda_i u_i u_i^T, \tag{6}$$

where N_g is the number of eigenvalues (excluding the largest one) that deviate from the bulk of the eigenvalue spectrum. Figure 4 shows the probability distribution of the elements of the three parts which constitute $\mathbf{C}$ for two different periods.

3.5 Time Evolution of the Market Mode

The market mode $\mathbf{C}^{market}$ is constructed from the eigenvector corresponding to the largest eigenvalue and contains information about correlation between stocks due to the global signal which often dominates the movement of stock prices in the market. Figure 5 shows the probability distribution of the elements of $\mathbf{C}^{market}$ for 85 different periods. The contour plot of the distribution shown on top of the figure illustrates

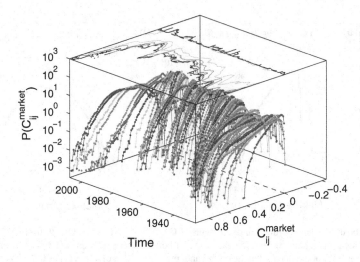

Fig. 5 Time evolution of probability density function of $\mathbf{C}^{market}$ elements over the period between December 31, 1925 and February 1, 2012. The *curves* lie on the *positive side* for the most part with very minimal extension on the *negative side*, indicative of the fact that the global signal mainly results in the correlated movement of all stocks. The contour plot on *top* highlights the varying level of skewness over the entire period spanning around 86 years

the heavily skewed nature of the distributions corresponding to some of the earliest periods and the more recent periods. This is possibly related to the high levels of volatility seen in the market during these periods. The skewness is positive indicating that large positive correlation exists between constituent stocks during the periods with high volatility.

3.6 Time Evolution of the Random Modes

The correlation matrix contribution from the random modes, $\mathbf{C}^{random}$, is constructed from eigenvectors corresponding to eigenvalues that occur below the upper bound λ_{max} given by the corresponding Wishart matrix. It models the spurious correlations that can appear between stocks as a result of stochastic fluctuations. Figure 6 shows the time evolution of $\mathbf{C}^{random}$ over the period under study. As expected, there is very little variation in the distribution of the random modes contribution over the different intervals. The distributions have a Gaussian nature and are centered about the origin, which is emphasized by the contour plot shown on top of the figure.

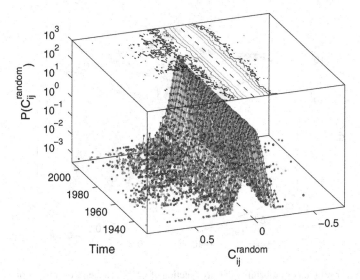

Fig. 6 Time evolution of probability density function of $\mathbf{C}^{random}$ elements over the period between December 31, 1925 and February 1, 2012. The *curves* are mostly symmetric around the origin, a fact that is emphasized by the contour plot on *top* of the figure. The shape of the *curve* shows negligible variation over the entire period spanning about 86 years

3.7 Time Evolution of the Intra-group Interaction Modes

The contribution to the correlation matrix from the modes arising from interaction between stocks, $\mathbf{C}^{group}$, is constructed from eigenvectors corresponding to the eigenvalues deviating from the random bulk (with the exclusion of the largest eigenvalue). It contains information relating to interaction between stocks that have non-stochastic origins, possibly arising from the stocks responding to information specific to the common industrial sector to which they belong. Figure 7 shows the time evolution of $\mathbf{C}^{group}$ over the period under study. The contour plot on top of the figure highlights intervals of large skewness, especially the recent ones that occur towards the end of the period under study. Higher incidence of large values of positive correlations indicates strongly correlated movements of stocks belonging to the same group.

We have considered the time variation of the skewed nature of the correlation matrix distribution in some detail, as the nature of asymmetry in distributions can inform us about important market movements (e.g., the alterations in the symmetry of the daily return distributions for an ensemble of stocks, quantified by measures of skewness, can identify market crashes and rallies [16]). In order to quantify the skewness of the distributions we have used Pearson's second skewness coefficient,

$$skewness = \frac{3(\mu - median)}{\sigma} \tag{7}$$

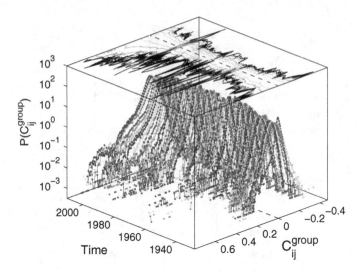

Fig. 7 Time evolution of probability density function of $\mathbf{C}^{group}$ elements over the period between December 31, 1925 and February 1, 2012. The *curves* are more skewed towards the *positive side* at the beginning and also towards the end of the period under study, while being relatively symmetric about the origin during the intervening period. The contour plot on *top* aids in showing the variation in shape of the distribution in the different intervals over the period of about 86 years

where μ and σ are the mean and standard deviation of the distribution respectively. Figure 8 shows the time evolution of Pearson's second skewness coefficient of $\mathbf{C}$, $\mathbf{C}^{market}$ and $\mathbf{C}^{group}$ over the 85 intervals spanning the period under study. Figure 8 (top) shows the variation of skewness for the elements of the entire correlation matrix $\mathbf{C}$ (solid line) and that for the market mode, $\mathbf{C}^{market}$ (dotted line). The skewness in $\mathbf{C}$ is very well emulated by that of $\mathbf{C}^{market}$, indicating that the skewness of the former is largely dictated by the latter. In those intervals where there is significant difference between the two, we observe negative skewness in the distribution of elements of correlation matrix contribution arising from interaction between stocks, $\mathbf{C}^{group}$ (Fig. 8, bottom). Although, in general, the magnitude of skewness of $\mathbf{C}^{group}$ elements is smaller compared to the skewness of $\mathbf{C}$, we do observe a significant increase in the former during recent times comprising the intervals that correspond to the 2008-9 financial crisis. Note that, the other period of major financial crisis, the Great Crash of 1929, is marked by the occurrence of peak value of skewness for the distribution of $\mathbf{C}$, as well as, the market mode, $\mathbf{C}^{market}$. The observation suggests that although both were cataclysmic events in NYSE, their essential nature were quite distinct, with the 1929 Crash being primarily dominated by global (or market-wide effects) while the 2008-9 Crisis was dominated by intra-sector dynamics.

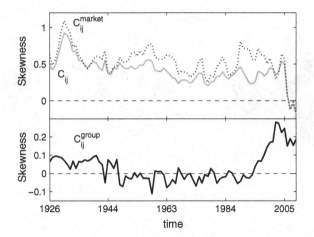

Fig. 8 Variation with time of the Pearson's second skewness coefficient for the distribution of elements of $\mathbf{C}$ (*top, solid line*), $\mathbf{C}^{market}$ (*top, dotted line*) and $\mathbf{C}^{group}$ (*bottom, solid line*) over the period of about 86 years. The *curves* for $\mathbf{C}$ and $\mathbf{C}^{market}$ have very similar trends for the most part indicative of the fact that $\mathbf{C}^{market}$ is the dominant contribution to $\mathbf{C}$. Unlike $\mathbf{C}^{market}$ and $\mathbf{C}$, the distributions of $\mathbf{C}^{group}$ elements show *negative* skewness during certain periods. Note that the 1929 Great Crash occurred during the peak value of skewness for $\mathbf{C}$ and $\mathbf{C}^{market}$ while the highest *positive* value for the skewness of $\mathbf{C}^{group}$ coincides with the 2008–2009 financial crisis

3.8 Percolation Inspired Approach to Clustering

Having split the correlation matrix into components and extracted the contribution arising from modes that correspond to interactions between stocks, one can now use $\mathbf{C}^{group}$ to explicitly construct a network of interactions among stocks. For this purpose we convert $\mathbf{C}^{group}$ into a matrix of z-scores, $\mathbf{Z}$, that quantify the statistical significance of the cross-correlation between a pair of stocks, by the operation

$$\mathbf{Z} = \frac{\mathbf{C}^{group} - \mu}{\sigma}, \tag{8}$$

where μ and σ are the mean and standard deviation of the elements of the matrix $\mathbf{C}^{group}$. In order to construct networks from the $\mathbf{Z}$ matrices we impose a threshold value Z_{th} on the z-scores to obtain corresponding adjacency matrices:

$$A_{ij} = \begin{cases} 0 \text{ if } Z_{ij} \le Z_{th}, \\ 1 \text{ if } Z_{ij} > Z_{th}. \end{cases} \tag{9}$$

The resulting adjacency matrix, $\mathbf{A}$, is symmetric (undirected) and binary (unweighted). To look for clustering between group of stocks we visualize the networks obtained for a particular $\mathbf{Z}$ matrix by imposing different values of z-score thresholds and using Frutcherman-Reingold layout in the graph visualization

(a)   (b)   (c)

(d)   (e)   (f)

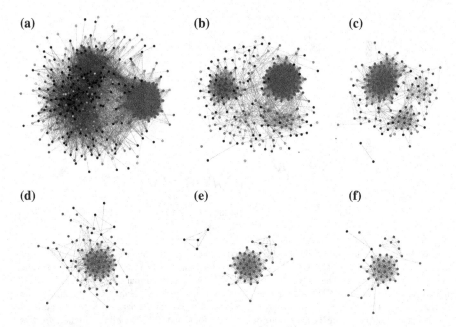

Fig. 9 Network of interaction among stocks obtained from the filtered correlation matrix $\mathbf{C}^{group}$ corresponding to the interval between Jan 30, 2006 and January 12, 2010. Only those matrix elements which have a z-score greater than a specified threshold are considered to construct a link between a pair of stocks. The z-score thresholds used in the different panels are **a**1, **b**2, **c**3, **d**4, **e**5 and **f**6. The nodes are shaded according to the industrial sectors they belong to. The largest cluster that survives till **f** corresponds to the sector termed Holding And Other Investment Offices (SIC id 67). The networks are generated using the Fruchterman-Reingold layout algorithm implemented in Gephi, an open source graph visualization software

software *Gephi*. Figure 9 shows how the network changes with increasing values of z-score thresholds. With increasing value of the threshold only clusters with very strong correlation between stocks, and hence high value of z-scores, survive.

We also analyze the nature of variation of the network topology as the z-score threshold value is changed. We take a "reverse-percolation" approach where we start from low value of z-score threshold and gradually increase the threshold value in small increments. As expected, the graph becomes sparser with increasing threshold. We quantify the change in topology of the network by means of graph measures such as the size of the Largest Connected Component (LCC), the ratio of the size of the second largest connected component (SCC) to that of LCC and the average clustering coefficient. For comparison with emirical networks, we use an ensemble of randomized networks obtained by random permutation of the weights of the empirical network keeping the symmetry of the adjacency matrix intact. Intervals that show higher degree of clustering between group of stocks tend to show significantly different behavior in terms of the above mentioned quantities when compared to their randomized surrogates. Although the LCC size shows very similar behavior with

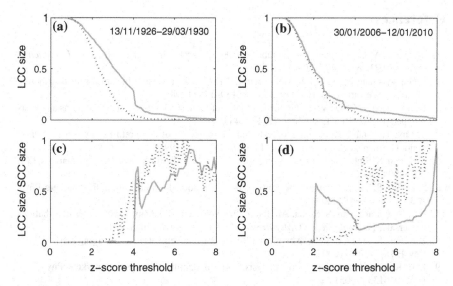

Fig. 10 Variation in graph properties of the networks of stocks resulting from the reverse-percolation approach as the z-score threshold is varied. The fraction of the nodes in the network belonging to the largest connected component (LCC) and the ratio of the sizes of the second largest connected component (SCC) to that of the LCC for two periods close to the beginning and the end of the period we are considering are shown. Connected component comprises nodes such that a connected path exists between any given pair of nodes belonging to it. The *solid curve* corresponds to the empirical data while the broken *curve* corresponds to the corresponding randomized networks (see text for details)

respect to the corresponding randomized networks (Fig. 10a, b), the ratio of SCC size to LCC size becomes significant much earlier than it does for the corresponding randomized surrogates in intervals with significant clustering between groups of stocks (Fig. 10d). On the other hand, no significant differences are observed with respect to the randomized surrogates when the clustering between groups of stocks is not prominent (Fig. 10c). In cases where the randomized surrogates and empirical networks exhibit very similar properties, we note that the characteristics of the empirical network can be mostly attributed to the distribution of the weights, i.e., the distribution of elements of the $\mathbf{C}^{group}$ matrix. In cases where there is significant difference, it is because of topological properties of the network that cannot be entirely captured by the distribution of the matrix elements.

Acknowledgments We would like to thank Wharton Research Data Service (WRDS) for providing access to the CRSP database. This work is supported in part by the IMSc Econophysics (XII Plan) Project.

References

1. S. Sinha, A. Chatterjee, A. Chakraborti, B.K. Chakrabarti, *Econophysics: An Introduction* (Wiley-VCH, Weinheim, 2011)
2. P. Gopikrishnan, M. Meyer, L.A.N. Amaral, H.E. Stanley, Inverse cubic law for the distribution of stock price variations. Eur. Phys. J. B **3**, 139–140 (1998)
3. D.W. Jansen, C.G. De Vries, On the frequency of large stock returns: putting booms and busts into perspective. Rev. Econ. Stat. **73**, 18–24 (1991)
4. T. Lux, The stable Paretian hypothesis and the frequency of large returns: an examination of major German stocks. Appl. Financ. Econ. **6**, 463–475 (1996)
5. R.K. Pan, S. Sinha, Self-organization of price fluctuation distribution in evolving markets. EPL **77**, 58004 (2007)
6. R.K. Pan, S. Sinha, Inverse-cubic law of index fluctuation distribution in Indian markets. Phys. A **387**, 2055–2065 (2008)
7. V.S. Vijayaraghavan, S. Sinha, Are the trading volume and the number of trades distributions universal?, *Econophysics of Order-driven Markets* (Springer, Milan, 2011)
8. R.K. Pan, S. Sinha, Collective behavior of stock price movements in an emerging market. Phys. Rev. E **76**, 046116 (2007)
9. D.H. Kim, H. Jeong, Systematic analysis of group identification in stock markets. Phys. Rev. E **72**, 046133 (2005)
10. http://www.statista.com/
11. C.R. Geisst, *Wall Street: A History* (Oxford University Press, New York, 1997)
12. http://www.crsp.com/
13. http://www.macrotrends.net/
14. L. Sandoval Junior, I.D.P. Franca, Correlation of financial markets in times of crisis. Phys. A **391**, 187–208 (2012)
15. A.M. Sengupta, P.P. Mitra, Distribution of singular values for some random matrices. Phys. Rev. E **60**, 3389–3392 (1999)
16. F. Lillo, R.N. Mantegna, Symmetry alteration of ensemble return distribution in crash and rally days of financial markets. Eur. Phys. B **15**, 603–606 (2000)

Bitcoin Dynamics: The *Inverse Square Law* of Price Fluctuations and Other Stylized Facts

Soumya Easwaran, Manu Dixit and Sitabhra Sinha

Abstract Analysis of time-series data of different markets have produced evidence for several stylized facts (universal features) including heavy tails characterized by power law exponents, which provide us tantalizing hints of the dynamics underlying such complex systems. It is especially important to see how these features evolve over time after the market is created and gradually develops. The recent advent of the digital currency, Bitcoin, and its growing popularity as an asset traded between agents over the last few years, provides us with an invaluable dataset for such a study. Similar to many financial markets, Bitcoin is de-centralized and its value is not controlled by a single institution, (e.g., a central bank). Here we have analyzed high-frequency Bitcoin trading data (with a resolution of one tick, i.e., a single trading event). We show that the distribution of price fluctuation (measured in terms of logarithmic return) has a heavy tail. The exponent of the tail implies that Bitcoin fluctuations follow an *inverse square law*, in contrast to the *inverse cubic law* exhibited by most financial and commodities markets. The distribution of transaction sizes and trading volume are seen to have Levy-stable distribution. Multi-scale analysis show the presence of long term memory effects in market behavior.

1 Introduction

Markets are fascinating examples of complex systems, with a large number of interacting agents acting according to strategic considerations and in response to a constant flow of external information [1]. In analogy with driven systems that self-organize into non-equilibrium steady states often characterized by power-law scaling,

S. Easwaran · S. Sinha (✉)
The Institute of Mathematical Sciences, CIT Campus, Taramani, Chennai 600113, India
e-mail: sitabhra@imsc.res.in

S. Easwaran
e-mail: soumyae@imsc.res.in

M. Dixit
Birla Institute of Technology and Science, Pilani 333031, India
e-mail: manudxt@gmail.com

© Springer International Publishing Switzerland 2015   121
F. Abergel et al. (eds.), *Econophysics and Data Driven Modelling of Market Dynamics*,
New Economic Windows, DOI 10.1007/978-3-319-08473-2_4

statistical physicists have sought to identify evidence of universal scaling behavior in markets [2]. One of the most robust of such "stylized facts" of market dynamics is the observation that the tails of the cumulative distribution of fluctuations in stock price (or market index) follow a power law, $P_c(x) \sim x^{-\alpha}$, with the exponent $\alpha \sim 3$ [3]. Initially reported for a small number of stocks from the S&P 100 list [4], this "inverse cubic law" of fluctuations [5] has subsequently been shown to hold for many different financial markets, including that of India [6]. Other distributions, such as that of trading volume, have also been occasionally claimed to have universal power-law scaling, although no consensus seems to have emerged as of now. It is thus important to see whether the type of scaling behavior associated with conventional financial market will also extend to markets in virtual financial assets that are rapidly becoming popular in the new century. In this article we analyze high-frequency price and trading data of Bitcoin, an electronic currency that has seen large rise in its trading value in the five years following its introduction. We report the existence of a novel scaling behavior in the price fluctuation that we refer to as the "inverse square law", as well as other stylized facts about the trading dynamics of the asset.

2 The Bitcoin

Bitcoin is a decentralized software-based crypto currency developed by Satoshi Nakamoto (pseudonym) in 2008, and first used on January 3, 2009. A cryptocurrency is a computer currency whose whose implementation is based on cryptography which is used to both validate transactions and generate new currency. Like most cryptocurrencies, the creation and transfer of bitcoins is based on an open source protocol which is independent of any central authority. In some respect Bitcoin is like cash in that both parties in a transaction are quasi-anonymous and are known only by their pseudonyms. The parties are identified with an address, a 27–34 alphanumeric character that comprises of a pair of public and private key. Addresses can be generated by any user at no cost and it is thus possible for a user to have more than one address. The ownership of the existing bitcoins is identified with the public key and outgoing payments are signed by an owner using the private key.

Bitcoin network operates on a peer to peer model without any regulatory oversight. All transactions in the bitcoin ecosystem are recorded in a public ledger which is called the blockchain. Thus although all transactions are between anonymous real-world entities, every transaction is globally visible. Since the blockchain is essentially the entire transaction history, each new payment is validated by checking its consistency with the blockchain and once it is verified by majority of peers in the network, is added to the blockchain. Anyone with an internet connection can download a client and become a part (node) of the network. Fraud is avoided by ensuring the majority of the participants agree on the blockchain. This is achieved by having a computationally intensive process as method for verifying transaction history. More nodes in the network mean a more secure network. In order to incentivize users to devote computational resources, new bitcoins are created periodically and distributed

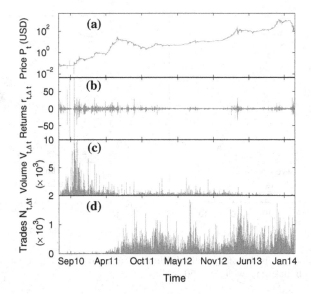

Fig. 1 Time evolution of the Bitcoin market (Sept 2010–Jan 2014). The time series of **a** price P_t, **b** normalized returns, $r_{t,\Delta t}$, **c** trading volume, $V_{t,\Delta t}$ and **d** the number of trades $N_{t,\Delta t}$ calculated over time-intervals $\Delta t = 5$ mins

among the participating nodes. Thus unlike fiat currency, new bitcoins are created and distributed by an algorithm at a fixed rate with no intervention from a central authority.

Description of the data set. The high frequency data that we analyze consists of the price and the transaction volume of Bitcoin. This data is obtained from the website of Bitcoin charts [7] which provides a simple API to download Bitcoin trading data from two exchanges: MtGox and Bitstamp. The files obtained using the API contains information about trade time, the price and amount of Bitcoins traded. The price is quoted against US Dollar and the unit for the transaction volume is 1 Bitcoin ($=10^8$ satoshi). The period over which we have collected data from MtGox exchange is Jan 11, 2010 to Feb 25, 2014, while for the Bitstamp exchange we have acquired data between Sep 11, 2011 to Jun 25, 2014. Most of the results reported here are based on our analysis of the data from the MtGox exchange.

3 Results

The price of Bitcoins, P_t, as well as, other features such as trading volume and number of trades, have shown remarkable rise and fall over the few years that it has been in existence (Fig. 1). The price fluctuations over a time scale Δt are measured in terms of the logarithmic return $R_{t,\Delta t} = log(P_{t+\Delta t}/P_t)$, that is normalized by

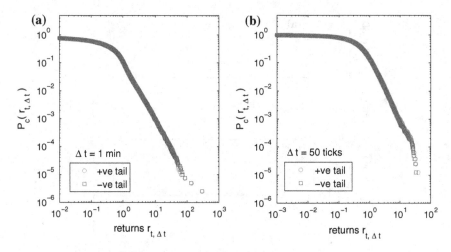

Fig. 2 Cumulative distribution of normalized returns calculated over **a** real time with time interval $\Delta t = 1$ min and **b** 'tick time' with $\Delta t = 50$ ticks

subtracting the mean value and dividing by the standard deviation (a measure of the volatility), $r_{t,\Delta t} = (R_{t,\Delta t} - \langle R_{t,\Delta t} \rangle)/\sigma_R$. Here $< \ldots >$ denotes time average over the given time period.

The distribution of r_t (Fig. 2a), as for most financial markets, show a large deviation from the Gaussian distribution. It is characterized by a heavy tail that follows a power law $P_c(r_{t,\Delta t}) \sim r_{t,\Delta t}^{-\alpha}$. However, the power-law exponent α is quite distinct from three that is seen for most markets (the well-known *inverse cubic law*). For a

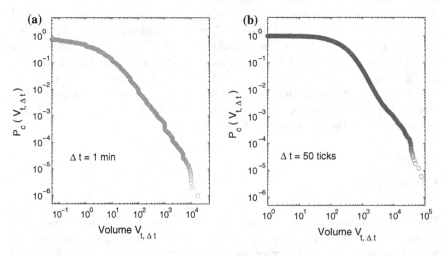

Fig. 3 Cumulative distribution of trading volume over **a** real time interval $\Delta t = 1$ min and **b** tick time interval $\Delta t = 50$ ticks

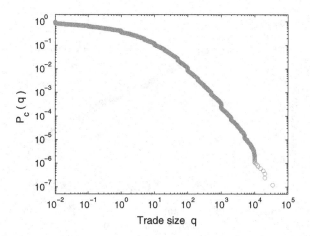

Fig. 4 Cumulative distribution of transaction sizes

time-interval of $\Delta t = 1$ min, Hill estimators of exponents for both the positive and negative tails yield $\alpha \simeq 2$. This *inverse square law* for fluctuations is marginally outside the Levy-stable regime, so that if returns are calculated over increasing values of Δt the Gaussian distribution can be recovered at large Δt.

While analysis of returns with the time interval $\Delta_t = 1$ minute is strongly suggestive of a power law tail for the price return distribution, we have also verified this using other time scales. We have also considered returns calculated using "tick"-time instead of real time, i.e., the time-scale is defined in terms of a fixed number of transactions. We again observe a power law (Fig. 2b) with Hill estimators of exponents for both positive and negative tails being $\alpha \simeq 2.2$ for $\Delta t = 50$ ticks. Thus the *inverse square* law of price fluctuations for the Bitcoin market seem to be remarkably robust with respect to the interval definition being used for measuring returns.

We observe that the trading volume distribution (i.e., the distribution of the total number of Bitcoins traded in a certain time interval) to be also characterized by a heavy tail. However, the distribution exponent is in the Levy-stable regime (Fig. 3). We also note that the distribution of the transaction size, i.e., the number of Bitcoins traded in each transaction, also happens to have a Levy-stable distribution (Fig. 4). One can expect that the trading volume distribution will have the same nature as the transaction size distribution, and in particular, that they have the same exponent if (i) the latter is Levy stable and (ii) successive transaction sizes are independent of each other. The power law exponent for transaction size has been calculated using the Hill estimator to be -1.42 ± 0.01 while the power law exponent for trading volume calculated for intervals $\Delta t = 50$ ticks and $\Delta t = 1$ min are -1.44 ± 0.08 and -1.36 ± 0.02, respectively.

Calculating the auto correlation properties of the the return time series, we see that $r(t)$ is uncorrelated, as expected from the efficient market hypothesis. However, the absolute values of the return $|r(t)|$, which measure the volatility, show a slow

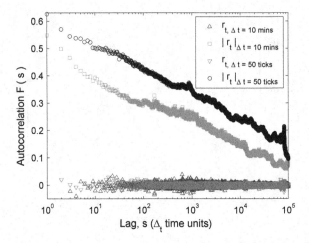

Fig. 5 The autocorrelation of normalized returns and volatility calculated over real time and tick time

logarithmic decay in their auto-correlation (Fig. 5), which is a signature of long-memory effects operating in actual market.

The distribution of real returns is also observed to exhibit multi-scaling behavior (Fig. 6). This multifractal scaling in the Bitcoin market is indicated by anomalous scaling of higher moments of price changes with time:

$$M_q(d) = < |r_d(t) - m_d|^q >_e \simeq A_q d^{\tau_q},$$

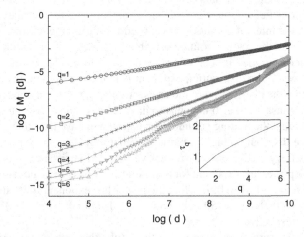

Fig. 6 The qth moments of absolute value of the fluctuations in return have a power-law scaling relation with respect to the time scale d. The inset shows the nonlinear variation of the corresponding power-law exponent, indicating multifractality

where m_d is the mean return calculated at scale d and $< \ldots >_e$ denotes the empirical average. The estimated τ_q has a concave shape indicating the multifractal nature of the returns.

4 Discussion

It is remarkable that the return distribution exponent not only departs significantly from the "inverse cubic law" reported for returns of many financial and commodities markets—and widely believed to be universal—but it is also pointing the occurrence of an "inverse square law" of markets, predicted by the model proposed in [8]. In this model, where N agents buy and sell an asset depending on the deviation of its instantaneous price from the long-term average (used as a surrogate for an estimate of the "fundamental" price of the asset), the return distribution exhibits a distribution with a heavy tail over a large range values of the parameter μ controlling the sensitivity of an agent to the magnitude of deviation of the price from its perceived fundamental value. It is of interest to note here that when agents are identical in terms of sensitivity to the price variation (i.e., every agent has the same μ (>50, say), the model predicts the inverse square law, whereas if the agents are heterogeneous (i.e., μ is distributed over an interval, say, $[10, 200]$), the usual inverse cubic law is obtained. Comparison between the model results and the Bitcoin market suggests that the latter may be distinct from conventional financial markets in that the agents trading in Bitcoin may be relatively more homogeneous in terms of their risk-sensitivity. Given that in the initial stages of the Bitcoin market, only a relatively small section of population, viz., technophiles, traded in it, the assumption of homogeneous agents is perhaps not unreasonable.

Acknowledgments We thank Frederic Abergel and Arnab Chatterjee for helpful discussions. This work is supported in part by the Department of Atomic Energy through the IMSc Econophysics (XII Plan) Project.

References

1. S. Sinha, A. Chatterjee, A. Chakraborti, B.K. Chakrabarti, *Econophysics: An Introduction* (Wiley, Weinheim, 2011)
2. R.N. Mantegna, Z. Palágyi, H.E. Stanley, Applications of statistical mechanics to finance. Phys. A **274**, 216–221 (1999)
3. T. Lux, The stable Paretian hypothesis and the frequency of large returns: an examination of major German stocks. Appl. Financ. Econ. **6**, 463–475 (1996)
4. D.W. Jansen, C.G. De Vries, On the frequency of large stock returns: putting booms and busts into perspective. Rev. Econ. Stat. **73**, 18–24 (1991)
5. P. Gopikrishnan, M. Meyer, L.A.N. Amaral, H.E. Stanley, Inverse cubic law for the distribution of stock price variations. Eur. Phys. J. B **3**, 139–140 (1998)

6. R.K. Pan, S. Sinha, Self-organization of price fluctuation distribution in evolving markets. EPL **77**, 58004 (2007)
7. http://api.bitcoincharts.com/v1/csv/
8. S.V. Vikram, S. Sinha, Emergence of universal scaling in financial markets from mean-field dynamics. Phys. Rev. E **83**, 016101 (2011)

Are Firms that Are Awarded More Patents More Productive?

**Shouji Fujimoto, Atushi Ishikawa, Takayuki Mizuno
and Tsutomu Watanabe**

Abstract What determines the productivity of firms? Firm productivity is improved by technology or efficiency. Patents are one critical indicator of a firm's technology or efficiency. In economics, production is generally described as a function of capital and labor called the production function. In this study, we analyze the relationship between patents and productivity through the production function. We classified firms by their capital and their number of employees and confirmed that patents are correlated with firm productivity that have large capital and many employees.

1 Introduction

Societies advance by technological innovation. In other words, economic growth is difficult to sustain without technological innovation. Nevertheless, quantitatively measuring technological innovation is difficult. Firm productivity is one index to

S. Fujimoto (✉) · A. Ishikawa
Faculty of Business Administration and Information Science,
Kanazawa Gakuin University, 10 Sue, Kanazawa, Ishikawa 920-1392, Japan
e-mail: fujimoto@kanazawa-gu.ac.jp

A. Ishikawa
e-mail: ishikawa@kanazawa-gu.ac.jp

S. Fujimoto · T. Mizuno · T. Watanabe
The Canon Institute for Global Studies, 11th Floor, Shin Marunouchi Building
1-5-1 Marunouchi, Chiyoda-ku, Tokyo 100-6511, Japan
e-mail: mizuno@nii.ac.jp

T. Mizuno
National Institute of Informatics, 2-1-2, Hitotsubashi, Chiyoda-ku, Tokyo 101-8430, Japan

T. Mizuno
Department of Informatics, Graduate University for Advanced Studies, Tokyo, Japan

T. Watanabe
Graduate School of Economics, The University of Tokyo, 7-3-1 Hongo, Bunkyo-ku,
Tokyo 113-0033, Japan
e-mail: watanabe@e.u-tokyo.ac.jp

© Springer International Publishing Switzerland 2015
F. Abergel et al. (eds.), *Econophysics and Data Driven Modelling of Market Dynamics*,
New Economic Windows, DOI 10.1007/978-3-319-08473-2_5

measure it. But what determines firm productivity? Even though this question is significant to quantify technological innovation, answering it is difficult because determining the productivity of firms requires various contributing factors. In this paper, we investigate patents to answer this question because they are one such contributory factor.

Although much research has addressed patents, a large portion of it has focused on networks [1–5], especially the structures of patent citation networks. Patent groups were classified by the network's structure to detect valuable patents by specifying which ones are network hubs.

Other research focused on the distribution of patents. The size distribution of the number of citations was also discussed [6]. By classifying patent applications into cities by country, the number was aggregated by cities [7]. This work confirmed that the distribution follows a power-law and compared the power-law indices of various countries.

In this paper, we employ a database that relates patents to the firms that own them to investigate the number of patents that belong to individual firms. First, we confirm that the distribution of the number of patent applications follows a power-law and that the citation counts obey a power-law distribution.

Second, we analyze the relation between the patents of firms and their productivity and conclude that the number of patent applicants of firms is related to their sales, even if their capital and labor inputs are constant. The number of applications is related to the total factor productivity (TFP), which appears in the Cobb-Douglas production function frequently used in economics.

2 Databases

In this paper, we employed the following databases. The European Patents Office (EPO) maintains a worldwide patent database called PATSTAT [8]. The Organization Economic Cooperation and Development (OECD) also has a worldwide patent database [9]. Bureau van Dijk [10] combined them and added to each patent an ID code called BvDID that enables us to classify patents by firms and to aggregate them. We can link the databases to another worldwide financial database (ORBIS) compiled by Bureau van Dijk using BvDID.

The number of applications of all kinds of patents and those owned by firms increased annually from 2000 to 2011 (Fig. 1) and from 1783 to 2011 (Fig. 2). Figure 2 also shows that the proportion of applicants owned by firms among all kinds of patents gradually increased. In such situations, we must investigate the features of the patents owned by firms.

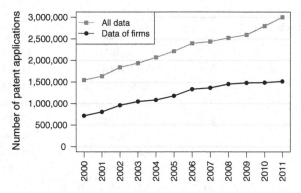

Fig. 1 Transitive graph of patent applications from 2000 to 2011

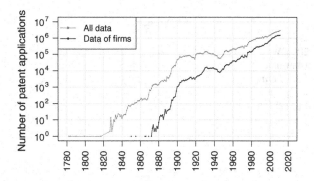

Fig. 2 Transitive graph of patent applications from 1782 to 2011

3 Distribution of Firm Patents

In this section, we observe several distributions of patents. Since an individual applies for patents several times a year at most, the distribution does not have a fat tail. In this paper, we investigate the number of patent applicants who work for firms. We denote the number of aggregated applications in each firm in some year as C. Table 1 shows the top 10 firms that applied for the most patents in 2010 and 2011. They generally belong to manufacturing or communication business industries. The number of applications by the top 10 firms is about 10,000. Note that the distribution of C in 2011 follows a power-law (Fig. 3). Figure 4 shows the annual distributions of C for a 12-year period. The power-law indices were stable in the 12 years between 1.12 and 1.17.

A citation count of patents is one index to measure a patent's value. A patent application must cite related patents as references. Therefore, frequently cited patents are important. In this paper, we denote the citation count of a patent as d (Fig. 5). By using the database described in Sect. 3, we can identify which firms own which

Table 1 Top 10 firms by patent applications in 2010 and 2011

Rank	2010	2011
1	20,867 (Samsung)	19,378 (Samsung)
2	12,670 (IBM)	12,751 (IBM)
3	12,089 (LGE)	11,594 (ZTE)
4	10,861 (Panasonic)	11,373 (LGE)
5	10,176 (Qualcomm)	10,356 (Panasonic)
6	9,525 (ZTE)	9,927 (Bosch)
7	9,422 (Bosch)	9,706 (Qualcomm)
8	8,465 (Sony)	9,087 (Canon)
9	8,365 (Canon)	8,154 (Sony)
10	7,125 (Microsoft)	7,124 (Microsoft)

Fig. 3 Cumulative distribution of patent applications in 2011. We estimated power-law index inside *vertical dashed gray lines*. *Solid gray line* is a guideline whose power-law index is 1.17

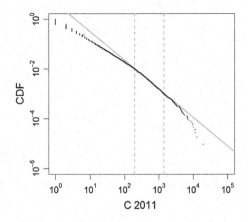

Fig. 4 Cumulative distribution of patent applications in 2000–2011. Power-law indices are between 1.12 and 1.17

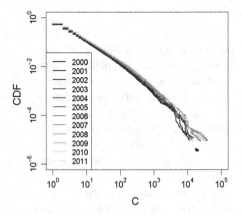

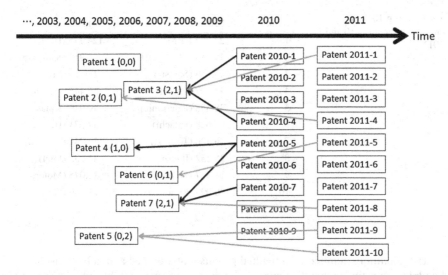

Fig. 5 Citation counts. Patent (d_{2010}, d_{2011}) was cited d_{2010} times in 2010 and d_{2011} times in 2011. For example, Patent 3 (2, 1) was cited by Patent 2010-1 and Patent 2010-4 in 2010 and was cited by Patent 2011-1 in 2011

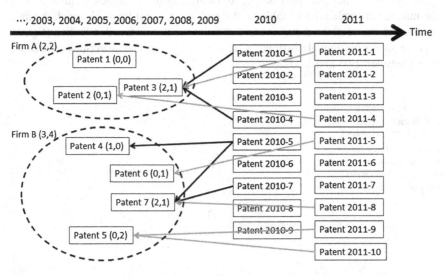

Fig. 6 Aggregation of citation counts for all firms. Firm (D_{2010}, D_{2011}) has patents that were cited D_{2010} times in 2010 and D_{2011} times in 2011. For example, Firm A (2, 2) = Patent 1 (0, 0) + Patent 2 (0, 1) + Patent 3 (2, 1)

patents. We can classify d into firms and aggregate them. Here the aggregated citation count in each firm is referred to as D (Fig. 6).

Table 2 Top 10 firms by citation counts in 2010 and 2011

Rank	2010	2011
1	127,277 (IBM)	124,775 (IBM)
2	48,353 (Microsoft)	51,912 (Samsung)
3	47,859 (Samsung)	48,619 (Microsoft)
4	42,497 (Canon)	46,445 (Canon)
5	38,358 (Panasonic)	40,046 (Panasonic)
6	37,628 (Hitachi)	37,571 (Hitachi)
7	36,326 (Toshiba)	36,241 (Toshiba)
8	35,252 (Intel)	35,235 (Intel)
9	31,317 (Sony)	32,075 (Motorola)
10	30,093 (Motorola)	31,839 (Sony)

The maximum number of individual patents that are cited annually is just a few hundred. We estimate that the power-law indices are between 2.98 and 4.08 the past 12 years. Therefore, the distribution of d lacks a fat enough tail.

Table 2 shows the top 10 firms whose applications are most frequently cited in 2010 and 2011. They generally belong to manufacturing industries. Compared with the number of applications (Table 1), the ranks of several firms changed. Against the distribution of d, the distribution of D follows a power-law distribution (Fig. 7). Figure 8 shows the annual distributions of D for 12 years. The power-law indices are stable between 1.05 and 1.09.

Fig. 7 Cumulative distribution of citation count in 2011. We estimated power-law index inside *vertical dashed gray lines*. *Solid gray line* shows that the power-law index is 1.05

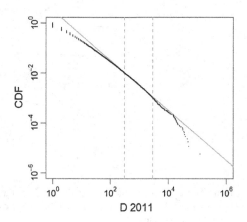

Fig. 8 Cumulative
distribution of citation
counts in 2000–2011.
Power-law indices range
from 1.05 to 1.09

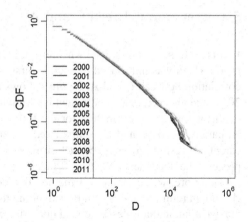

4 Cobb-Douglas Production Function and Total Factor Productivity

Identifying the variables that represent the technology of firms is crucial. In economics, total factor productivity (TFP) A is defined in the Cobb-Douglas production function [11]:

$$Y = F(K, L) = AK^{\alpha}L^{\beta}. \tag{1}$$

TFP is referred to as technology. Y is a variable for total production output, K is a variable for capital input, and L is a variable for labor input. Parameters α and β are the output elasticities of capital and labor. TFP A in Eq. (1) represents an efficiency that cannot be measured in terms of K and L.

In previous studies [12–14], we reported that the Cobb-Douglas production function can be interpreted as the following multiple regression:

$$\log Y_i = \alpha \log K_i + \beta \log L_i + \log a + \log R_i. \tag{2}$$

Here index i signifies an individual firm. We have adopted sales (in one thousands U.S. dollars), plant assets (in one thousands U.S. dollars), and the number of employee to measure the total production Y, the capital K, and the labor L respectively. α, β, and a are fitting parameters that are determined by multiple linear regression analysis in the region where Y, K, and L follow a power-law distribution.

R_i in multiple regression (2) is the residue. TFP A_i defined by aR_i is one of the variable which constitute Y_i. The variable A_i dose not depend on capital input K_i and labor input L_i. It is considered that A_i contains some input values concerning with technological efficiency which cannot explained by K_i and L_i to constitute Y_i.

5 Correlation Between C and A

In this section, we investigate whether total factor productivity A can be explained by C, which was introduced in Sect. 3. First, direct measurement cannot detect the correlation between C and A. Figure 9 shows the scatter-plots between C and A in 2011, and the correlation coefficient is estimated by -0.08. The correlation was not confirmed by direct estimation for the following reason. To determine each firm's A, parameters α, β, and a are first fixed by multiple regression (2). With them, we calculated residue R_i or A_i of each firm. When we used multiple regression in this process, the dependence of A on C fades.

Therefore, we adopt another method that detects the correlation between C and A without multiple regression. By extending the Cobb-Douglas production function (1), we assume that sales Y depend not only on K and L but also on C:

$$Y = AK^\alpha L^\beta = BC^\gamma K^\alpha L^\beta. \tag{3}$$

In Sect. 3, since we confirmed that C follows power-law distributions as well as K, L, and Y, this extension is natural. If we observe Eq. (3) in the data analysis, by comparing Eqs. (1) and (3), we conclude that

$$A = BC^\gamma. \tag{4}$$

Here B is residue. Our strategy confirms the correlation between C and Y.

Figures 10, 11, 12, 13, 14, and 15 show scatter-plots among C, K, L, and Y. Variables C and Y are correlated, and the correlation coefficient is estimated to be 0.48 (Fig. 10). This correlation conceivably comes from the correlations between C and K and between C and L through the extended Cobb-Douglas function (1). In fact, the correlation coefficients between C and K and between C and L are estimated by 0.45 and 0.51, respectively (Figs. 11 and 12). In this case, the correlation between C and Y is suspected of having spurious correlations. We have to estimate a partial correlation

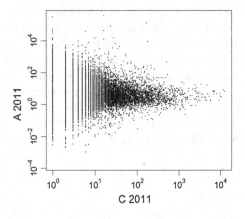

Fig. 9 Scatter-plot of patent applications C versus TFP A of all firms in 2011. Correlation coefficient between $\log C$ and $\log A$ is -0.08

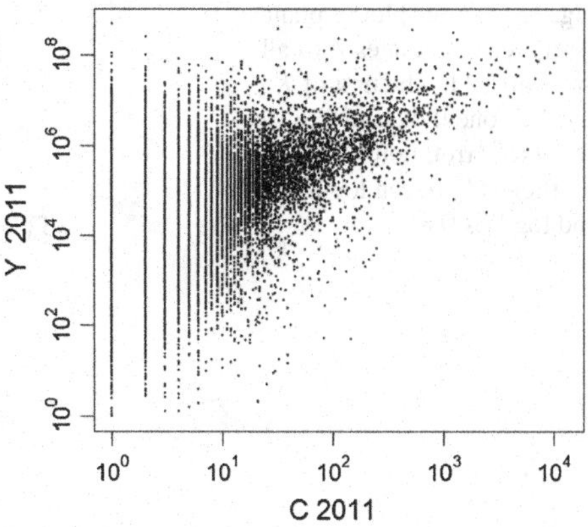

Fig. 10 Scatter-plot of
patent applications C versus
sales Y of all firms in 2011.
The unit of Y is one thousand
U.S. dollars. Correlation
coefficient between $\log C$
and $\log Y$ is 0.48

Fig. 11 Scatter-plot of
patent applications C versus
plant assets K of all firms in
2011. The unit of K is one
thousand U.S. dollars.
Correlation coefficient
between $\log C$ and $\log K$ is
0.45

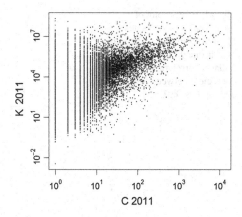

Fig. 12 Scatter-plot of
patent applications C versus
employees L of all firms in
2011. Correlation coefficient
between $\log C$ and $\log L$ is
0.51

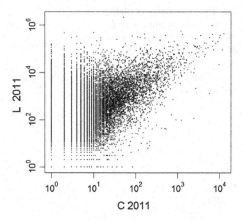

Fig. 13 Scatter-plot of plant assets K versus sales Y of all firms in 2011. The unit of K and Y is one thousand U.S. dollars. Correlation coefficient between $\log K$ and $\log Y$ is 0.85

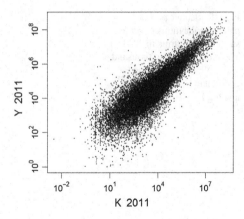

Fig. 14 Scatter-plot of employees L versus sales Y of all firms in 2011. The unit of Y is one thousand U.S. dollars. Correlation coefficient between $\log L$ and $\log Y$ is 0.91

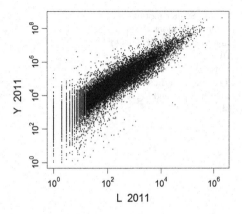

Fig. 15 Scatter-plot of plant assets K versus employees L of all firms in 2011. The unit of K is one thousand U.S. dollars. Correlation coefficient between $\log K$ and $\log L$ is 0.86

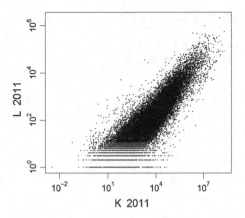

coefficient, which is a residue removed from the contributions of K and L. The value can be calculated as 0.038 using the correlation coefficients among C, K, L, and Y.

Consequently, with this approach, no true correlation between C and Y was confirmed for the following reason. In the estimation of the partial correlation, the correlations among C, K, L, and Y are calculated in all the regions. On the other hand, the correlation between C and Y must be detected in the large-scale region; it was not detected in the non-large-scale region. In this method, therefore, no correlation was detected that only exists in the large-scale region.

To avoid this problem, we restrict parameters K and L in the narrow ranges and assume that K and L are fixed and that Eq. (1) can be approximately reduced to

$$Y = \text{Const. } C^{\gamma}. \tag{5}$$

By restricting K and L in the narrow ranges of the large scales, the correlation between C and Y must emerge. For instance, Figs. 16 and 17 depict a scatter-plot between the C and Y of firms, between K and L, which are restricted in $10^2 \leq K < 10^3$ and $10^0 \leq L < 10^1$, and between $10^4 \leq K < 10^5$ and $10^2 \leq L < 10^3$. Hear, the unit of Y and K is one thousand U.S. dollars.

Table 3 shows the correlation coefficients between the C and Y of firms. The firms' K and L are confined into nine ranges of K and six ranges of L. In the table, the values in the parentheses are the amount of the data in each region. In Table 3, no correlation between C and Y was observed in the bins that contain small-sized firms. However, correlation was observed in the bins that contain large-sized firms. As a result, the C and Y of firms with large-scale K and L are correlated. This means that C and A of the large-scale firms are correlated, as we expected.

Fig. 16 Scatter-plot between C and Y of firms restricted in $10^2 \leq K < 10^3$ and $10^0 \leq L < 10^1$. The unit of K and Y is one thousand U.S. dollars. Amount of data is 1,356. Correlation coefficient between $\log C$ and $\log Y$ is 0.02

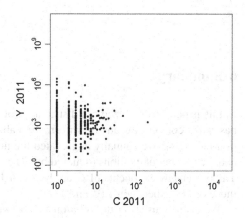

Fig. 17 Scatter-plot
between C and Y of firms
restricted in $10^4 \leq K < 10^5$
and $10^2 \leq L < 10^3$. The unit
of K and Y is one thousand
U.S. dollars. Amount of data
is 6308. Correlation
coefficient between $\log C$
and $\log Y$ is 0.22

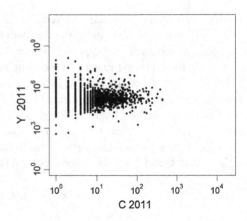

Table 3 Table of correlation coefficients between $\log C$ and $\log Y$ of firms K and L that are confined to nine ranges of K and six ranges of L. The unit of K and Y is one thousand U.S. dollars. Values in parentheses are amount of data in each region

	$\log L$	0	1	2	3	4	5
$\log K$		1	2	3	4	5	6
0	1	0.00(627)	−0.03(131)				
1	2	−0.03(1,414)	−0.02(1,114)	0.08(39)			
2	3	0.02(1,356)	0.01(4,043)	0.06(528)	−0.18(9)		
3	4	0.14(490)	0.05(6,157)	0.20(4,054)	0.05(91)		
4	5	0.27(50)	0.13(1,150)	0.22(6,308)	0.17(1,603)	0.17(10)	
5	6		−0.14(24)	0.10(764)	0.17(2,317)	−0.01(369)	
6	7			−0.13(26)	0.23(308)	0.18(498)	0.19(37)
7	8				0.45(34)	0.20(94)	0.40(48)
8	9						0.76(4)

6 Summary

In this paper, we analyzed the distribution of patents using worldwide patent databases that contains exhaustive information about ownership by firms. By classification into firms, we annually aggregated the number of applications (denoted by C) and the number of citations (denoted by D). C and D obey a power-law distribution. The power-law exponents of C are between 1.12 and 1.17 from 2000 to 2011, and those of D are between 1.05 and 1.09.

We also focused on the relationship between the number of patents owned by firms and firm productivity. By assuming that sales (denoted by Y) depend on C of the Cobb-Douglas type, we investigated the relation between Y and C. Although a correlation seemingly exists between Y and C in each firm, we confirmed that this correlation is spurious. Next we classified firms into logarithmically equal size bins

K and L and estimated the correlation coefficient between C and Y in each bin. Although no correlation between C and Y was observed in the bins that contain small-sized firms, we confirmed the correlation of the bins that contain large-sized firms. Under the assumption that Y depends on C by the Cobb-Douglas type, the correlation between C and total factor productivity A is also correlated. From these results, in large-sized firms, C is considered a factor that constitutes A. Variable C contains some technological features that improve the productivity of large-sized firms.

In this study, we proposed one statistical method to measure the contribution of patents owned by firms to the production or technology of the firms. Variables may exist that represent the influence in a more appropriate manner than C. We also adopted citation count D as well as C and investigated the similar correlation between D and Y. The correlation between D and Y did not change drastically from the correlation between C and Y. This is probably because patents are published to defend patent rights. Patent citations differ from the citations of scientific papers. We employed sales Y to represent production. However, sales contain various outputs that are not directly related to patents. For example, if we adopt patent income instead of sales, which represent production, we can observe more clearly the correlation between patents and production.

We discussed the relation between C and Y in large-scale firms. This does not necessarily mean that C influences Y. To establish the influence's presence, we must consider the relation between the changes of C and Y at intervals of several years. This is a future problem.

Acknowledgments The authors thank the Saha Institute of Nuclear Physics, where this work was presented for "ECONOPHYS-KOLKATA VIII" as well as Prof. H. Aoyama for many useful discussions and comments about spurious correlations. This work was supported in part by a Grant-in-Aid for Scientific Research (C) (No. 24510212) from the Ministry of Education, Culture, Sports, Science and Technology, Japan. T. M. was supported in part by a Grant-in-Aid for Young Scientists (B) (No.24710156).

References

1. X. Li, H. Chen, Z. Huang, M.C. Roco, Patent citation network in nanotechnology (1976–2004). J. Nanopart. Res. **9**, 337–352 (2007)
2. S.B. Chang, K.K. Lai, S.M. Chang, Exploring technology diffusion and classification of business methods: using the patent citation network. Technol. Forecast. Soc. Change **76**, 107–117 (2009)
3. J.C. Wang, C.H. Chiang, S.W. Lin, Network structure of innovation: can brokerage or closure predict patent quality? Scientometrics **84**, 735–748 (2010)
4. B. Gress, Properties of the USPTO patent citation network: 1963–2002. World Pat. Inf. **32**, 3–21 (2010)
5. S.W. Hung, A.P. Wang, Examining the small world phenomenon in the patent citation network: a case study of the radio frequency identification (RFID) network. Scientometrics **82**, 121–134 (2010)

6. G. Silverberg, B. Verspagen, The size distribution of innovations revisited: an application of extreme value statistics to citation and value measures of patent significance. J. Econom. **139**, 318–339 (2007)
7. D.R.J. O'Neale, S.C. Hendy, Power law distributions of patents as indicators of innovation. PROS ONE (2012). doi:10.1371/journal.pone.0049501
8. PATSTAT, http://www.epo.org/searching/subscription/raw/product-14-24.html
9. OECD Patent Statistics, http://www.oecd-ilibrary.org/science-and-technology/data/oecd-patent-statistics_patent-data-en
10. Bureau van Dijk Co., http://www.bvdinfo.com/Home.aspx
11. C.W. Cobb, P.H. Douglas, A theory of production. Am. Econ. Rev. **18**, 139–165 (1928)
12. S. Fujimoto, A. Ishikawa, T. Mizuno, T. Watanabe, A new method for measuring tail exponents of firm size distributions. Econom.: Open-Access, Open-Assess. E-J. **5**, 1–20 (2011)
13. A. Ishikawa, S. Fujimoto, T. Mizuno, T. Watanabe, The emergence of different tail exponents in the distributions of firm size variables. Phys. A **392**, 2104–2113 (2013)
14. A. Ishikawa, S. Fujimoto, T. Mizuno, T. Watanabe, Analytical derivation of power laws in firm size variables from Gibratfs law and quasi-inversion symmetry: a geomorphological approach. J. Phys. Soc. Jpn. **83**, 034802 (2014)

Nonlinear Dynamics of Stock Markets During Critical Periods

Kousik Guhathakurta

Abstract Stock market crashes have always been a subject of intimate study in financial economics literature. Starting from [1] to [2], the reasons, nature and impact of stock market crashes have been analysed in a various ways by various authors ranging from econometric to behavioural and physics based models to explain the phenomena. Mostly, these recent works have shown an analogy between crashes and phase transition. Using a technique evolved from nonlinear dynamics and physics, it is possible to graphically represent the dynamic evolution of a system. This technique known as Recurrence Plot (RP) can detect critical phases in the system and changes in the same. Inspired by this several authors used this technique to try and detect bubbles and crashes including [28]. The present work extends the findings of the same work. Using the recurrence statistics, we show that it is possible to detect critical periods in advance for all the cases where there was a known bubble building up in the market. RP alone can not predict cashes but definitely, this tool may be used to identify changes in market dynamics and can serve as a warning bell.

1 Introduction

Stock market crashes have always been a subject of intimate study in financial economics literature. Starting from [1] to [2], the reasons, nature and impact of stock market crashes have been analysed in a various ways by various authors. Some authors like [3] focus on micro-behaviour of traders and information asymmetry to explain crashes. Shiller [4] tries to provide a behavioural finance perspective to stock market crashes. Li and Xue [5] identify a Bayesian investors belief evolution when facing a structural break in economy and links it to the bubbles and crashes. There has been some significant investigations of the major crashes like the 1987 black

K. Guhathakurta (✉)
Indian Institute of Management Indore,
Prabandh Shikhar, Rau-Pithampur Road, Indore 453556, Madhya Pradesh, India
e-mail: kousikg@gmail.com

© Springer International Publishing Switzerland 2015   143
F. Abergel et al. (eds.), *Econophysics and Data Driven Modelling of Market Dynamics*,
New Economic Windows, DOI 10.1007/978-3-319-08473-2_6

Monday [4, 6–9], or the tech bubble [10]. Bates [2] tries to capture the stochastic volatility and the substantial outliers observed in U.S. stock market returns over the past 85 years and thus providing a clue to stock market crash risk. There has been a host of empirical studies on financial market crashes as well [11–16].

While all these works were based on traditional econometric modelling, there has been considerable research work towards modelling financial crashes based on analogies from physics, most suggesting that, close to a crash the market behaves like a thermodynamic system which undergoes phase transition. Some propose a picture of stock market crashes as critical points in a system with discrete scale invariance. The critical exponent is then complex, leading to log-periodic fluctuations in stock market indices. This picture is in the spirit of the known earthquake-stock market analogy and of recent work on log-periodic fluctuations associated with earthquakes [17, 18]. Some has shown that stock market crashes are caused by the slow build up of long-range correlations between traders, leading to a collapse of the stock market in one critical instant. A crash is interpreted as a critical point [19, 20].

Mostly, these recent works have shown an analogy between crashes and phase transition [21–24]; as in earthquakes, log periodic oscillations have been found before some crashes [19, 25], and then it was proposed that an economic index increases as a complex power law. These works have significance for understanding crash from policy perspective. Using standard econometric models it is not possible to detect the bubble in advance or predict a crash even in approximate terms. However, using a technique evolved from nonlinear dynamics and physics, it is possible to graphically represent the dynamic evolution of a system. This technique known as Recurrence Plot (RP) can detect critical phases in the system and changes in the same. Inspired by this, Fabretti and Ausloos [26] first show that using recurrence plot and its quantification one can detect endogenous crashes. Guhathakurta et al. [27–29] further established the tool as a crash detector by using the techniques to distinguish between endogenous and exogenous crashes. Bastos [30] also reinforced their findings.

The present work extends the original findings of Guhathakurta et al. [29] by analysing eight different financial crashes at different times. The purpose is to identify the critical phases as also to understand the dynamics of the stock market during such periods. Using recurrence statistics, we show that it is possible to detect critical periods in advance for all the cases where there was a known bubble building up in the market. We can therefore, consider Recurrence Analysis as a valid tool to detect bubbles in the market. Using this technique we also detect the average length of the bubble and see whether we can predict the same or not. We also offer an economic intuition for the reason of change in dynamic characteristics of market during the critical phases.

The rest of the paper is presented in the following way. The next section gives back ground of the crashes that we have selected. Section 3 mentions the data source and software used. Section 4 deals with the methodology. Next section discusses the results and their interpretation. In the concluding section we summarise our findings and discuss the implications of the same for policy makers and investors.

2 The Stock Market Crashes Under Study

A stock market crash takes place when prices in the stock market plunges deep within a short period of time due to overselling. This may occur as a fall out of an unreasonable rise of stock prices in the market (when the rise is not justified by economic fundamentals) when the market 'corrects' itself by bringing the price down. We call such unreasonable price rise in stock market 'bubbles' and the crashes that follow them as endogenous crashes. Sometimes a crash is caused by an external event like failure of major corporate bodies or political turmoil or even terrorist attacks. In such a case we designate the crash as exogenous crash. The problem of distinguishing the crashes is that often they are a mix of both in that an external shock triggers a crash in a market where already a 'bubble' was present. In our study we have tried to identify certain crashes where no 'bubbles' were reported and examined them alongside some known 'endogenous' crashes. The purpose was to establish the robustness of our tool. If the tool is robust then it should not detect any bubble in all the exogenous crashes while it will definitely detect bubbles in endogenous crashes. We can then use this analysis to detect endogenous crashes and bubbles. We now present a brief description of each of the stock market crashes that we have chosen for our study.

2.1 Black Monday (Oct 19, 1987) Crash

October 19, 1987 the date which is named as "Black Monday" was the date on which investor saw one of the biggest fall in global stock markets. Intra-day S&P 500 Index declined by more than 20 % and the Hong Kong market fell by a staggering 45 % by month end.

Several factors are attributed to this crash. These included market overvaluation, high US trade deficit data and rising interest rates globally, housing committee plans to tax mergers events heavily, continued federal budget deficits. However, the most important factor that came out of analysis was "Programme Trading" strategies which were meant to provide "portfolio insurance" by applying limit loss function which automatically sell shares when prices fell in order to avoid big losses. Excessive selling pressure drove prices down which created a vicious circle of selling. Kyle [31] argued that unusual price volatility during the crash was due to insufficient liquidity provided by long term investors to balance the trade initiated by portfolio insurance programme. Gennotte [32], presented a model of supply shocks, showing portfolio insurance program leading to sudden permanent price revisions. Since no evidence of a sustained rally in market was there, and it was mainly a temporally localised event we classify this crash as an exogenous one.

2.2 Japanese Bubble (1986–1991)

The Japanese asset price bubble from 1986 to 1991, resulted in price inflation of real estate and stock prices and according to Paul Krugman, this bubble burst left Japan lurking in liquidity trap. The research paper of the biggest contributor of this bubble was loose monetary policy during 1980s which resulted in the continual lowering of the interest rate. The interest rate, was much lower in Japan than its global counterparts, which caused Japan's stock prices to go skyrocketed [33]. The high rate of land price inflation in the late 1980s was also attributed to high money growth [33]. This crash as we can see, is obviously an endogenous crash.

2.3 The Friday the 13th Mini-crash (Oct 13, 1989)

The Friday the 13th mini-crash is referred to the stock market crash that occurred on Friday, October 13, 1989. The Dow Jones Industrial Average closed 190.58 points or 6.91 % down, to 2,569.26. The NASDAQ Composite fell by 14.90 points, or 3.09 %, to 467.30, and the S&P 500 Index fell 21.74 points, or 6.12 %, to 333.65.

The news of disruption of a $6.75 billion leveraged buyout deal for UAL Corporation, created the trigger for crash with the collapse of the junk bond market. Indices started plunging just moment after this news. We should consider this as an exogenous crash.

2.4 October 27, 1997 Mini-crash

The October 27, 1997 mini-crash is the global stock market crash triggered by an Asian economic crisis. Dow Jones Industrial Average suffered its eighth biggest point loss since its creation in 1896. The crash started overnight in Asia as Hong Kong's Hang Seng Index plummeted 6 % and then contagiously spread to the European markets where London's FTSE 100 Index fell 98.90 points, or just about 2 %, to 4,871.30. The Frankfurt DAX index fell sharply as well, the Dow hit its first trading curb halt when it fell 350 points. Several stock market analysts attributed this crash as a "correction" to the overheated markets, which saw 100 % upside in only 30 months. With this evidence, one may consider it as an example of endogenous crash.

2.5 11 Sept NYSE, 2001 Crash

The NYSE crash in the aftermath of 11 Sept, 2001 tragedy led to a sharp decline in stock markets. Investors reaching to safe heavens gold prices spiked upwards, from 215.50 to 287 an ounce in London trading. Gas prices in the United States also briefly shot up. The United States dollar fall sharply against the Euro, British pound, and

Japanese yen. The next day, European stock markets fell sharply, including declines of 4.6 % in Spain, 8.5 % in Germany, and 5.7 % on the London Stock Exchange, 9.2 % drop in Brazil, 5.2 % drop in Argentina, and 5.6 % decline in Mexico, before trading was halted. By all accounts, 9/11 crash is definitely an exogenous crash.

2.6 Stock Market Downturn of 2002, DJIA

The stock market downturn of 2002 is the sharp drop in stock prices during 2002 in stock exchanges across the United States, Canada, Asia, and Europe. After recovering from lows reached in the aftermath of September 11 attacks, indices fell steadily starting in March 2002, with sharp declines in July and September leading 6 year lows.

This crash can be viewed as part of a correction that began in 2000, After a decade long bull market had led to unrealistically high stock valuations (Cleveland Federal Committee Report). Collapse of Enron & many internet companies (Webvan, Exodus Communications, and Pets.com went bankrupt due to unsustainable high stock valuations. Accounting scandals of Arthur Andersen, Adelphia, Enron, and WorldCom may also be noted.) tarnished investor confidence as numerous large corporations were forced to restate earnings. As evidence suggests, this is an endogenous crash.

2.7 2010 Flash Crash

The May 6, 2010 Flash Crash, was a United States stock market crash on Thursday May 6, 2010 in which the Dow Jones Industrial Average plummeted by about 1,000 points/9 %. It was the second largest point swing, 1,010.14 points, and the biggest one-day point decline, 998.5 points, on an intraday basis in Dow Jones Industrial Average history.

One major work found that the crash was caused due to overreliance on computer systems and high-frequency trading [34]. Other significant contributors identified included the use of market and stop-loss orders, a decrease in market maker trading activity, and order routing issues among securities exchanges. High frequency traders contribution to higher trading volumes was mistaken for liquidity by Fundamental Traders. But at the time of re-balancing their positions, High Frequency Traders competed for liquidity and amplified price volatility. Here also we do not find any evidence of bubble and hence we consider this crash as exogenous.

2.8 China 2007 Crash

The global stock market plunge of February 27, 2007 started from China which destroyed a market value of about hundreds of billions. After rumours of interest rate rising to curb inflation and clamp down on speculative trading with borrowed money, the SSE Composite Index of the Shanghai Stock Exchange fell by 9 %, the largest drop in 10 years. After the Chinese market crash, the Dow Jones Industrial Average in the United States dropped by 416 points, or 3.29 % from 12,632 to 12,216.

According to research the overvaluation of listed state owned companies was the most important reason for the correction of Chinese stock market [35]. The three large state owned banks- Bank of China (BOC), China Construction Bank (CCB) and ICBC had been heavily recapitalised by the government. With such massive support from government, the balance sheets of these three banks looked healthy; However it skyrocketed their shares to unrealistic values. By the end of 2007, rocketing share prices catapulted five of the Chinese listed enterprises, including Petro China, ICBC, Sinopec, China Life Insurance and China Mobile to ranks of the worlds largest 10 companies in terms of market value. However, the profitability and productivity of this Chinese enterprise were well below those of their global counterparts. It was overvalued by at least six times compared with ExxonMobil in terms of their price to earnings (P/E) ratios. This is our final sample of endogenous crash.

3 Data & Software

As explained in the section before our analysis covers eight different stock market crashes which occurred in different stock exchanges across the world. The choice of the crashes were such that four of the crashes could be identified as endogenous, i.e., preceded by a bubble while four others, the crashes were reported to be more caused by external shock than because of a bubble building up beforehand. The reason for this choice was to check for the robustness of our method. If the method is robust then we should be able to distinguish between endogenous and exogenous crashes. The data for analysis was the closing value of the respective stock indices around the crash period (Table 1).

Table 1 Details of Stock market crashes under study

Crash details	Index	Type
Black monday (Oct 19, 1987) crash	Hong Kong AOI	Exogenous
Japanese bubble (1986–91)	NIKKEI	Endogenous
The friday the 13th mini-crash (Oct 13, 1989)	Dow Jones Industrial Average (DJIA)	Exogenous
October 27, 1997 mini-crash	HangSeng	Endogenous
11 Sept NYSE, 2001 crash	DJIA	Exogenous
Stock market downturn of 2002, DJIA	DJIA	Endogenous
2010 flash crash	DJIA	Exogenous
China 2007 crash	CSI300	Endogenous

The data is taken from 3 years before the crash to 3 years post crash. All the analysis was carried out on MATLAB platform.

4 Theoretical Background of Empirical Tools

4.1 Recurrence Plot

4.1.1 Recurrence Analysis

Natural processes can have a distinct recurrent behavior, e.g., periodicity (as seasonal or Milankovich cycles), but also irregular cyclicities (as ElNiño/Southern Oscillation). Recurrence of states $\mathbf{x_i} \in \mathbb{R}^m$ (with m the dimension of the phase space), in the meaning that states are arbitrary close after some time, is a fundamental property of deterministic dynamical systems.

Eckmann et al. have introduced a tool which visualizes the recurrence of states $\mathbf{x}_i$ in phase space [36]: the recurrence plot. A recurrence plot (RP) is a visualisation of state-space dynamics that shows all those times at which a state of the dynamical system recurs:

$$R_{i,j} = \Theta(\varepsilon - \parallel \mathbf{x}_i - \mathbf{x}_j \parallel), \quad \mathbf{x}_i \in \mathbb{R}^m, \quad i, j = 1, \ldots, N, \quad (1)$$

where $\mathbf{R}$ is the recurrence matrix, N is the number of considered states x_i, ε is a threshold distance, $\parallel \cdot \parallel$ a norm, and $\Theta(\cdot)$ the Heaviside function. A recurrence of a state at time i at a different time j is, thus, marked within a two-dimensional squared matrix with ones and zeros. Both axes of the recurrence matrix are time axes. This representation is called recurrence plot (RP). RPs has shown to be useful for analysing short and non-stationary data [37].

In our study we apply the RP in order to reveal the characteristics of the dynamics of the economic time series under investigation. For an economic time series, the patterns over time tell us whether the series is disrupted, non-stationary or nonlinear in nature. The plot is time versus time graph where the recurrence distances represent by colour codes indicate the state of the system evolving over time.

4.1.2 Embedding Parameters

If only one observable is available, the phase space can be reconstructed using time-delay embedding [37]. Thus, we need to choose an appropriate value for the time delay d and the embedding dimension m. Several methods have been developed to best estimate m and d. Frequently used methods are the Average Mutual Information Function (AMI) for the time delay [38] and the False Nearest Neighbors (FNN) method for the embedding dimension [39]. As for the embedding delay, we choose such a value where the mutual information has its first minimum or changes its scaling behavior, and for the embedding dimension, we use such a value for m where the number of false nearest neighbours in the phase space vanishes.

4.1.3 Structures in Recurrence Plots

The initial purpose of RPs was the visual inspection of recurrences of phase space trajectories. The view on RPs gives hints about the time evolution of these trajectories. RPs exhibit characteristic large scale *typology* and small scale patterns (*texture*). The typology offers a global impression which can be characterized as homogeneous, periodic, drift, and disrupted [37]. Small scale structures are single dots, diagonal lines as well as vertical and horizontal lines (the combination of vertical and horizontal lines obviously forms rectangular clusters of recurrence points). For a recurrence analysis, the diagonal and vertical line structures are important.

A diagonal line $R_{i+k,j+k} = 1$ (for $k = 1, \ldots, l$, where l is the length of the diagonal line) occurs when a segment of the trajectory runs parallel to another segment, i.e., the trajectory visits the same region of the phase space at different times. The length of this diagonal line is determined by the duration of such similar local evolution of the trajectory segments and can give an idea about its divergence behavior, i.e., the faster the trajectory segments diverge, the shorter are the diagonal lines.

A vertical (horizontal) line $R_{i,j+k} = 1$ (for $k = 1, \ldots, v$, where v is the length of the vertical line) marks a time length in which a state does not change or changes very slowly. It seems, that the state is trapped for some time. This is a typical behavior of laminar states (intermittency).

These small scale structures are the base of a quantitative analysis of the RPs. In our case, we need to observe whether the plot reveals any significant change in the dynamic state of the time series under study during the period of reported bubbles. Though the visual interpretation of RPs requires some experience, their quantification offers a more objective way for the investigation of the considered system. A detailed discussion on the application and interpretation of RPs and the various structures in a RP can be found in [37].

4.2 Quantification of Recurrence Plots (Recurrence Quantification Analysis) With Confidence Intervals

A quantification of recurrence plots (Recurrence Quantification Analysis, RQA) was developed in order to distinguish between different appearances of RPs [40, 41]. Measures which base on diagonal structures are able to find chaos-order transitions, whereas measures based on vertical (horizontal) structures are able to find chaos-chaos transitions (laminar phases) [37].

Using the histogram of diagonal line lengths, we define the fraction of recurrence points forming diagonal lines as a measure called *determinism DET*,

$$DET = \frac{\sum\limits_{l=l_{\min}}^{N} l\,P(l)}{\sum\limits_{l=1}^{N} l\,P(l)}, \tag{2}$$

where $P(l)$ is the histogram of the diagonal lines of exactly length l, and $l_{\min}$ is a minimal length a diagonal structure should have to be counted as a line. Processes with uncorrelated or weakly correlated, stochastic or irregular chaotic behaviour cause none or very short diagonals, hence, small *DET*. In contrast, regular deterministic processes lead to longer diagonals and less isolated recurrence points, resulting in higher values of *DET*. This measure can also be interpreted as characterizing the predictability of the system.

The *average diagonal line length*

$$L = \frac{\sum\limits_{l=l_{\min}}^{N} l\,P(l)}{\sum\limits_{l=l_{\min}}^{N} P(l)} \tag{3}$$

gives the average time that two segments of the trajectory are close to each other, and can be interpreted as the mean prediction time.

Analogously to the definition of the determinism in Eq. (2), we can use the histogram of the vertical lines of exactly length v, and define the fraction of recurrence points forming vertical structures in the RP as the *laminarity LAM*

$$LAM = \frac{\sum\limits_{v=v_{min}}^{N} v\,P(v)}{\sum\limits_{v=1}^{N} v\,P(v)}. \tag{4}$$

The computation of *LAM* is realized for those v that exceed a minimum length v_{min} in order to decrease the influence of the tangential motion (time-continuous systems that are discretized with sufficiently high sampling rate and an appropriately large threshold ε result in a large amount of recurrences coming from succeeding states $x_i, x_{i+1}, x_{i+2}, \ldots$). *LAM* represents the occurrence of laminar states in the system without describing the length of these laminar phases. In particular, *LAM* decreases if the RP consists of more isolated recurrence points than vertical structures.

The average length of vertical structures is given by

$$TT = \frac{\sum\limits_{v=v_{min}}^{N} vP(v)}{\sum\limits_{v=v_{min}}^{N} P(v)}, \tag{5}$$

and is called *trapping time*. As in the case of *LAM*, the computation of *TT* requires the consideration of a minimal length v_{min} as well. The trapping time estimates the mean time that the system will abide at a specific state, i.e., how long the state will be trapped.

Both *LAM* and *TT* have been proven to be useful for describing the dynamics of discrete systems and studying chaos-chaos transitions. RQA consists of further measures which are not used in this study. RQA as the whole is a very powerful technique for quantifying differences in the dynamics of complex systems and has meanwhile found numerous applications, e.g., in astrophysics, biology, engineering, geo- and life sciences, or protein research [41].

5 Analysis of Empirical Results

The Recurrence Plot of any time series reveals the dynamic patterns of the time evolution of the data set. We have used a un-thresholded version of the Recurrence Plot. For disrupted systems like stock markets, un-thresholded RP reveals clearer picture, depicting the change in the recurrence distance matrix. The time series representing stock market indices are non-stationary and their evolution always shows change in their determinism over different epochs as depicted by the RPs. If we look carefully at the plots of respective Recurrence Plots of the different data set, we can see that the colour bands changed over the entire regime indicating a transition in dynamic state. The colour code gives us a clue about the level of determinism in the system. The colour band at the bottom of each RP depicts the range of recurrences which in turn indicate the level of determinism in the system. As we move from left to right (from red to magenta, the level of determinism increases). This means that the light red and yellow colours indicate lesser recurrence distance pointing to a random or stochastic system, while the green and blue indicates a high recurrence distance pointing out to a high degree of determinism. The significance of this is that whenever we see a dark band (Blue and Magenta) emanating we can understand the dynamic system is no longer exhibiting pure stochastic behaviour but a trend is emerging instead. With respect to stock market data, we can take this as a bubble setting in. With this frame work if we look at the Recurrence Plot of the respective time series representing the different stock index data around the Crash dates, we can find out whether the RP can display such dark bands before the rash dates. The recurrence quantification analysis was done by dividing the entire time period into several epochs of 100 days each. So the values of the RQA statistics in the couple of epochs before the crash are of significance to us.

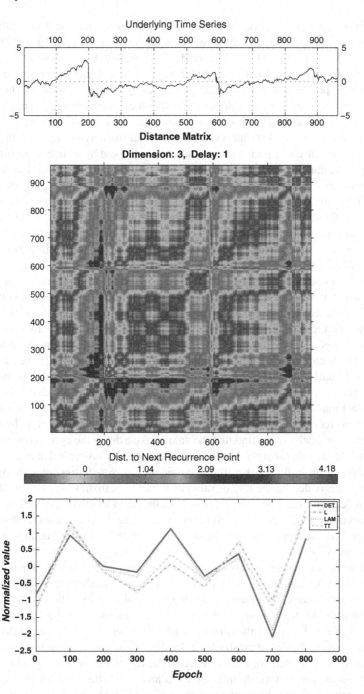

Fig. 1 Recurrence plot of time series representing the daily close of HongKong AOI index (*top*). One cannot clearly discern any *dark band* emanating. RQA values of the same time series (*bottom*). We can see that the same fluctuates randomly around the crash period

We now try to see what picture emerges from the recurrence analysis of the eight crashes under study. We start with the famous Black Monday (Oct 19, 1987) Crash. The index under study is Hong Kong AOI. We know that this crash was triggered by an external shock and no known endogenous bubble was building up in the market over a long period. If the recurrence plot picks up change in trend, it should not display any significant change in dynamics. In other words no significant change in colour of the map should be observed before or after the crash. The recurrence plot of the time series (see Fig. 1) reflects an exactly similar story. The graph is a time versus time plot with the entire period under study being denoted by number of trading days staring from one. If we move along the x-axis, we find that there is a thin dark line before the crash point at around the 200th point. But there is no significant change that is persisting for long enough a period to suggest a trend emerging in the market, which should have been present, had this been a bubble. If we look at the values of RQA statistics (bottom half of Fig. 1), we can see that the values fluctuate randomly and there is no distinctive increase in determinism of the system in period before crash. We now move forward to examine an endogenous crash.

The Japanese bubble, as explained in Sect. 2.2, was a known stock market bubble that persisted over a long period. In this case the recurrence plot should be able to detect a change in dynamic state of the system as we move through the time evolution of the data series. Like we have discussed in the introductory paragraph of this section, a dark band should emerge some trading days before the day of the crash. As we carefully inspect the recurrence plot (top half of Fig. 2), we can easily detect such dark band emerging at least 200 trading days before the day of the crash, which is equivalent to a period almost 9–10 months. This is our first vindication of the fact that as because during a bubble, the stock market goes thorough a phase change, the recurrence plot can detect the same well in advance. When we look at the quantification analysis, we find that the values take a dip in the epoch containing the crash while they rise sharply in epoch before that. This suggests that as the bubble was building up in the market, the dynamic system was following a strong trend which was revealed by the rise in value of the RQA statistics which signified high level of determinism. The crash effect breaks the trend which is captured by the fall in values of the RQA statistics in the epoch containing the crash. This reinforces our findings from the recurrence plot. Next we turn our attention once again to a known exogenous crash.

The next stock market crash under study is the one infamously known as the 'Friday the 13th mini-crash' (Oct 13, 1989). This was another crash caused chiefly by an external shock, viz., the news of disruption of a $6.75 billion leveraged buy-out deal for UAL Corporation (refer Sect. 2.3). If we examine the recurrence plot (top half of Fig. 3) of the time series, we find that there is no distinctive dark band emerging before the day of crash (day 250). Thus in case of exogenous crash, where no bubble was reported we find once again the recurrence plot not picking up any wrong signal. The recurrence quantification analysis(bottom half of Fig. 3) do not pick up any significant change in the determinism of the system during the preceding epoch and the epoch containing the crash day. We now investigate another reported endogenous crash.

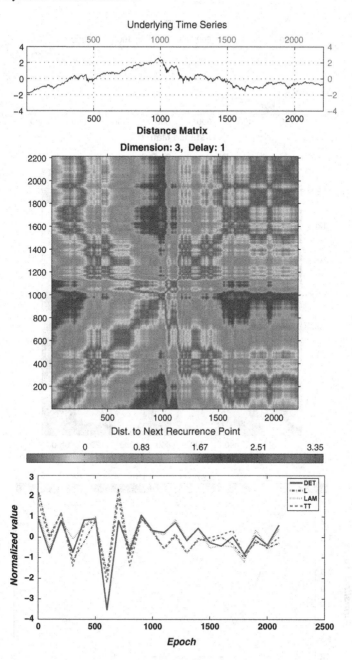

Fig. 2 Recurrence plot of time series representing the daily close of NIKKEI index (*top*). Here, we can clearly detect a *dark band* emerging about 200 trading days before the crash point (from around 800 to 1,000 on X-axis). RQA values of the same time series (*bottom*) shows that there is a rise in determinism in the epoch before the crash and a sudden drop in same as we approach the crash

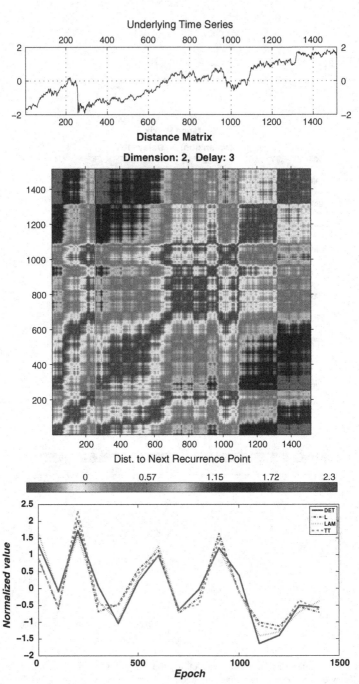

Fig. 3 Recurrence plot of time series representing the daily close of DJIA index (*top*) to examine the 'Friday the 13th mini-crash'. There is no distinguishable change in character in RP near the crash. RQA values of the same time series (*bottom*) which also supports the findings in RP

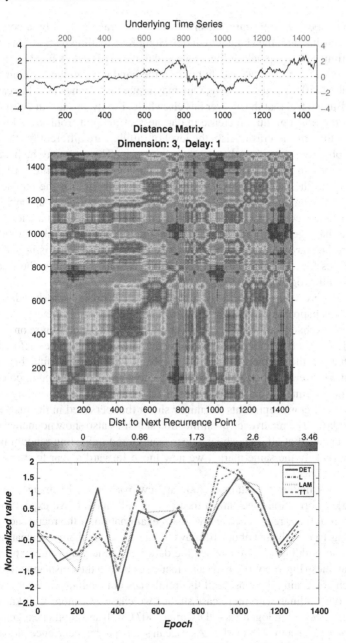

Fig. 4 Recurrence plot of time series representing the daily close of HangSeng index (*top*). The October, 1997 mini crash being an endogenous crash, the RP does detect a trend as shown by the *dark band* before the crash (800th trading day). However, it occurs for a short period, say about 50 trading days. This may be due to the fact that the extraordinary rally may have begun quite late. RQA values of the same time series (*bottom*) also fluctuate randomly with no clear indication around the crash period

The October 27, 1997 mini-crash was a global crash caused by a correction of markets starting with the HangSeng index Hong Kong falling by 6%. On examining the recurrence plot (top half of Fig. 4) we find that there is a dark band just before the crash. However, in this case we do not find the band to continue for long. In fact it is only about 50 trading days long. One may argue that RP has failed to capture the bubble. But we feel that the answer lies elsewhere. Firstly, on careful observation we find that there is a gradual increase in darkness of the plot from say 100 days, i.e., three months prior to crash. Also, the level of rallying are different for the different bubbles observed across markets in different times. The case may be here that the herd behaviour of the market peaked only about a month before the crash day. The very fact that the market recovered also fast also tells us that the bubble was not too deep. We can argue that not only does recurrence plot picks up bubbles it can distinguish between soft and hard rally in the markets. The RQA statistics reveal that the values have general rising trend about 400 days before the crash(at 800th trading day from the start)while it dips down just before the crash. This is in line with our expectations for a bubble. Our next set of observations is based on the study of the mother of all exogenous crashes.

If ever there was any decidedly exogenous stock market crash, when nothing unusual was happening in the market prior to the crash, it was the crash of the New York stock exchange after the 9/11 attacks on the World Trade Centre on September 11, 2001. We put the time series containing the DJIA closing values before and after the crash under the scanner of recurrence analysis. Expectedly we find the recurrence plot (Fig. 5) revealing a graph where there is no significant change in colour in the period approaching the crash. The very thin dark band which incidentally starts only on the date of crash represents the minor shock that persisted in the market for the next few days. The recurrence quantification statistics also show no increasing trend near the crash. If at all, they show a decreasing trend indicating random behaviour in the market. In the same market we now move forward a year to study a known endogenous crash.

The Stock market downturn of 2002 at New York Stock Exchange was a result of a market correction mechanism to previous bull market. We expect to detect a bubble through our recurrence analysis. An examination of the recurrence plot (top half of Fig. 6) reveals that about 100 days before the crash there is a clear indication of a bubble building up. We can see the dark band emerging quite explicitly. The RQA statistics (Fig. 6 bottom half) also increases before the crash and takes a dip in the epoch containing the crash. All this reinforces our findings in other crashes. To validate our findings with more case studies we chase two more well known stock market crashes one endogenous—the China 2007 Crash and another exogenous—the 2010 Flash crash at the US market. As we can see from the recurrence plot and RQA (Fig. 7), in case of the China crash, recurrence analysis shows a dark band in RP and rise and fall of RQA statistics around the crash to indicate a strong presence of bubble. On the other hand the same graphs for the Flash Crash (Fig. 8) shows no such signs. Our analysis thus establishes the capacity of the recurrence analysis to detect critical regimes and change in dynamics of stock markets operating in different times and places.

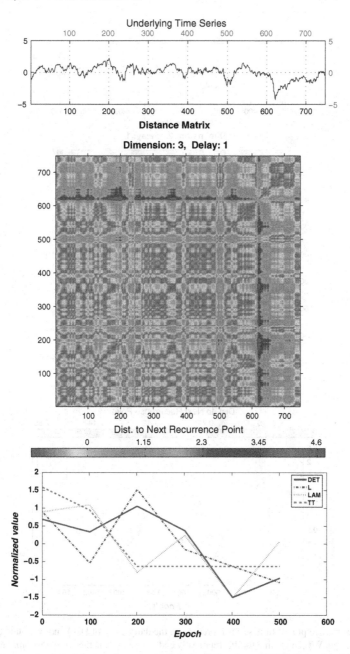

Fig. 5 Recurrence plot of time series representing the daily close of DJIA index (*top*). The 9/11 crash being another exogenous crash, the RP once again does not detect any significant change in trend around the crash. RQA values of the same time series (*bottom*) also fluctuate randomly with no clear indication around the crash period

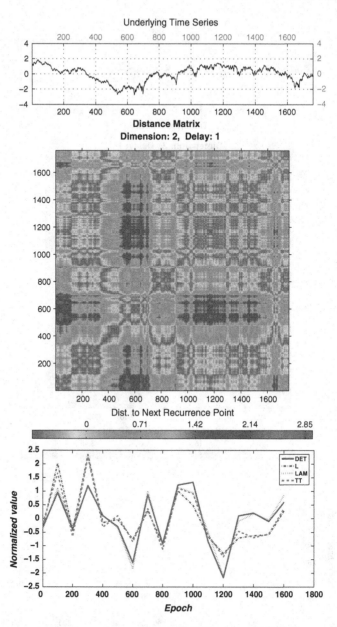

Fig. 6 Recurrence plot of time series representing the daily close of DJIA index around the 2002 downturn (*top*). We can see that for this reportedly endogenous crash, there is a *dark band* emerging about 100 trading days before the critical point. RQA values of the same time series (*bottom*) also take a dip just before the crash period, indicating a drop in determinism and change in dynamics

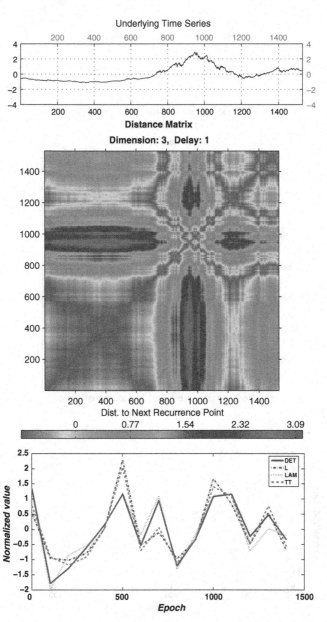

Fig. 7 Recurrence plot of time series representing the daily close of CSI300 index (*top*). We can see that for this reportedly endogenous crash, there is a *dark band* emerging about 100–110 trading days before the critical point. RQA values of the same time series (*bottom*) also take a dip just before the crash period, indicating a drop in determinism and change in dynamics

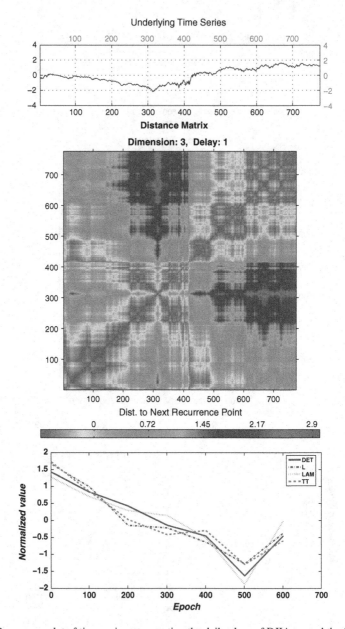

Fig. 8 Recurrence plot of time series representing the daily close of DJIA around the flash crash (*top*). We can observe that there is hardly any significant indication of a bubble in the map. RQA values of the same time series (*bottom*) also do not indicate any dynamics that suggest a trend emerging before the crash

Overall, our findings show that in all the endogenous cases, where a bubble was reported, viz., The Japanese Crash, October 27, 1997 mini-crash, The China Crash and the 2002 Dow Jones downturn, we see that a dark band is emerging. We can also see that the average length of such bands is lasting about 100 trading days which points out to about four months of active trading. This indicates that in all the major crashes the bubble always gathered momentum three to four months before the crash occurred. The corresponding RQA statistics also reinforces our findings. The determinism of the system increases before the crash, while it drops suddenly just before the crash occurs. In case of The Black Monday, Flash crash, Friday the 13th and 9/11 crash, no such distinguishable change surface in the Recurrence Plot. The RQA statistics also vary quite randomly, thus indicating no emergent behaviour around the crash point.

The presence of dark band indicates an increasing recurrence distance signifying formation of trend in the time signal. Phenomenological interpretation of the same may be emergence of herd behaviour. More specifically, as [42] points out, speculative bubbles are motivated by precipitating factors which lead to public opinion about markets or that have an immediate impact on demand, and by amplification mechanisms that take the form of price-to-price feedback. This explains the change in recurrence which indicates a long recurrence which can happen because of the feedback effects. The findings of [43–45] which find strong relation between higher moments and crashes also indicates prior to crashes there is an unusual rally of prices. If that be the case, then also, the recurrence plot which is a delay embedded vector of the original time series will have points that will tend to have larger recurrence distances during such periods of price rally. The larger distances will be reflected in the dark colour of the region in the map.

6 Conclusions

Our study reveals that for all the known endogenous crashes, where evidences of bubble were beyond doubt, the recurrence plot of the related data always showed a pattern. In all the cases, a dark band indicating a trend covered a period of three to four months before the crash. This reinforces the findings of [29]. Another interesting finding was the fact that in all the cases, the bubble length varied between three to four months suggesting that a trend if it continues for more than three months is likely to lead to a critical point resulting in a crash. The statistics yielded by recurrence quantification analysis also suggests that the degree of determinism of the data set increased rapidly before the critical point, i.e., the day of the crash. Corresponding to the epoch of 100 days before the crash there was a marked change in the values of the statistics. This definitely corroborates our findings from the recurrence plot. We also noted that for the cases where there were no known evidence of any bubble, the recurrence analysis clearly showed that there was no remarkable change in either the plot or the RQA values. Thus recurrence plot can be confidently used in identifying bubble. In case the bubble continues for three months one can expect a crash. However, more evidence is required before one can precisely predict a crash. Our present work shows that though the bubble is detected, for each of the crash the bubble length is different. Therefore, RP alone can not predict cashes and also our findings raise doubt about the ability of the phase transition model as proposed by [20] to precisely predict the crashes. But definitely, this tool may be used to identify changes in market dynamics and can serve as a warning bell.

Future works may evolve around modelling bubbles as regime changes. Rigourous econometric models which support a dynamics indicated by the recurrence plot may be built to try and capture the essential statistics of the time series.

References

1. I. Fisher, *The Stock Market Crash-and After* (Macmillan Company, New York, 1930)
2. D.S. Bates, US stock market crash risk, 19262010. J. Financ. Econ. **105**(2), 229–259 (2012)
3. G. Berlevi, P. Veronesi, Rational panics and stock market crashes. J. Econ. Theory **110**, 234–263 (2003)
4. R.J. Shiller, Fashions, fads and bubbles in financial markets, in *Knights, Raiders and Targets: The Impact of the Hostile Takeover*, ed. by Je Coffeein (Oxford University Press, Oxford, 1987)
5. C. LI, Hui Xue, A Bayesian's bubble. J. Financ. **64**(6), 2665–2701 (2009)
6. H.N. Seyhun, Overreaction or fundamentals: some lessons from insiders' response to the market crash of 1987. J. Financ. **45**(5), 1363–1388 (1990)
7. L. Harris, The October 1987 S&P 500 stock-futures basis. J. Financ. **44**(1), 77–99 (1989)
8. D.S. Bates, The crash of '87: was it expected? The evidence from options markets. J. Financ. **46**(3), 1009–1044 (1991)
9. M.L. Mitchell, J.M. Netter, Triggering the 1987 stock market crash: antitakeover provisions in the proposed house ways and means tax bill. J. Financ. Econ. **24**(1), 37–68 (1989)
10. J.M. Griffin, S. Topaloglu, Who drove and burst the tech bubble? J. Financ. **66**(4), 1251–1290 (2011)
11. Z. Ding, C.W. Granger, R.F. Engle, A long memory property of stock market returns and a new model. J. Empir. Financ. **1**(1), 83–106 (1993)
12. R.J. Barro, The stock market and investment. Rev. Financ. Stud. **3**(1), 115–131 (1990)
13. Beni Lauterbach, U.R.I. BEN-ZION, Stock market crashes and the performance of circuit breakers: empirical evidence. J. Financ. **48**(5), 1909–1925 (1993)
14. F.M. Longin, The asymptotic distribution of extreme stock market returns. J. Bus. **69**(3), 383–408 (1996)
15. T. Choudhry, Stock market volatility and the crash of 1987: evidence from six emerging markets. J. Int. Money Financ. **15**(6), 969–981 (1996)
16. J.K. Yang, I. Min, Stock market integration and financial crises: the case of Asia. Appl. Financ. Econ. **13**(7), 477–486 (2003)
17. J.A. Feigenbaum, P.G. Freund, Discrete scale invariance in stock market before crashes. Int. J. Mod. Phys. B **10**, 3737–3745 (1996)
18. J.A. Feigenbaum, P.G. Freund, Discrete scale invariance and the second black monday. Int. J. Mod. Phys. B **12**, 57 (1998)
19. D.A. Sornette, Crashes as critical points. Int. J. Theor. Appl. Financ. B **10**, 3737–3745 (1996)
20. D. Sornette, A. Johansen, J.P. Bouchaud, Stock market crashes, precursors and replicas. Int. J. Phys. I Fr. 167–175 (1996)
21. D. Sornette, A. Johansen, Critical crashes. RISK **39** (1999)
22. A. Johansen, D. Sornette, Financial anti-bubbles: log periodicity in gold and Nikkei collapses. Int. J. Mod. Phys. C **10**, 563–575 (1999)

23. N. Vandewalle, M. Ausloos, P. Boveroux, A. Minguet, How the financial crash of October 1997 could have been predicted. Eur. Phys. J. B **4**, 139–141 (1998)
24. N. Vandewalle, P. Boveroux, A. Minguet, M. Ausloos, The crash of October 1987 seen as a phase transition. Phys. A **255**, 201–210 (1998)
25. A. Johansen, D. Sornette, The Nasdaq crash of April 2000: yet another example of logperiodicity in a speculative bubble ending in a crash. Eur. Phys. J. B **17**, 319–328 (2000)
26. A. Fabretti, M. Ausloos, Recurrence plot and recurrence quantification analysis techniques for detecting a critical regime: examples from financial market indices. Int. J. Mod. Phys. C **16**(5), 671–706 (2005)
27. K. Guhathakurta, B. Bhattacharya, A. Roychowdhury, Analysing financial crashes using recurrence plot- a comparative study on selected financial markets, in *Forecasting Financial Markets*, ed. by R.P. Pradhan (Allied Publishers, New Delhi, 2009), pp. 22–29
28. K. Guhathakurta, B. Bhattacharya, A. Roychowdhury, An examination of critical periods of stock price movements using recurrence plot, in *Essays in Finance*, ed. by B. Bhattacharya, M. Roy (Allied Publishers, New Delhi, 2010), pp. 98–113
29. K. Guhathakurta, B. Bhattacharya, A. Roychowdhury, Using recurrence plot analysis to distinguish between endogenous and exogenous stock market crashes. Phys. A **389**(9), 1874–1882 (2010)
30. J.A. Bastos, J. Caiado, Recurrence quantification analysis of global stock markets. Phys. A **390**(7), 1315–1325 (2011)
31. R.S. Kyle, Informed speculation with imperfect competition. Rev. Econ. Stud. **56**(3), 317–355 (1989)
32. G. Gennotte, Market liquidity, hedging, and crashes. Am. Econ. Rev. 999–1021 (1990)
33. K. Sato, Bubbles in Japan's stock market: a macroeconomic analysis. Working Paper Series, Center on Japanese Economy and Business, Graduate School of Business, Columbia University, Working Paper No. 95 (1995)
34. A. Kirilenko, M. Samadi, A.S. Kyle, T. Tuzun, *The Flash Crash: The Impact of High Frequency Trading on an Electronic Market*, (Working paper, 2011)
35. Shujie Yao, Dan Luo, The economic psychology of stock market bubbles in China. World Econ. **32**(5), 667–691 (2009)
36. J.P. Eckmann, S.O. Kamphorst, D. Ruelle, Recurrence plots of dynamical systems. Europhys. Lett. (EPL) **4**(9), 973 (1987)
37. N. Marwan, M. Carmen Romano, M. Thiel, J. Kurths, Recurrence plots for the analysis of complex systems. Phys. Rep. **438**(5), 237–329 (2007)
38. A.M. Fraser, H.L. Swinney, Independent coordinates for strange attractors from mutual information. Phys. Rev. A **33**(2), 1134 (1986)
39. M.B. Kennel, R. Brown, H.D. Abarbanel, Determining embedding dimension for phase-space reconstruction using a geometrical construction. Phys. Rev. A **45**(6), 3403 (1992)
40. J. Zbilut, C. Webber, Embeddings and delays as derived from quantification of recurrence plots. Phys. Lett. A **171**(3–4), 199–203 (1992)
41. N. Marwan, A historical review of recurrence plots. Eur. Phys. J. Spec. Top. **164**(1), 3–12 (2008)
42. R. Shiller, *Irrational Exuberance* (Princeton University Press, Princeton, 2000)
43. G. Bakshi, D. Madan, What is the probability of a stock market crash, University of Maryland (unpublished, September 2, 1998)
44. G. Bakshi, D. Madan, *Crash Discovery in Stock and Options Markets* (Unpublished P aper, University of Maryland, 1999)
45. G. Bakshi, D. Madan, G. Panayotov, Deducing the implications of jump models for the structure of stock market crashes, rallies, jump arrival rates, and extremes. J. Bus. Econ. Stat. **28**(3), 380–396 (2010)

Probabilistic Flows of Inhabitants in Urban Areas and Self-organization in Housing Markets

Takao Hishikawa and Jun-ichi Inoue

Abstract We propose a simple probabilistic model to explain the spatial structure of the rent distribution of housing market in city of Sapporo. Here we modify the mathematical model proposed by Gauvin et al. [1]. Especially, we consider the competition between two distances, namely, the distance between house and center, and the distance between house and office. Computer simulations are carried out to reveal the self-organized spatial structure appearing in the rent distribution. We also compare the resulting distribution with empirical rent distribution in Sapporo as an example of cities designated by ordinance. We find that the lowest ranking agents (from the viewpoint of the lowest 'willing to pay') are swept away from relatively attractive regions and make several their own 'communities' at low offering price locations in the city.

1 Introduction

Collective behaviour of interacting animals such as flying birds, moving insects or swimming fishes has attracted a lot of attentions by scientists and engineers due to its highly non-trivial properties. Several remarkable attempts have even done to figure out the mechanism of the collective phenomena by collecting empirical data of flocking of starlings with extensive data analysis [2], by computer simulations of realistic flocking based on a simple algorithm called BOIDS [3–5]. Applications of such collective behavior of animals also have been proposed in the context of engineering [6].

Apparently, one of the key factors to emerge such non-trivial collective phenomena is 'local interactions' between agents. The local interaction in the microscopic level causes non-trivial structures appearing in the macroscopic system. In other

T. Hishikawa · J. Inoue (✉)
Graduate School of Information Science and Technology, Hokkaido University,
N14-W-9, Kita-ku, Sapporo 060-0814, Japan
e-mail: jinoue@cb4.so-net.ne.jp; j_inoue@complex.ist.hokudai.ac.jp

T. Hishikawa
e-mail: hishikawa@complex.ist.hokudai.ac.jp

© Springer International Publishing Switzerland 2015
F. Abergel et al. (eds.), *Econophysics and Data Driven Modelling of Market Dynamics*,
New Economic Windows, DOI 10.1007/978-3-319-08473-2_7

words, there is no outstanding leader who designs the whole system, however, the spatio-temporal patterns exhibited by the system are 'self-organized' by local decision making of each interacting ingredient in the system.

These sorts of self-organization by means of local interactions between agents might appear not only in natural phenomena but also in some social systems including economics. For instance, decision making of inhabitants in urban areas in order to look for their houses, the resulting organization of residential street (area) and behavior of housing markets are nice examples for such collective behavior and emergent phenomena. People would search suitable location in their city and decide to live a rental (place) if the transaction is approved after negotiation in their own way on the rent with buyers. As the result, the spatio-temporal patterns might be emerged, namely, both expensive and cheap residential areas might be co-existed separately in the city. Namely, local decision makings by ingredients—inhabitants—determine the whole structure of the macroscopic properties of the city, that is to say, the density of residents, the spatial distribution of rent, and behavior of housing markets.

Therefore, it is very important for us definitely to investigate which class of inhabitants chooses which kind of locations, rentals, and what is the main factor for them to decide their housings. The knowledge obtained by answering the above naive questions might be useful when we consider the effective urban planning. Moreover, such a simple but essential question is also important and might be an advanced issue in the context of the so-called spatial economics [7].

In fact, constructing new landmarks or shopping districts might encourage inhabitants to move to a new place to live, and at the same time, the resulting distribution of residents induced by the probabilistic flow of inhabitants who are looking for a new place to live might be important information for the administrator to consider future urban planning. Hence, it could be regarded as a typical example of 'complex systems' in which macroscopic information (urban planning) and microscopic information (flows of inhabitants to look for a new place to live) are co-evolved in relatively long time scale under weak interactions.

To investigate the macroscopic properties of the system from the microscopic viewpoint, we should investigate the strategy of decision making for individual person. However, it is still extremely difficult for us to tackle the problem by making use of scientifically reliable investigation. This is because there exists quite large person-to-person fluctuation in the observation of individual behaviour. Namely, one cannot overcome the individual variation to find the universal fact in the behaviour even though several attempts based on impression evaluation or questionnaire survey have been done extensively. On the other hand, in our human 'collective' behaviour instead of individual, we sometimes observe several universal facts which seem to be suitable materials for computer scientists to figure out the phenomena through sophisticated approaches such as agent-based simulations or multivariate statistics accompanying with machine learning technique.

In a mathematical housing market modelling recently proposed by Gauvin et al. [1], they utilized several assumptions to describe the decision making of each inhabitant in Paris. Namely, they assumed that the intrinsic attractiveness of a city depends on the place and there exists a single peak at the center. They also used the assumption that each inhabitant tends to choose the place where the other inhabitants having the similar or superior income to himself/herself are living. In order to find the best possible place to live, each buyer in the system moves from one place to the other according to the transition (aggregation) probability described by the above two assumption and makes a deal with the seller who presents the best condition for the buyer. They concluded that the resulting self-organized rent distribution is almost consistent with the corresponding empirical evidence in Paris. However, it is hard for us to apply their model directly to the other cities having plural centers (not only a single center as in Paris).

Hence, here we shall modify the Gauvin's model [1] to include the much more detail structure of the attractiveness by taking into account the empirical data concerning the housing situation in the city of Sapporo. Sapporo is the fourth-largest city in Japan by population, and the largest city on the northern Japanese island of Hokkaido. Sapporo is also recognized as one of big cities designated by ordinance and it has ten wards (we call 'ku' for 'ward' in Japanese), namely, *Chuo (Central)*, *Higashi (East)*, *Nishi (West)*, *Minami*, *Kita (North)*, *Toyohira*, *Shiraishi*, *Atsubetsu*, *Teine* and *Kiyota* as shown in Fig. 1.

We also consider the competition between two distances, namely, the distance between house and center, and the distance between house and office. Computer simulations are carried out to reveal the self-organized structure appearing in the

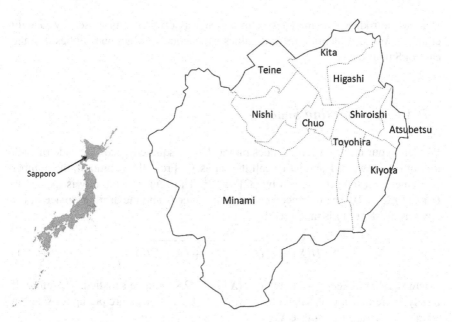

Fig. 1 Sapporo is the largest city on the northern Japanese island of Hokkaido. Sapporo is also recognized as one of big cities designated by ordinance and it has ten wards (we call 'ku' for 'ward' in Japanese), namely, *Chuo (Central)*, *Higashi (East)*, *Nishi (West)*, *Minami*, *Kita (North)*, *Toyohira*, *Shiraishi*, *Atsubetsu*, *Teine* and *Kiyota*

rent distribution. Finally, we compare the resulting distribution with empirical rent distribution in Sapporo as an example of cities designated by ordinance. We find that the lowest ranking agents (from the viewpoint of the lowest 'willing to pay') are swept away from relatively attractive regions and make several their own 'communities' at low offering price locations in the city.

This paper is organized as follows. In the next Sect. 2, we introduce the Gauvin's model [1] and attempt to apply it to explain the housing market in city of Sapporo, which is one of typical cites designated by ordinance in Japan. In Sect. 3, we show the empirical distribution of averaged rent in city of Sapporo and compare the distribution with that obtained by computer simulations in the previous Sect. 2. In Sect. 4, we will extend the Gauvin's model [1] in which only a single center exists to much more generalized model having multiple centers located on the places of ward offices. In Sect. 5, we definitely find that our generalized model can explain the qualitative behavior of spatial distribution of rent in city of Sapporo. In the same section, we also show several results concerning the office locations of inhabitants and its effect on the decision making of inhabitant moving to a new place. The last Sect. 5.4 is devoted to summary and discussion.

2 The Model System

Here we introduce our model system which was originally proposed by Gauvin et al. [1]. We also mention the difficulties we encounter when one applies it to the case of Sapporo city.

2.1 A City—Working Space—

We define our city as a set of nodes on the $L \times L$ square lattice. The side of each unit of the lattice is 1 and let us call the set as Ω. From the definition, the number of elements in the set is given by $|\Omega| \equiv L^2$. The center of the city is located at $O \equiv (L/2, L/2)$. The distance between the center O and the arbitrary place in the city, say, $X \equiv (x, y)$ is measured by

$$D(X) = \sqrt{(x - L/2)^2 + (y - L/2)^2} \tag{1}$$

where we should keep in mind that $D(X) \leq L/2$ should be satisfied. Therefore, if totally $\mathcal{N}$ rentals are on sale in the city, $N \equiv \mathcal{N}/L^2$ houses are put up for sale 'on average' at an arbitrary place X.

Fig. 2 The mark showing where here is in city of Sapporo (urban district). We can easily find it on the *top* of traffic signal as 'South 5 West 4', which means that here is south by 5 blocks, west by 4 blocks from the origin ('Odori Park', center)

2.1.1 Why Do We Choose City of Sapporo?

As we will see later, our modelling is applicable to any type of city. However, here we choose our home town, city of Sapporo, as a target city to be examined.

For the above setting of working space, city of Sapporo is a notable town. This is because the roadways in the urban district are laid to make grid plan road [8]. Hence, we can easily specify each location by the two-dimensional vector, say, 'S5-W4' which means that south by 5 blocks, west by 4 blocks from the origin ('Odori Park', center). Those labels are usually indicated by marks on the top of the traffic signals (see Fig. 2). These kinds of properties might help us to collect the empirical data sets and compare them with the outputs from our probabilistic model.

2.2 Agents

We suppose that there co-exist three distinct agents, namely, 'buyers', 'sellers' and 'housed'. The total number of these agents is not constant but changing in time. For instance, in each (unit) time step $t = 1$, Γ (≥ 1) agents visit to the city as 'new buyers', whereas α percentage of the total housed persons let the house go for some amount of money and they become 'new sellers'. When the 'new sellers' succeeded in selling their houses, they leave the city to move to the other cities. The situation is illustrated by a cartoon in Fig. 3.

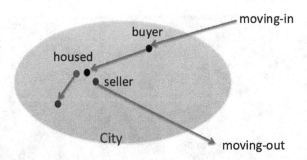

Fig. 3 Three kinds of agents in our model system and those typical behaviors. A newcomer as a '*buyer*' looks for the housing in the city (say, Sapporo). The '*housed*' who is living his/her own house becomes a '*seller*' when he/she would move to the other *city* (e.g., Tokyo or Osaka). Each *seller* presents the possible rent to the *buyers*. Once the *seller* accepts the offer price and makes a contract with the *buyer*, the *buyer* becomes a '*housed*' at the place and the *seller* immediately leaves from the *city* (Sapporo)

2.2.1 Ranking of Agents

Each of agent is categorized (ranked) according to their degree of 'willing to pay' for the housing. Let us define the total number of categories by K. Then, the agent belonging to the category $k \in \{0, \ldots, K-1\}$ can pay P_k (measured by a currency unit, for instance *Japanese yen*) for the housing. Hence, we can give the explicit ranking to all agents when we put the price P_k in the following order

$$P_0 < P_1 < \cdots < P_{K-1}. \tag{2}$$

In this paper, the price P_k of 'willing to pay' is given by

$$P_k = P_0 + k \frac{\Delta}{K-1}, \quad k = 0, \ldots, K-1, \tag{3}$$

namely, as the difference between the highest rent P_0 and the lowest P_{K-1} leads to a gap Δ, each category (person) is put into one of the Δ/K intervals. In our computer simulations which would be given later on, the ranking of each agent is allocated randomly from $\{0, \ldots, K-1\}$.

2.3 Attractiveness of Locations

A lot of persons are attracted by specific areas close to the railway (subway) stations or big shopping districts for housing. As well-known, especially in city of Sapporo, *Maruyama*-area which is located in the west side of the Sapporo railway station has been appealed to, in particular, high-ranked persons as an exclusive residential district. Therefore, one can naturally assume that each area in the city possesses its own attractiveness and we might regard theattractiveness as time-independent quantity.

However, the attractiveness might be also dependent on the categories (ranking) of agents in some sense. For instance, the exclusive residential district is not attractive for some persons who have relatively lower income and cannot afford the house. On the other hand, some persons who have relatively higher income do not want to live the area where is close to the busy street or the slum areas.

Taking into account these two distinct facts, the resulting attractiveness for the area X should consists of the part of the intrinsic attractiveness $A^0(X)$ which is independent of the categories $k = 0, \ldots, K - 1$ and the another part of the attractiveness which depends on the categories. Hence, we assume that attractiveness at time t for the person who belongs to the category k at the area X, that is, $A_k(X, t)$ is updated by the following spatio-temporal recursion relation:

$$A_k(X, t + 1) = A_k(X, t) + \omega(A^0(X) - A_k(X, t)) + \Phi_k(X, t) \tag{4}$$

where $A^0(X)$ stands for the time-independent intrinsic attractiveness for which any person belonging to any category feels the same amount of charm.

For example, if the center of the city $O = (L/2, L/2)$ possesses the highest intrinsic attractiveness A^0_{max} as in Paris [1], we might choose the attractiveness $A^0(X)$ as a two-dimensional Gaussian distribution with mean $O = (L/2, L/2)$ and the variance R^2 as

$$A^0(X) = \frac{A^0_{max}}{\sqrt{2\pi R^2}} \exp\left[-\frac{\{(x - L/2)^2 + (y - L/2)^2\}}{2R^2}\right] \tag{5}$$

where the variance R^2 denotes a parameter which controls the range of influence from the center. We should notice that the Eq. (4) has a unique solution $A^0(X)$ as a steady state when we set $\Phi_k(X, t) = 0$.

Incidentally, it is very hard for us to imagine that there exist direct interactions (that is, 'communications') between agents who are looking for their own houses. However, nobody doubts that children's education (schools) or public peace should be an important issue for persons (parents) to look for their housing. In fact, some persons think that their children should be brought up in a favorable environment with their son's/daughter's friends of the same living standard as themselves. On the other hand, it might be rare for persons to move to the area where a lot of people who are lower living standard than themselves. Namely, it is naturally assumed that people seek for the area as their housing place where the other people who are higher living standard than themselves are living, and if possible, they would like to move such area.

Hence, here we introduce such 'collective effects' into the model system by choosing the term $\Phi_k(X, t)$ as

$$\Phi_k(X, t) = \varepsilon \sum_{k' \geq k} v_{k'}(X, t) \tag{6}$$

where $v_k(X, t)$ stands for the density of housed persons who are in the category k at the area X at time t. Namely, from Eqs. (4) and (6), the attractiveness of the area X for the people of ranking k, that is, $A_k(X, t)$ increases at the next time step $t+1$ when persons who are in the same as or higher ranking than k start to live at X. It should bear in mind that the intrinsic part of attractiveness $A^0(X)$ is time-independent, whereas $\Phi_k(X, t)$ is time-dependent through the flows of inhabitants in the city. Thus, the $A_k(X, t)$ could have a different shape from the intrinsic part $A^0(X)$ due to the effect of the collective behavior of inhabitants $\Phi_k(X, t)$.

2.4 Probabilistic Search of Locations by Buyers

The buyers who have not yet determined their own house should look for the location. Here we assume that they move to an arbitrary area 'stochastically' according to the following probability:

$$\pi_k(X, t) = \frac{1 - \exp(-\lambda A_k(X, t))}{\sum_{X' \in \Omega} \{1 - \exp(-\lambda A_k(X', t))\}} \tag{7}$$

Namely, the buyers move to the location X to look for their housing according to the above probability. The situation is shown as a cartoon in Fig. 4. We easily find from Eq. (7) that for arbitrary λ, the area which exhibits relatively high attractiveness is more likely to be selected as a candidate to visit at the next round. Especially, in the limit of $\lambda \to \infty$, the buyers who are looking for the locations visit only the highest attractiveness location

$$X_k = \mathrm{argmax}_X A_k(X, t). \tag{8}$$

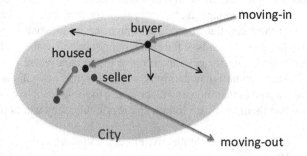

Fig. 4 In each time step, a *buyer* visits a place X with a probability (7). By repeating the transition processes, a *buyer* explores the suitable and desirable location in the *city* from place to place

On the other hand, for $\lambda \ll 1$, Eq. (7) leads to

$$\pi_k(X, t) = \frac{A_k(X, t)}{\sum_{X' \in \Omega} A_k(X', t)},\tag{9}$$

and the probability for buyers to visit the place X at time t is proportional to the attractiveness corresponding to the same area $A_k(X, t)$.

Here we should mention that when we regard the 'location' as 'company', the present model system is described by similar aggregation probability to the probabilistic labor market proposed by Chen et al. [9–11] where the parameter γ corresponds to the λ in the Gauvin's model [1].

2.5 Offering Prices by Sellers

It is not always possible for buyers to live at the location X where they have selected to visit according to the probability $\pi_k(X, t)$ because it is not clear whether they can accept the price offered by the sellers at X or not. Obviously, the offering price itself depends on the ranking k of the sellers. Hence, here we assume that the sellers of ranking k at the location X offer the buyers the price:

$$P_k^o(X) = P^0 + [1 - \exp(-\xi \overline{A}(X, t))]P_k\tag{10}$$

for the rental housing, where $\overline{A}(X, t)$ means the average of the attractiveness $A_k(X, t)$ over the all categories $k = 0, \ldots, K - 1$, that is to say,

$$\overline{A}(X, t) \equiv \frac{1}{K} \sum_{k=0}^{K-1} A_k(X, t),\tag{11}$$

and ξ is a control parameter. In the limit of $\xi \to \infty$, the offering price by a seller of ranking k is given by the sum of the basic rent P^0 and the price of 'willing to pay' for the sellers of ranking k, namely, P_k as $P_k^o = P^0 + P_k$.

For the location X in which any transaction has not yet been approved, the offering price is given as

$$P_k^o(X) = P^0 + [1 - \exp(-\xi \overline{A}(X, t))]P^1\tag{12}$$

because the ranking of the sellers is ambiguous for such location X.

2.6 The Condition on Which the Transaction is Approved

It is needed for buyers to accept the price offered by sellers in order to approve the transaction. However, if the offering price is higher than the price of 'willing to pay' for the buyer, the buyer cannot accept the offer. Taking into account this limitation, we assume that the following condition between the buyer of the ranking k and the seller of the ranking k' at the location X should be satisfied to approve the transaction.

$$P_k > P^o_{k'}(X) \tag{13}$$

Thus, if and only if the above condition (13) is satisfied, the buyer can own the housing (the seller can sell the housing).

We should keep in mind that there is a possibility for the lowest ranking people to fail to own any housing in the city even if they negotiate with the person who also belongs to the lowest ranking. To consider the case more carefully, let us set $k = k' = 0$ in the condition (13), and then we have $P_0 > P^0/\exp[-\xi \overline{A}(X, t)]$. Hence, for the price of 'willing to pay' P_0 of the lowest ranking people, we should determine the lowest rent P^0 so as to satisfy $P^0 < \exp[-\xi \overline{A}(X, t)]P_0$. Then, the lowest ranking people never fail to live in the city.

We next define the actual transaction price as interior division point between P_k and $P^o_{k'}(X)$ by using a single parameter β as

$$P_{tr} = (1 - \beta)P_{k'}(X) + \beta P_k. \tag{14}$$

By repeating these three steps, namely, probabilistic searching of the location by buyers, offering the rent by sellers, transaction between buyers and sellers for enough times, we evaluate the average rent at each location X and draw the density of inhabitants, the spatial distribution of the average rent in two-dimension. In following, we show our preliminary results.

2.7 Computer Simulations: A Preliminary

In Fig. 5, we show the spatial density distribution $\rho(r)$ of inhabitants by the Gauvin's model [1] having a single center in the city. The horizontal axis r of these panels stands for the distance between the center O and the location X, namely, $r = D(X)$. We set the parameters appearing in the model as $L = 100$, $\alpha = 0.1$, $K = 10$, $P_0 = 15,000$, $\Delta + P_0 = 120,000$ ($\Delta = 105,000$), $\omega = 1/15$, $R = 10$, $\varepsilon = 0.0022$, $\lambda = 0.01$, $\zeta = 0.1$, $P^0 = 9,000$, $P^1 = 200,000$, $\beta = 0.1$, $\Gamma = L^2/K$, $T = 100$ ($\equiv$ Total number of updates (4)). It should be noted that the definition of density is given by

$$\rho(r) \equiv \frac{(\text{\# of inhabitants of ranking } k \text{ at the location } r)}{(\text{Total \# of inhabitants at the location } r)}. \tag{15}$$

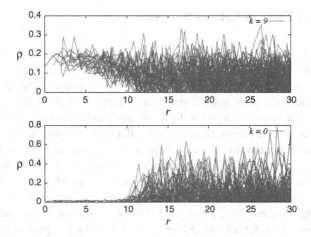

Fig. 5 The spatial density $\rho(r)$ of inhabitants obtained by the Gauvin's model [1] having a single center in the city. The horizontal axis r stands for the distance between the center $O = (L/2, L/2)$ and the location $X = (x, y)$, namely, $r = D(X)$. The *upper panel* is the result for the highest raining people ($k = K - 1$), whereas the *lower panel* shows the result for the lowest ranking inhabitants ($k = 0$). We easily recognize that the persons who are belonging to the lowest ranking cannot live the area close to the center

From the lower panel, we easily recognize that the persons who are belonging to the lowest ranking cannot live the area close to the center in the city. Thus, we find that there exists a clear division of inhabitants of different rankings.

Intuitively, these phenomena might be understood as follows. The persons of the highest ranking ($k = K - 1$) can afford to accept any offering price at any location. At the same time, the effect of aggregation induced by $\Phi_{K-1}(X, t)$ in the searching probability $\pi_{K-1}(X, t)$ and the update rule of $A_{K-1}(X, t)$ in (4) is the weakest among the K categories. As the result, the steady state of the update rule (4) is not so deviated from the intrinsic attractiveness, namely, $A_{K-1}(X, t) \simeq A^0(X)$. Hence, people of the highest ranking are more likely to visit the locations where are close to the center $X = O$ and they live there. Of course, the density of the highest ranking persons decreases as r increases.

On the other hand, people of the lowest ranking ($k = 0$) frequently visit the locations where various ranking people are living due to the aggregation effect of the term $\Phi_0(X, t)$ to look for their housing. However, it is strongly dependent on the offering price at the location whether the persons can live there or not. Namely, the successful rate of transaction depends on which ranking buyer offers the price at the place X. For the case in which the 'housed' people changed to sellers at the location where is close to the center, the seller is more likely to be the highest ranking person because he/she was originally an inhabitant owing the house near the center. As the result, the price offered by them might be too high for the people of the lowest ranking to pay for approving the transaction. Therefore, the lowest raining people might be driven away to the location where is far from the center.

3 Empirical Data in City of Sapporo

Apparently, the above simple modelling with a single center of city is limited to the specific class of city like Paris. Turning now to the situation in Japan, there are several major cites designated by ordinance, and city of Sapporo is one of such 'mega cities'. In Table 1, we show several statistics in Sapporo in 2010. From this table, we recognize that in each year, 63,021 persons are moving into and 57,587 persons are moving out from Sapporo. Hence, the population in Sapporo is still increasing by approximately six thousand in each year. As we already mentioned, Sapporo is the fourth-largest city in Japan by population, and the largest city on the northern Japanese island of Hokkaido. Sapporo is recognized as one of big cities designated by ordinance and it has ten wards (what we call 'ku' in Japanese), namely, *Chuo, Higashi, Nishi, Minami, Kita, Toyohira, Shiraishi, Atsubetsu, Teine* and *Kiyota* as shown in Table 1 (for details, see [8] for example). Hokkaido prefectural office is located in *Chuo*-ku and the other important landmarks concentrate in the wards. Moreover, as it is shown in Table 1, the highest and the lowest rents for the 2DK-type (namely, a two-room apartment with a kitchen/dining area) flats in *Chuo*-ku are both the highest among ten wards. In this sense, *Chuo*-ku could be regarded as a 'center' of Sapporo. However, the geographical structure of rent distribution in city of Sapporo is far from the symmetric one as given by the intrinsic attractiveness $A^0(X)$ having a single center (see Eq. 5). In fact, we show the rough distribution of average rents in city of Sapporo in Fig. 6 by making use of the empirical data collected from [12]. From this figure, we clearly confirm that the spatial structure of rents in city of Sapporo is not symmetric but apparently asymmetric.

From this distribution, we also find that the average rent is dependent on wards, and actually it is very hard to simulate the similar spatial distribution by using the Gauvin's model [1] in which there exists only a single center in the city. This is

Table 1 Statistics in city of Sapporo for the number of persons who were moving-into and -out, the lowest and highest rents (the unit is Japanese yen) of 2DK-type flats in city of Sapporo

Wards	# of moving-in	# of moving-out	Lowest (yen)	Highest (yen)
Chuo (*Central*)	12,132	10,336	19,000	120,000
Kita (*North*)	8,290	7,970	15,000	73,000
Higashi (*East*)	7,768	7,218	20,000	78,500
Shiraishi	6,857	6,239	25,000	67,000
Atsubetsu	4,003	3,736	33,000	57,000
Toyohira	7,854	7,037	20,000	69,000
Kiyota	2,560	2,398	30,000	55,000
Minami (*South*)	3,824	3,794	23,000	58,000
Nishi	6,315	5,788	20,000	80,000
Teine	3,418	3,071	20,000	68,000
Total	63,021	57,587	—	—

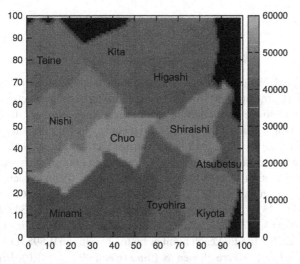

Fig. 6 The spatial distribution of the averaged rent in city of Sapporo by using the empirical housing data collected from [12]. In Sapporo, ten wards *Chuo* (*Central*), *Higashi* (*East*), *Nishi* (*West*), *Minami* (*South*), *Kita*, *Toyohira*, *Shiraishi*, *Atsubetsu*, *Teine* and *Kiyota* exist (Color figure online)

because in a city designated by ordinance like Sapporo, each ward formulates its own community, and in this sense, each ward should be regarded as a 'small city' having a single (or multiple) center(s). It might be one of the essential differences between Paris and Sapporo.

4 An Extension to a City Having Multiple Centers

In the previous Sect. 3, we found that the Gauvin's model [1] having only a single center is not suitable to explain the empirical evidence for a city designated by ordinance such as Sapporo where multiple centers as wards co-exist.

Therefore, in this section, we modify the intrinsic attractiveness $A^0(X)$ to explain the empirical evidence in city of Sapporo. For this purpose, we use the label $l = 1, \ldots, 10$ to distinguish ten words in Sapporo, namely, *Chuo* (*Central*), *Higashi* (*East*), *Nishi* (*West*), *Minami* (*South*), *Kita* (*North*), *Toyohira*, *Shiraishi*, *Atsubetsu*, *Teine* and *Kiyota* in this order, and define $\boldsymbol{B}_l = (x_{B_l}, y_{B_l}), l = 1, \ldots, 10$ for each location where each ward office is located. Then, we shall modify the intrinsic attractiveness in terms of the $\boldsymbol{B}_l$ as follows.

$$A^0(\boldsymbol{X}) = \sum_{l=1}^{10} \frac{\delta_l}{\sqrt{2\pi} R_l} \exp\left[-\frac{\{(x - x_{B_l})^2 + (y - y_{B_l})^2\}}{2R_l^2}\right] \quad (16)$$

where

$$\delta_1 + \cdots + \delta_{10} = 1 \tag{17}$$

should be satisfied. Namely, we would represent the intrinsic attractiveness $A^0(X)$ in city of Sapporo by means of a two-dimensional mixture of Gaussians in which each mean corresponds to the location of the ward office. R_l ($l = 1, \ldots, 10$) denotes a set of parameters which control the spread of the center. In our computer simulations, we set $R_l = 5$ for all $l = 1, \ldots, 10$ in our intrinsic attractiveness (16).

Here we encounter a problem, namely, we should choose each weight δ_l, $l = 1, \ldots, 10$. For this purpose, we see the number of estates (flats) in each wards. From a real-estate agents in Sapporo [13], we have the statistics as *Chuo (Central)* (9,598), *Higashi (East)* (6,433), *Nishi (West)* (5,830), *Minami (South)* (1,634), *Kita (North)* (4,671), *Toyohira* (4,893), *Shiraishi* (5,335), *Atsubetsu* (1,104), *Teine* (2,094), *Kiyota* (962). Hence, by dividing each number by the maximum 9,598 for *Chuo*-ku, the weights δ_l, $l = 1, \ldots, 10$ are chosen as *Chuo (Central)* ($\delta_1 \propto 1.0$), *Higashi (East)* ($\delta_2 \propto 0.67$), *Nishi (West)* ($\delta_3 \propto 0.61$), *Minami (South)* ($\delta_4 \propto 0.17$), *Kita (North)* ($\delta_5 \propto 0.49$), *Toyohira* ($\delta_6 \propto 0.51$), *Shiraishi* ($\delta_7 \propto 0.56$), *Atsubetsu* ($\delta_8 \propto 0.12$), *Teine* ($\delta_9 \propto 0.22$), *Kiyota* ($\delta_{10} \propto 0.1$). Of course, we normalize these parameters so as to satisfy the condition (17).

5 Computer Simulations

In this section, we show the results of computer simulations.

5.1 Spatial Structure in the Distribution of Visiting Times

In Fig. 7 (left), we show the distributions of the number of persons who checked the information about the flat located at X and the number of persons who visited the place X according to the transition probability (7) in the right panel. From this figure, we find that the locations (flats) X where people checked on the web site [13] most frequently concentrate around each ward office. From this fact, our modelling in which we choose the locations of multiple centers as the places of wards might be approved. Actually, from the right panel of this figure, we are confirmed that the structure of spatial distribution is qualitatively very similar to the counter-empirical distribution (right panel).

In order to investigate the explicit ranking dependence of the housing-search behavior of agents, in Fig. 8 (the upper panels), we plot the spatial distribution of the number of visits for the lowest ranking $k = 0$ (left) and the highest ranking $k = 9$ (right) agents. We also plot the corresponding spatial distributions of the number of the transaction approvals in the lower two panels. From this figure, we

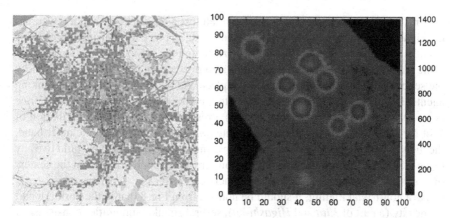

Fig. 7 The distributions of the number of persons who checked the information about the flat located at X on the web site (*left*, from [13]) and the number of persons who visited the place X according to the transition probability (7) (*right*) in our artificial society (Color figure online)

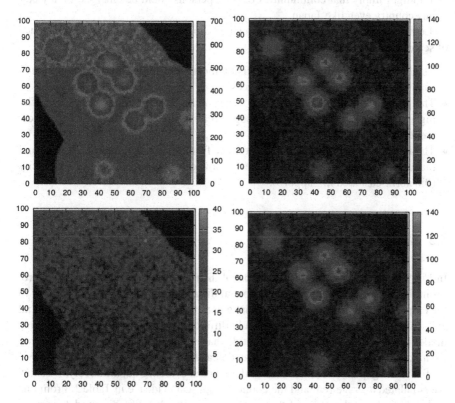

Fig. 8 The *upper two panels* show the spatial distribution of the number of visits consisting of the lowest ranking $k = 0$ (*left*) and the highest ranking $k = 9$ (*right*) agents. The corresponding spatial distributions of the number of the transaction approvals are shown in the *lower panels* (Color figure online)

confirm that the lowest ranking agents visit almost whole area of the city, whereas the highest ranking agents narrow down their visiting place for housing search. This result is naturally understood as follows. Although the lowest ranking agents visit some suitable places to live, they cannot afford to accept the offer price given by the sellers who are selling the flats at the place. As the result, such lowest ranking agents should wander from place to place to look for the place where the offer price is low enough for them to accept. That is a reason why the spatial distribution of visit for the lowest ranking agents distributes widely in the city. On the other hand, the highest ranking agents posses enough 'willing to pay' P_9 and they could live any place they want. Therefore, their transactions are easily approved even at the centers of wards with relatively high intrinsic attractiveness $A^0(X)$.

As a non-trivial finding, it should be noticed from Fig. 8 that in the northern part of the city (a part of *Kita* and *Higashi*-ku), several small communities consisting of the lowest ranking persons having their 'willing to pay' P_0 emerge. In our modelling, we do not use any 'built-in' factor to generate this sort of non-trivial structure. This result might imply that communities of poor persons could be emerged in any city in any country even like Japan.

Let us summarize our findings from simulation concerning ranking dependence of search-approvals by agents below.

- The lowest ranking agents ($k = 0$) visit almost all of regions in city even though such places are highly 'attractive places'.
- The highest ranking agents ($k = 9$) visit relatively high attractive places. The highest ranking agents are rich enough to afford to accept any offering price, namely, # of contracts $\simeq$ # of visits.
- The lowest ranking agents are swept away from relatively attractive regions and make several their own 'communities' at low offering price locations in the city (the north-east area in Sapporo).

5.2 The Rent Distribution

In Fig. 9 (left), we plot the resulting spatial distribution of rent in city of Sapporo. From this figure we confirm that the spatial distribution is quantitatively similar to the empirical evidence. We also find that a complicated structure—a sort of spatial anisotropy—emerges and it is completely different from the result by the Gauvin's model [1]. In particular, we should notice that relatively high rent regions around *Chuo*-ku appear. These regions are located near *Kita, Higashi, Nishi* and *Shiraishi*.

To see the gap between our result and empirical evidence quantitatively, we show the simulated average rent and the counter-empirical evidence in Table 2. From this table, we find that the order of the top two wards simulated by our model, namely, *Chuo* and *Shiraishi* coincides with the empirical evidence, and moreover, the simulated rent itself is very close to the market price. However, concerning the order of the other wards, it is very hard for us to conclude that the model simulates the empirical

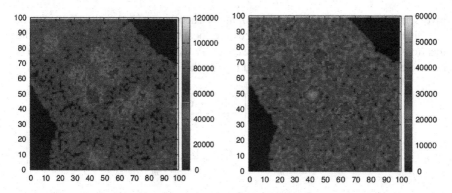

Fig. 9 The resulting spatial distributions of rent in city of Sapporo. In the *left panel*, we do not consider the effect of office locations on buyer's decision, whereas in the *right panel*, the effect is taken into account (Color figure online)

Table 2 The left list shows the ranking (order) and the market prices

Ranking	Market price (yen)	Ranking	Simulated average rent (yen)
Chuo (Central)	54,200	Chuo (Central)	50,823
Shiraishi	51,100	Shiraishi	43,550
Nishi (West)	48,200	Higashi (East)	44,530
Kiyota	45,800	Kita (North)	43,516
Teine	42,900	Nishi (West)	43,093
Kita (North)	41,700	Toyohira	42,834
Higashi (East)	40,100	Minami (South)	39,909
Atsubetsu	39,700	Teine	39,775
Toyohira	39,600	Kiyota	37,041
Minami (South)	34,400	Atsubetsu	36,711

The simulated average rent for each ward and the ranking are shown in the right list. The unit of price is Japanese yen [13]

data. Of course, the market price differences in those wards are very small and it is very difficult to simulate the correct ranking at present. Thus, the modification of our model to generate the correct ranking and to obtain the simulated rents which are much closer to the market prices should be addressed our future study.

5.3 On the Locations of Offices

In the previous sections, we modified the intrinsic attractiveness so as to possess the multiple peaks at the corresponding locations of the ward offices by (16). However, inhabitants must go to their office every weekday, and the location of office might give some impact on the decision making of each buyer in the city. For a lot of inhabitants in Sapporo city, their offices are located within the city, however, the locations are distributed. Hence, here we specify the ward in which his/her office is located by the label $m = 1, \ldots, 10$ and rewrite the intrinsic attractiveness (16) as

$$
A_m^0(X) = \sum_{l \neq m}^{10} \frac{\delta_l}{\sqrt{2\pi} R_l} \exp\left[-\frac{\{(x - x_{B_l})^2 + (y - y_{B_l})^2\}}{2R_l^2} \right]
$$
$$
+ \frac{(\delta_m + \eta)}{\sqrt{2\pi} R_m} \exp\left[-\frac{\{(x - x_{B_m})^2 + (y - y_{B_m})^2\}}{2R_m^2} \right].
$$

Namely, for the buyer who has his/her office within the ward m, the ward m might be a 'special region' for him/her and the local peak appearing in the intrinsic attractiveness is corrected by η. If he/she seeks for the housing close to his/her house (because the commuting cost is high if the office is far from his/her house), the correction η takes a positive value. On the other hand, if the buyer wants to live the place located far from the office for some reasons (for instance, some people want to vary the pace of their life), the correction η should be negative. To take into account these naive assumptions, we might choose η as a snapshot from a Gaussian with mean zero and the variance $\sigma^2 (< \delta_m)$. From this type of corrections, the buyer, in particular, the buyer of the highest ranking ($k = K - 1$) might feel some 'frustration' to make their decision, which is better location for them between *Chuo*-ku as the most attractive ward and the ward m where his/her office is located under the condition $\delta_1 \simeq \delta_m + \eta$. For the set of weights δ_m, $m = 1, \ldots, 10$, we take into account the number of offices in each ward, that is, *Chuo* (23,506), *Kita* (8,384), *Higashi* (8,396), *Shiraishi* (7,444), *Toyohira* (6.652), *Minami* (3,418), *Nishi* (6,599), *Atsubetsu* (2,633), *Teine* (3,259), *Kiyota* (2,546). Then, we choose each δ_m by dividing each number by the maximum of *Chuo*-ku as *Chuo* ($\delta_1 \propto 1.00$), *Higashi* ($\delta_2 \propto 0.36$), *Nishi* ($\delta_3 \propto 0.28$), *Minami* ($\delta_4 \propto 0.15$), *Kita* ($\delta_5 \propto 0.36$), *Toyohira* ($\delta_6 \propto 0.28$), *Shiraishi* ($\delta_7 \propto 0.32$), *Atsubetsu* ($\delta_8 \propto 0.11$), *Teine* ($\delta_9 \propto 0.14$), *Kiyota* ($\delta_{10} \propto 0.11$). For the bias parameter η, we pick up the value randomly from the range:

$$
|\eta| < \delta_m, \tag{18}
$$

instead of the Gaussian.

The resulting spatial distribution is shown in the right panel of Fig. 9. From this panel, we are clearly confirmed that the spatial structure of rents distributes more widely in whole city than that without taking into account the office location (see the left panel in Fig. 9 for comparison). We should notice that the range of simulated rent in the city is remarkably reduced from $[0, 120, 000]$ to $[0, 60,000]$ due to the diversification of values to consider the location of their housing.

5.4 On the Effective Time-Scale of Update Rule

Until now, we did not make a mention of time scale in the spatio-temporal update rule
(4) in the attractiveness $A_k(X, t)$. However, it might be important to for us consider
how long the time in our model system (artificial society) goes on for the minimum
time step $t \rightarrow t + 1$, especially when we evaluate the necessary period of time
to complete the accumulation of community after new-landmarks or shopping mall
come out. To decide the effective time-scale for $t \rightarrow t + 1$, we use the information
about the number of persons moving-into city of Sapporo through the year. Let us
define the number from the empirical data by C. Then, we should remember that
in our simulation, we assumed that in each time step ($t \rightarrow t + 1$), Γ newcomers
visit the city. Hence, the actual time τ for the minimum time step $t \rightarrow t + 1$ in our
artificial society is effectively given by

$$\tau = 365 \times \frac{\Gamma}{C} \text{ [days].} \tag{19}$$

Therefore, by using our original set-up $\Gamma \equiv L^2/K = (100 \times 100)/10 = 1,000$ and
by making use of the data listed in Table 1, we obtain $C = 63,021$, and substituting
the value into (19), we finally have $\tau = (1,000 \times 365)/63,021 = 5.79$ [days] for
$t \rightarrow t + 1$. This means that approximately 579 days have passed when we repeat
the spatio-temporal update rule (4) by $T = 100$ times. This information might be
essential when we predict the future housing market, let us say, after constructing the
Hokkaido Shinkansen (a rapid express in Japan) railway station, related landmarks
and derivative shopping mall.

6 Summary and Discussion

In this paper, we modified the Gauvin's model [1] to include the city having multiple
centers such as the city designated by ordinance by correcting the intrinsic attractiveness $A^0(X)$. As an example for such cities, we selected our home town Sapporo
and attempted to simulate the spacial distribution of averaged rent. We found that our
model can explain the empirical evidence qualitatively. Especially, we found that the
lowest ranking agents (from the viewpoint of the lowest 'willing to pay') are swept
away from relatively attractive regions and make several their own 'communities' at
low offering price locations in the city.

However, we should mention that we omitted an important aspect in our modelling.
Namely, the spatial resolution of working space and probabilistic search by buyer
taking into account their office location. In following, we will make remarks on those
two issues.

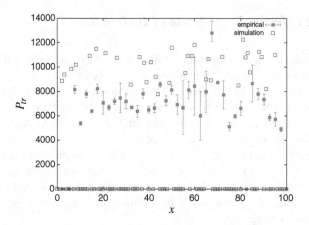

Fig. 10 The resulting rent distribution along the *Tozai*-subway line obtained by the quasi-one-dimensional model. The correspondence between x-axis and the location of each subway station is explicitly given as follows: *Bus-center mae* ($x = 86$), *Ōdori* ($x = 76$), *Nishi 11-Choume* ($x = 54$), *Nishi 18-Choume* ($x = 36$), *Maruyama Kouen* ($x = 22$), and *Nishi 28-Choume* ($x = 11$)

6.1 The 'Quasi-One-dimensional' Model

The problem of spatially low resolution of working space might be overcame when we model the housing market focusing on *Chuo*-ku instead of whole (urban) part of Sapporo. In the modelling, we restrict ourselves to the 'quasi-one-dimensional' working space and this approach enables us to compare the result with the corresponding empirical data. Although it is still at the preliminary level, we show the resulting rent distribution along the *Tozai*-subway line which is running across the center of Sapporo (*Chuo*-ku) from the west to the east as in Fig. 10. Extensive numerical studies accompanying with collecting data with higher resolution are needed to proceed the present study and it should be addressed as our future work.

6.2 Probabilistic Search Depending on the Location of Office

Some of buyers might search their housing locations by taking into account the place of their office. Here we consider the attractiveness $B_{k,i}(X, t)$ of office location X for i ranking k at time t. Namely, we should remember that the attractiveness for the place to live is updated as

$$A_k(X, t + 1) = A_k(X, t) + \omega(A^0(X) - A_k(X, t)) + \varepsilon \sum_{k' \geq k} v'_k(X, t) \qquad (20)$$

depending on the intrinsic attractiveness of place to live, whereas the attractiveness of the place X for agent i of ranking k who takes into account the location of their office place $Y_{k,i}$ is also defined accordingly and it is governed by

$$B_{k,i}(X, t+1) = B_{k,i}(X, t) + \varpi(B_{k,i}^0(|X - Y_{k,i}|) - B_{k,i}(X, t)) + \bar{\varepsilon} \sum_{k' \geq k} v_{k'}(X, t) \quad (21)$$

where $B_{k,i}^0(|X - Y_{k,i}|)$ is 'intrinsic attractiveness' of the location X for the agent i whose office is located at $Y_{k,i}$ and it is given explicitly by

$$B_{k,i}^0(|X - Y_{k,i}|) = \frac{1}{\sqrt{2\pi}Q} \exp\left[-\frac{(X - Y_{k,i})^2}{2\pi Q^2}\right], \quad (22)$$

and $\varpi, \bar{\varepsilon}$ are parameters to be calibrated using the empirical data sets.

Then, agent i looks for the candidates X_A, X_B to live according to the following probabilities

$$\pi_k^{(A)}(X, t) = \frac{1 - \exp(-\lambda A_k(X, t))}{\sum_{X' \in \Omega}\{1 - \exp(-\lambda A_k(X', t))\}} \quad (23)$$

$$\pi_{k,i}^{(B)}(X, t) = \frac{1 - \exp(-\lambda B_{k,i}(X, t))}{\sum_{X' \in \Omega}\{1 - \exp(-\lambda B_{k,i}(X', t))\}} \quad (24)$$

If the both X_A, X_B are approved, transaction price for each place is given by

$$P_{\text{tr}}^{(A)} = (1 - \beta)P_{k'}(X_A) + \beta P_k \quad (25)$$

$$P_{\text{tr}}^{(B)} = (1 - \beta)P_{k'}(X_B) + \beta P_k \quad (26)$$

The finial decision X_F is

$$X_F = \arg\min\{P_{\text{tr}}^{(A)} + \rho_m l_A, , P_{\text{tr}}^{(B)} + \rho_m l_B\}, \; F = \{A, B\} \quad (27)$$

where $\rho_m l_A, \rho_m l_B$ are travel costs between X_A, X_B and office (see Fig. 11).

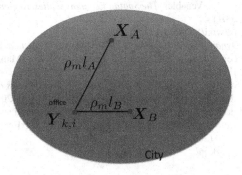

Fig. 11 Each buyer decides the final location $X_F = \arg\min\{P_{\text{tr}}^{(A)} + \rho_m l_A, P_{\text{tr}}^{(B)} + \rho_m l_B\}, F = \{A, B\}$

Therefore, the agents might prefer relatively closer place to the office to the attractive place to live when the distance between the attractive place to live and the office is too far for agents to manage the cost by the commuting allowance.

Acknowledgments One of the authors (JI) thanks Jean-Pierre Nadal in École Normale Supérieure for fruitful discussion on this topic and useful comments on our preliminary results at the international conference *Econophysics-Kolkata VII*. The discussion with Takayuki Mizuno, Takaaki Onishi, and Tsutomu Watanabe was very helpful to prepare this manuscript. This work was financially supported by Grant-in-Aid for Scientific Research (C) of Japan Society for the Promotion of Science(JSPS) No. 22500195, Grant-in-Aid for Scientific Research (B) No. 26282089, and Grant-in-Aid for Scientific Research on Innovative Area No. 2512001313. Finally, we would like to acknowledge the organizers of *Econophys-Kolkata VIII* for their hospitality during the conference, in particular, Frederic Abergel, Hideaki Aoyama, Anirban Chakraborti, Asim Ghosh and Bikas K. Chakrabarti.

References

1. L. Gauvin, A. Vignes, J.-P. Nadal, Modeling urban housing market dynamics: can the socio-spatial segregation preserve some social diversity? J. Econ. Dyn. Controls. **37**(7), 1300–1321 (2013)
2. M. Ballerini, N. Cabibbo, R. Candelier, A. Cavagna, E. Cisbani, I. Giardina, V. Lecomte, A. Orlandi, G. Parisi, A. Procaccini, M. Viale, V. Zdravkovic, Interaction ruling animal collective behaviour depends on topological rather than metric distance. Evidence from a field study. Proc. Nat. Acad Sci. USA **105**, 1232–1237 (2008)
3. C.W. Reynolds, Flocks, herds, and schools: a distributed behavioral model. Comput. Graph. **21**, 25 (1987)
4. M. Makiguchi, J. Inoue, Numerical study on the emergence of anisotropy in artificial flocks: a BOIDS modelling and simulations of empirical findings. *Proceedings of the Operational Research Society Simulation Workshop 2010 (SW10), CD-ROM*, (2010), pp. 96–102 (the preprint version, arxiv:1004 3837)
5. M. Makiguchi, J. Inoue, Emergence of anisotropy in flock simulations and its computational analysis. Trans. Soc. Instrum. Control Eng. **46**(11), 666–675 (2010). (in Japanese)
6. R. Olfati-Saber, Flocking for multi-agent dynamic systems: algorithms and theory. IEEE Trans. Autom. Control **51**(3), 401–420 (2006)
7. M. Fujita, P. Krugman A.J. Venables, *The Spatial Economy: Cities, Regions, and International Trade*. MIT Press (2001)
8. http://en.wikipeia.org/wiki/Sapporo
9. H. Chen, J. Inoue, *Dynamics of Probabilistic Labor Markets: Statistical Physics Perspective*. Lecture Notes in Economics and Mathematical Systems, vol. 662 (Managing Market Complexity, Springer, 2012), pp. 53–64
10. H. Chen, J. Inoue, Statistical mechanics of labor markets. *Econophysics of Systemic Risk and Network Dynamics*. New Economic Windows (Springer,(Italy-Milan), 2013), pp.157-171
11. H. Chen, J. Inoue, Learning curve for collective behavior of zero-intelligence agents in successive job-hunting processes with a diversity of Jaynes-Shannon's MaxEnt principle. Evol. Inst. Econ. Rev. **10**(1), 55–80 (2013)
12. http://www.souba-adpark.jp/
13. http://www.homes.co.jp/
14. http://rnk.uub.jp/

Part II
Miscellaneous

Performance of Inequality Indices

Anurag Chaplot and Rituparna Sen

Abstract It is well known that unequal income distribution, yielding poverty, can be a serious social and economic problem. Therefore, reliable inequality analysis of total population and the sub-populations is required in order to make efficient policy decisions. Among many inequality measures, Gini coefficient is the most popular and widely used. Other interesting measures used in this study are Zenga index and K-index. Non-parametric estimators of the indices are presented and analyzed in terms of the bias, efficiency and normality in order to study the small sample bias of the indices. Confidence intervals for these indices are studied for simple random samples using normal approximation, bootstrap percentile and bootstrap-t intervals. The methods are applied to real data on socioeconomic indicators for different states of India

1 Introduction

Economic inequality is the state of unequal distribution of income, assets or wealth among the individuals, group of individuals or countries. Large strands of research work has mainly focused on the issue of the income inequality, its determinants and its effects on the economy. Numerous indices have been developed to quantify the inequality of which the most widely used is Gini index along with Pietra index, Atkinson index and Zenga index. For a review of inequality measures in economics, see Cowell [4]. Ghosh et al. [10] developed the K-index and studied the inequalities in societies, academic institutions and science journals using Gini index and K-index. The definitions of all of these indices are based on the Lorenz curve.

The authors would like to thank the Indian Statistical Institute for funding this research.

A. Chaplot
Department of Humanities and Social Sciences IIT Kharagpur, Kharagpur, India
e-mail: anuragchaplot@gmail.com

R. Sen (✉)
Indian Statistical Institute, Chennai, India
e-mail: rsen@isichennai.res.in

© Springer International Publishing Switzerland 2015
F. Abergel et al. (eds.), *Econophysics and Data Driven Modelling of Market Dynamics*,
New Economic Windows, DOI 10.1007/978-3-319-08473-2_8

Recent works have concentrated more on the issue of estimation of these indices from sample data. Deltas [7] studied the small sample bias of the Gini index and established that it exhibits small sample downward bias. The paper also suggested some measures to deal with the bias. Statistical investigations, like Glasser [12] and Yitzhaki [21], focused largely on variance estimation for Gini measure. Qin et al. [18] studied the confidence intervals on Gini index using asymptotic theory and bootstrapping.

Greselin et al. [13] studied the Zenga index, developed some non-parametric estimators and estimated the confidence intervals on the index. Jedrzejczak [14] studied the asymptotic properties of the Gini index and Zenga index and the small sample properties through simulations. They presented some non-parametric estimators of the indices and analyzed properties such as bias, efficiency and normality.

The current study proposes a natural non-parametric estimator for K-index and derives its asymptotic distribution using the theory of estimating equations. Using this distribution approximate confidence intervals are obtained. These are compared in simulations to bootstrap confidence intervals for small sample properties. The real data analysis is carried out for K-index together with other indices like Gini and Zenga index for comparative purposes.

The rest of the paper is organized as follows. In Sect. 2 we define the different measures of inequality. We present the point estimators and interval estimators in Sect. 3. In particular, we derive the asymptotic distribution of the k-estimator and propose a large sample, as well as bootstrap confidence interval for this measure. We conduct thorough simulation studies regarding coverage probability and lengths of confidence intervals under different data generating mechanisms in Sect. 4. We present several real data examples of socio-economic variables and their inequality over the different states of India in Sect. 5.

2 Different Measures of Inequality

Lorenz curve is the graphical representation of the proportion of the total income that is cumulatively earned by the bottom $x\%$ of the population, and was introduced by Lorenz [16] to represent inequality of wealth distribution. The 45 degree line represents the line of perfect equality and the Lorenz curve lies beneath it. It is the most widely used measure for inequality.

Definition 1 (*Lorenz curve*) is given as

$$L(p) = \mu^{-1} \int\limits_0^p F^{-1}(t)dt \tag{1}$$

where $p = F(y)$ is the cumulative distribution function of income, μ denotes the expected value of the random variable Y and $F^{-1}(p)$ is the distribution pth quantile.

The Gini index was developed by Gini [11] as a summary measure of income inequality in society. A Gini value of 0 represents perfect equality in income distribution whereas a value of 1 represents maximal inequality. It is defined in terms of Lorenz curve as the ratio of the area that lies between the line of equality and the Lorenz curve over the total area under the equality line, or equivalently it is twice the area between the equality line and the Lorenz curve.

Definition 2 (*Gini index*) is given as

$$G = 2 \int_0^1 (p - L(p)) dp = \frac{2}{\mu} \int_0^\infty y F(y) dF(y) - 1 \tag{2}$$

K-index is a recent index developed by Ghosh etal [10] in view of high social inequalities in Kolkata. The corresponding values of k-index are $k = 0.5$ when $g = 0$ and $k = 1$ when $g = 1$.

Definition 3 (*k-index*) is given by the intersection of the Lorenz curve with the diagonal orthogonal to the equality line. It is the value of p such that

$$p + L(p) = 1 \tag{3}$$

The Zenga curve Z(p) is the ratio of the mean income of the poorest 100p % in the distribution to that of the rest of the $100(1 - p)$ % richest in the distribution, see Zenga [23].

Definition 4 (*Zenga index*) is defined in terms of Lorenz curve as

$$Z = \int_0^1 Z(p) dp$$

$$\text{where} \quad Z(p) = 1 - \frac{L(p)}{p} \frac{(1-p)}{(1-L(p))} \tag{4}$$

3 Estimation of Inequality Indices

In most common cases, we have a sample of n independent observations from an underlying distribution F_Y, for which we are interested in the inequality measure. Hence there is a need to find an estimator of the inequality measure based on the sample. This if often done by replacing the theoretical distribution F_Y by the empirical distribution function F_n that is a discrete cdf putting mass $1/n$ at each observation point.

All the inequality indices are defined in terms of the Lorenz curve, which in turn, is defined in terms of the quantile function. The quantile function is not uniquely defined for discrete distributions. Also, all interpretations of inequality indices assume that the Lorenz curve is strictly increasing. However, for discrete distributions, depending on how quantiles are defined, it may be a step function.

In the following we present point estimators and interval estimators for the inequality measures. The results for k-estimator are derived in this paper while those for Gini and Zenga index already exist in the literature and are presented for comparative purposes. In Sect. 3.1 we present the non-parametric point estimators. In Sect. 3.2 we derive the large sample properties of this estimator and thereby derive an asymptotic confidence interval. We present the bootstrap confidence intervals in Sects. 3.3 and 3.4, since the asymptotic intervals may not be precise in small samples. For a review of bootstrap re-sampling methods, see Efron and Tibshirani [8].

3.1 Point Estimates

Let $y_1, \ldots, y_n$ be an iid sample from F_Y and $F_n(u) = n^{-1} \sum_{j=1}^{n} I(y_j <= u)$ is the empirical distribution based on the sample, where I(.) is the indicator function. Following Sen [19], and Fei et al. [9], a point estimator of Gini index based on the sample is given by:

$$\hat{G} = \frac{1}{n\bar{y}} \sum_{i=1}^{n} \{2F_n(y_i) - 1\} y_i \tag{5}$$

A bias-corrected estimator $\tilde{G}$ was proposed by Davidson [6].

The estimator of k index is given by

$$\hat{K} = \frac{1}{n} \left(k_1 + \frac{S_{(-k_1)}/S_n - k_1/n}{y_{(k_1+1)}/S_n + 1/n} \right) \tag{6}$$

where S_n is the sum of all y_i, $y_{(i)}$ is the ith order statistic, $S_{(-i)}$ is sum of observation greater than $y_{(i)}$, $S_{(i)}$ is sum of observations less than and including $y_{(i)}$ and

$$k_1 = \max\{i : S_{(i)}/S_n < 1 - i/n\}$$

The commonly used nonparametric estimator of the Zenga index was introduced by Aly and Hervas [1] and can be expressed by the following equation:

$$\hat{Z} = 1 - \frac{1}{n\bar{y}} \left(y_{1:n} + \sum_{j=1}^{n-1} y_{\left[\frac{\sum_{i=1}^{j} y_{i:n}}{\bar{y}} \right]:n} \right) \tag{7}$$

where $y_{i:n}$ is the ith order statistics in a sample of size n.

3.2 Normal Approximation Confidence Intervals

A confidence interval for G based on asymptotic normal approximation has been derived by Qin et al. [18], as

$$(\hat{G} - z_{\alpha/2}\frac{\hat{\sigma}_1}{\sqrt{n}} \; , \; \hat{G} + z_{\alpha/2}\frac{\hat{\sigma}_1}{\sqrt{n}})$$

$$\text{where} \quad \hat{\sigma}_1{}^2 = \frac{1}{(n-1)\bar{y}^2} \sum_{i=1}^{n}(u_i - \bar{u})^2$$

$$\text{with} \quad u_i = 2h_1(y_i) - (\hat{G}+1)y_i$$

$$\text{and} \quad \hat{h}_1(u) = uF_n(u) + \frac{1}{n}\sum_{j=1}^{n} y_j I(y_j \geq u)$$

We derive a similar result for the K-index. We prove the following theorem in the Appendix B.

Theorem 1 *Let $y_1, \ldots, y_n$ be an iid sample from F_Y, such that $0 < E(Y_1^2) < \infty$, and $\hat{K}$ is as defined in Eq. (6). Then as $n \to \infty$,*

$$\sqrt{n}(\hat{K} - K) \Rightarrow N(0, \sigma_1^2) \tag{8}$$

where

$$\sigma_1^2 = \frac{K\lambda + K(1-K)(\xi-\gamma)^2}{(\xi+\mu)^2} \tag{9}$$

where $\mu = E(Y)$, $\xi = F^{-1}(K)$, $\gamma = E(Y \mid Y < \xi)$ and $\lambda = \text{Var}(Y \mid Y < \xi)$.

The variance is estimated by the corresponding sample quantities, that is,

$$\hat{\sigma}_1{}^2 = \frac{\hat{K}\hat{\lambda} + \hat{K}(1-\hat{K})(\hat{\xi}-\hat{\gamma})^2}{(\hat{\xi}+\hat{\mu})^2}$$

$$\hat{\mu} = \bar{y}$$

$$\hat{\xi} = y_{([n\hat{K}]+1)}$$

and $\hat{\gamma}$ and $\hat{\lambda}$ are the sample mean and sample variance of $y_{(1)}, \ldots, y_{([n\hat{K}])}$. A $(1-\alpha)$ percent confidence interval for K is given by

$$(\hat{K} - z_{\alpha/2}\frac{\hat{\sigma}_1}{\sqrt{n}}, \; \hat{K} + z_{\alpha/2}\frac{\hat{\sigma}_1}{\sqrt{n}}) \tag{10}$$

where $z_{\alpha/2}$ is the upper $\alpha/2$ quantile from standard normal distribution.

The corresponding confidence intervals of Zenga index are given in Greselin et al. [13].

3.3 Bootstrap Percentile Confidence Interval

A bootstrap sample $y_1^\star, \ldots, y_n^\star$ is taken from the original sample $y_1, \ldots, y_n$ by simple random sampling with replacement. $\hat{K}^\star$ is computed for this bootstrap sample. This procedure is repeated B times to obtain the sampling distribution of $\hat{K}^\star$. The $(1 - \alpha)$ level bootstrap-percentile confidence interval for K is given by

$$(P_{1-\alpha/2}, P_{\alpha/2})$$

where P_α is the 100αth percentile of the sampling distribution of $\hat{K}^\star$, i.e., $P_\alpha = \hat{K}_{[\alpha B]}^\star$.

3.4 Bootstrap-t Confidence Interval

The bootstrap-t confidence interval on K is constructed as

$$(\hat{K} - T_{1-\alpha/2}\frac{\hat{\sigma}}{\sqrt{n}}, \ \hat{K} - T_{\alpha/2}\frac{\hat{\sigma}}{\sqrt{n}}) \tag{11}$$

where T_α is the 100αth percentile of the sampling distribution of $(\hat{K}^\star - \hat{K})/(\sigma_1^\star/\sqrt{n})$ where $\hat{K}^\star$ and $\sigma_1^\star/\sqrt{n}$ are the estimator of K and associated standard error based on a bootstrap sample $y_1^\star, \ldots, y_n^\star$.

4 Simulation Study

A simulation study was conducted to investigate properties of the estimators for the three inequality coefficients.

In the experiment four different probability distributions were used as population models

1. Two-parameter lognormal distribution
2. Three-parameter Dagum distribution
3. Singh-Maddala
4. Weibull distribution

These are the most commonly used distributions for modelling prices, wages, incomes etc., see Chotikapanich [3]. The lognormal distribution is a classical model of income and wage size distributions. The Dagum model, also known also as Burr type-III distribution, has proved sufficient goodness-of-fit in many applications. It is a flexible distribution that can approximate both income distributions and wealth distributions, depending on parameters (for details see: Dagum [5], Kleiber and Kotz [15]). The Singh-Maddala distribution, introduced by Singh and Maddala [20], comes from a generalization of the Beta distribution of the second order and is very widely used in modelling household income. Although the Weibull distribution is more widely used in survival analysis, it is also common to model income distributions, in cases where the log-normal is not a good fit. The lognormal and Weibull distributions are special and limiting cases of the generalized Gamma distribution. Dagum and Singh-Maddala both generalized Beta distributions of Type 2 (see McDonald et al. [17]).

We start with actual income data coming from the Philippines National Statistics Office survey on Family Income and Expenditure conducted in 1997. This dataset is used, for example, in Zeileis [22]. The reason for using this data set is purely to get realistic values of parameters to carry out simulation studies. Different distributions, mentioned above, were fitted to the income data and their parameters were estimated using non-parametric methods.

The theoretical values of each index were calculated for those distributions under the estimated parameters. The expressions for these theoretical values are given in the Appendix A. In cases where theoretical values were not available, we performed numerical integration.

Random samples of different sizes were generated for each of the distributions and the indices were estimated. The number of repetitions of Monte Carlo experiment was $N = 10,000$.

We report the sample bias, variance and skewness for estimators of the different inequality measures in Tables 1, 2, 3 and 4 for different sample sizes under the different probability distributions. We make the following observations:

- The theoretical values of the indices depend on the chosen model. Although the same dataset is used, the indices are not robust to the choice of models.
- For small sample sizes, the value of the indices differed significantly from their theoretical values. But all the bias terms go down to zero as the sample size increases.
- The standard deviation of the K-index is the smallest.
- The variance of Gini and K indices decrease with increase in sample size. But this is not the case for the Zenga index.
- The skewness of Gini and K-index are very small under the lognormal and Weibull model, but quite large under the other models. In fact, in the later case, they increase with sample size. The skewness of the Zenga index is always moderate.

We studied confidence intervals for the different measures of inequality based on simulations from the different probability distributions with varying sample sizes. We obtained the intervals using normal approximation, bootstrap percentile and bootstrap-t methods described in Sect. 3. In Tables 5 and 6 we present the average length, the empirical coverage probability of 95 % confidence intervals, as well as the non-coverage probability in the lower tail and the upper tail of each index. We observe the following:

- As the sample size increases, each of the intervals get shorter and the coverage probability gets closer to 95 %. Hence, the accuracy of the intervals increases with the sample size.

Table 1 Characteristics of empirical distributions of different inequality measures under lognormal model

Sample size	Gini index estimators				k-index estimators			Zenga index estimators		
	Bias(G)	Bias(G)	Std. deviation	Coeff. of skewness	Bias(k)	Std. deviation	Coeff. of skewness	Bias($\hat{Z}$)	Std. deviation	Coeff. of skewness
Actual	0.385179				0.639			0.72427		
100	0.0050	0.0012	0.0287	0.1914	0.0064	0.0115	0.1546	0.0088	0.0442	1.3418
200	0.0024	0.0005	0.0206	0.1336	0.0032	0.0081	0.0908	0.0044	0.0462	1.3732
300	0.0013	0.0001	0.0170	0.1597	0.0020	0.0066	0.1400	0.0030	0.0468	1.3950
400	0.0012	0.0002	0.0146	0.1296	0.0016	0.0057	0.1071	0.0024	0.0473	1.4097
500	0.0010	0.0002	0.0132	0.0926	0.0013	0.0051	0.0836	0.0019	0.0477	1.4198
1,000	0.0005	0.0001	0.0093	0.1013	0.0007	0.0036	0.0959	0.0010	0.0483	1.4467
5,000	0.0000	0.0000	0.0041	0.0839	0.0002	0.0016	0.0648	0.0001	0.0490	1.4783

Table 2 Characteristics of empirical distributions of different inequality measures under Dagum model

Sample size	Gini index estimators			k-index estimators			Zenga index estimators		
	Bias($\hat{G}$)	Std. deviation	Coeff. of skewness	Bias($\hat{k}$)	Std. deviation	Coeff. of skewness	Bias($\hat{Z}$)	Std. deviation	Coeff. of skewness
Actual	0.450478			0.661			0.7673		
100	0.0138	0.0634	1.4182	0.0078	0.0248	1.9039	0.0158	0.0382	1.2340
200	0.0086	0.0500	1.4760	0.0042	0.0188	1.7436	0.0097	0.0416	1.3617
300	0.0061	0.0432	1.6074	0.0027	0.0161	1.9748	0.0069	0.0429	1.4222
400	0.0048	0.0402	1.8893	0.0020	0.0149	2.3381	0.0055	0.0438	1.4537
500	0.0042	0.0354	1.9924	0.0016	0.0131	2.5760	0.0047	0.0442	1.4785
1,000	0.0018	0.0289	2.555	0.0005	0.0102	3.0904	0.0023	0.0453	1.5310
5,000	0.0008	0.0141	2.5240	−0.0001	0.0049	2.7440	0.0008	0.0465	1.5922

Table 3 Characteristics of empirical distributions of different inequality measures under Singh-Maddala model

Sample size	Gini index estimators			k-index estimators			Zenga index estimators		
	Bias($\hat{G}$)	Std. deviation	Coeff. of skewness	Bias($\hat{k}$)	Std. deviation	Coeff. of skewness	Bias($\hat{Z}$)	Std. deviation	Coeff. of skewness
Actual	0.4517926			0.66			0.770807		
100	0.0152	0.0664	1.2582	0.0079	0.0257	1.5608	0.0175	0.0428	1.2609
200	0.0095	0.0526	1.5650	0.0040	0.0200	1.9936	0.0105	0.0460	1.3037
300	0.0063	0.0467	1.8555	0.0023	0.0177	2.5824	0.0072	0.0474	1.3440
400	0.0055	0.0419	1.9090	0.0018	0.0157	2.8408	0.0062	0.0481	1.3702
500	0.0041	0.0402	2.2029	0.0011	0.0152	3.4889	0.0049	0.0486	1.3874
1,000	0.0028	0.0309	2.2081	0.0004	0.0112	2.6992	0.0032	0.0496	1.4303
2,000	0.0010	0.0234	3.2418	−0.0003	0.0086	5.7682	0.0016	0.0503	1.4551

Table 4 Characteristics of empirical distributions of different inequality measures under Weibull model

Sample size	Gini index estimators				k-index estimators				Zenga index estimators			
	Bias($\widehat{G}$)	Bias($\widetilde{G}$)	Std. deviation	Coeff. of skewness	Bias($\widehat{k}$)	Std. deviation	Coeff. of skewness	Bias($\widehat{Z}$)	Std. deviation	Coeff. of skewness
Actual	0.40596				0.645			0.7644		
100	0.0035	−0.0006	0.0250	−0.0290	0.0053	0.0105	0.0418	0.0067	0.0803	0.9670
200	0.0020	0.0000	0.0175	−0.0308	0.0025	0.0072	0.0149	0.0038	0.0811	0.9954
300	0.0014	0.0001	0.0144	−0.0058	0.0014	0.0059	0.0235	0.0027	0.0813	1.0060
400	0.0008	−0.0002	0.0124	0.0130	0.0008	0.0051	0.0476	0.0018	0.0814	1.0069
500	0.0007	−0.0001	0.0111	−0.0027	0.0005	0.0045	0.0044	0.0016	0.0815	1.0116
1,000	0.0004	0.0000	0.0079	−0.0155	−0.0001	0.0032	−0.0089	0.0009	0.0817	1.0151
2,000	0.0002	0.0000	0.0056	0.0421	−0.0004	0.0022	0.0762	0.0005	0.0817	1.0178

Table 5 Simulation results for 95 % confidence interval under lognormal and Dagum models

n	CI	L	CP	U	AL	L	CP	U	AL
		Gini				k-index			
Lognormal									
20	NA	7.3	87.9	4.8	0.207	1.1	96.7	2.3	0.113
	BCp	24	72.8	3.2	0.194	2.8	87.7	9.5	0.086
	BCt	20.2	70.2	9.6	0.146	1.6	85	13.4	0.079
40	NA	4.2	91.0	4.8	0.157	0.6	97.6	1.8	0.083
	BCp	11.4	85.2	3.4	0.149	2.8	89.6	7.6	0.063
	BCt	12.4	77.8	9.8	0.112	2.7	84.6	12.7	0.057
60	NA	4.5	91.3	4.2	0.134	1.1	96.8	2.1	0.068
	BCp	10.5	86.5	3.0	0.129	2.5	90.3	7.2	0.053
	BCt	11.4	79.5	9.1	0.098	2.3	87.9	9.8	0.049
80	NA	2.3	93.4	4.3	0.118	0.7	98	1.3	0.059
	BCp	6.4	90.3	3.3	0.114	1.5	92.8	5.7	0.046
	BCt	8.6	82.7	8.7	0.086	1.6	88.7	9.7	0.043
100	NA	3.2	92.7	4.1	0.106	0.8	97.6	1.6	0.053
	BCp	5.0	92	3.0	0.103	2.3	91.5	6.2	0.042
	BCt	8.5	82.7	8.8	0.078	2.5	89.4	8.1	0.039
Dagum									
20	NA	5	76.8	18.2	0.233	6	78	16	0.103
	BCp	18.2	67.7	14.1	0.233	6.6	71.7	21.7	0.105
	BCt	8.0	65.1	26.9	0.204	4.6	65.3	30.1	0.092
40	NA	3.8	79.8	16.4	0.216	10.4	74.7	14.9	0.075
	BCp	12.4	73.7	13.9	0.205	4.9	77.1	18	0.091
	BCt	5.6	70.9	23.5	0.183	3.7	73.8	22.5	0.084
60	NA	1.7	79.8	18.5	0.204	9.5	75.2	15.3	0.062
	BCp	7.4	76.6	16	0.184	3	80.5	16.5	0.079
	BCt	3.6	69.2	27.2	0.161	2.7	76.6	20.7	0.074
80	NA	2.1	81.3	16.6	0.199	10.2	74.7	15.1	0.054
	BCp	6.5	78.9	14.6	0.176	1.8	82.4	15.8	0.073
	BCt	3.2	72.4	24.4	0.151	1.7	79.7	18.6	0.071
100	NA	1.3	81.6	17.1	0.181	10.1	73.6	16.3	0.048
	BCp	5.4	80	14.6	0.168	2.2	82.6	15.2	0.068
	BCt	2.9	72.6	24.5	0.142	1.5	79.4	19.1	0.067

- Among the three methods used for interval estimation, the one with high coverage probability also has high average length. Therefore, nothing can be said as to which interval estimation method is better, on the whole.
- The bootstrap t-interval is, in most cases, shorter than the bootstrap percentile interval.

Table 6 Simulation results for 95 % confidence interval under Singh-Maddala and Weibull models

n	CI	L	CP	U	AL	L	CP	U	AL
		Gini				k-index			
Singh-Maddala									
20	NA	3.8	75.9	20.3	0.232	7.4	76.4	16.2	0.104
	BCp	17.3	66.6	16.1	0.234	5.3	69.6	25.1	0.105
	BCt	8.6	61.8	29.6	0.203	3.5	64.2	32.3	0.093
40	NA	2.7	76.8	20.8	0.217	8.5	73.6	17.9	0.075
	BCp	11.9	70.3	17.8	0.206	3.4	73.9	22.7	0.089
	BCt	4.6	66.1	29.3	0.189	2.6	67.8	29.6	0.082
60	NA	2	79.8	18.2	0.209	8.5	76.1	15.4	0.062
	BCp	6.8	77.3	15.9	0.193	2.1	79.9	18	0.081
	BCt	2.2	70.6	27.2	0.169	2.1	76.4	21.5	0.077
80	NA	1.8	78.7	19.5	0.200	10.3	73	16.7	0.054
	BCp	6.0	76.1	17.9	0.179	2.3	80.7	17	0.073
	BCt	3.1	69.9	27	0.154	1.6	78.2	20.2	0.072
100	NA	1.7	81.8	16.5	0.197	10.4	71.7	17.9	0.049
	BCp	5.9	78.5	15.6	0.173	2.1	82	15.9	0.069
	BCt	3.1	71.7	25.2	0.147	1.4	80	18.6	0.068
Weibull									
20	NA	8.8	89.3	1.9	0.205	0	98.9	1.1	0.142
	BCp	18.1	81.2	0.7	0.204	3.3	90.6	6.1	0.087
	BCt	21.2	73.9	4.9	0.143	3.1	87.4	9.5	0.079
40	NA	5.2	92.4	2.4	0.178	0.1	99.7	0.2	0.103
	BCp	10.1	88.4	1.5	0.148	2.8	92.4	4.8	0.062
	BCt	14.6	79.7	5.7	0.108	2.5	90.2	7.3	0.056
60	NA	2	91.7	2.7	0.165	0	99.9	0.1	0.085
	BCp	8.3	89.7	2.0	0.122	2.3	93.5	4.2	0.051
	BCt	11.9	81.3	6.8	0.090	2.7	91.5	5.8	0.046
80	NA	1.8	94.4	1.9	0.157	0	99.9	0.1	0.074
	BCp	6.8	91.8	1.4	0.107	2.6	93.6	3.8	0.044
	BCt	12.6	81.9	5.5	0.080	2.6	92.4	5	0.040
100	NA	4.1	93.3	2.6	0.152	0	99.8	0.2	0.066
	BCp	6.7	91.2	2.1	0.095	2.6	94.3	3.1	0.039
	BCt	10.7	83.5	5.8	0.072	2.9	92.4	4.7	0.036

- For the Gini index, the bootstrap intervals are always shorter than the Normal approximation interval.
- For the K-index, the bootstrap intervals are always shorter than the Normal approximation interval for the lognormal and Weibull distributions, but not the other two cases.

- In case of lognormal and Weibull distributions, for sufficiently high sample size, all three methods of confidence intervals for both indices have quite high coverage probability. This is not the case for the other two distributions, where, even for sample size 100, all methods have coverage probability around 70–80 %. The problem is more pronounced in the upper tail.

It is seen that the behavior of the intervals depends on whether the generating distribution is of generalized beta or generalized gamma type. This leaves scope for further investigation in this direction.

5 Real Data Examples

We choose to study the different states and union territories of India with respect to some measures of development. The source of the data is Directorate of Economics Statistics of respective State Governments aggregated from the database of the Planning Commission of India. In particular, we study the following quantities

1. Net State Domestic Product—annual data from 2001–2002 to 2012–2013
2. Percentage of Population Below Poverty line—data for the years 1973–1974, 1977–1978, 1983–1984, 1987–1988, 1993–1994, 1999–2000, 2004–2005, almost every 5 years.
3. Infant Mortality Rate—Annual data from 1999 to 2012
4. Literacy Rate—data every 10 years from 1951 to 2011, which are the census years.

The point estimates of different indices and their confidence intervals using normal approximation, percentile bootstrap and t-bootstrap are presented in Figs. 1, 2, 3 and 4.

Here are some features that are common to all the data sets:

- We observe that the bias corrected Gini index is always higher than the simple Gini index.
- The confidence intervals by all three methods are very close for the Gini and Zenga indices. However, the asymptotic interval for the K-index is much wider than the corresponding bootstrap intervals in all data sets except the Net State Domestic Product.
- As we expected from the simulation study, the sample variance of the K-index is much lower than the other two indices. Hence we get much narrower confidence intervals for the former.
- For the Gini and Zenga-indices, the normal approximation intervals are always lower than the Bootstrap percentile intervals. This points to the positive skewness of the sampling distribution of the indices as seen in the simulations. The normal approximation produces a symmetric interval, which is not valid for skew distributions.

Here are some features common to all indices:

Fig. 1 Inequality in Net
State Domestic Product. *Top*
Gini Index, *Middle* K-index,
Bottom Zenga Index. For
each plot, the *solid curve* is
the index, *dashed lines* are
confidence intervals obtained
by Normal Approximation
(*blue*), Bootstrap Percentile
(*red*) and Bootstrap-t (*green*)
methods. The magenta line
of the *top* figure is the
bias-corrected Gini index
(Color figure online)

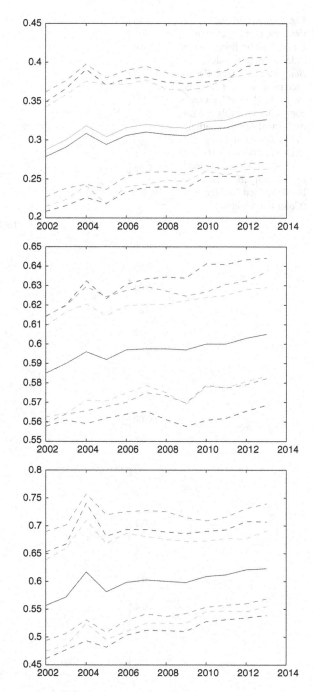

Fig. 2 Inequality in
Percentage of Population
Below Poverty line. *Top* Gini
Index, *Middle* K-index,
Bottom Zenga Index. For
each plot, the *solid curve* is
the index, *dashed lines* are
confidence intervals obtained
by Normal Approximation
(*blue*), Bootstrap Percentile
(*red*) and Bootstrap-t (*green*)
methods. The magenta line
of the *top* figure is the
bias-corrected Gini index
(Color figure online)

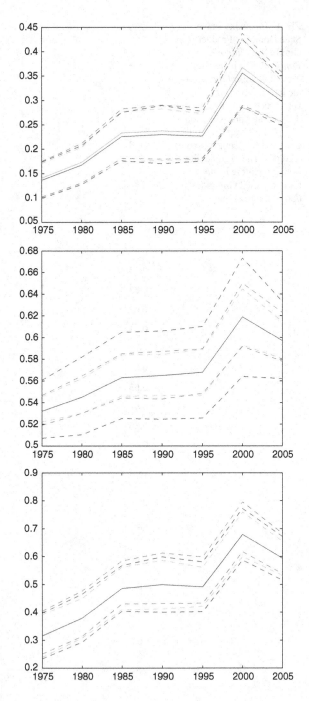

Fig. 3 Inequality in Infant Mortality Rate. *Top* Gini Index, *Middle* K-index, *Bottom* Zenga Index. For each plot, the *solid curve* is the index, *dashed lines* are confidence intervals obtained by Normal Approximation (*blue*), Bootstrap Percentile (*red*) and Bootstrap-t (*green*) methods. The magenta line of the *top* figure is the bias-corrected Gini index (Color figure online)

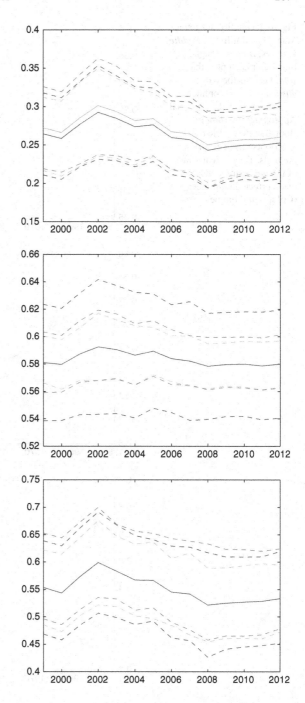

Fig. 4 Inequality in Literacy
Rate. *Top* Gini Index, *Middle*
K-index, *Bottom* Zenga
Index. For each plot, the
solid curve is the index,
dashed lines are confidence
intervals obtained by Normal
Approximation (*blue*),
Bootstrap Percentile (*red*)
and Bootstrap-t (*green*)
methods. The magenta line
of the *top* figure is the
bias-corrected Gini index
(Color figure online)

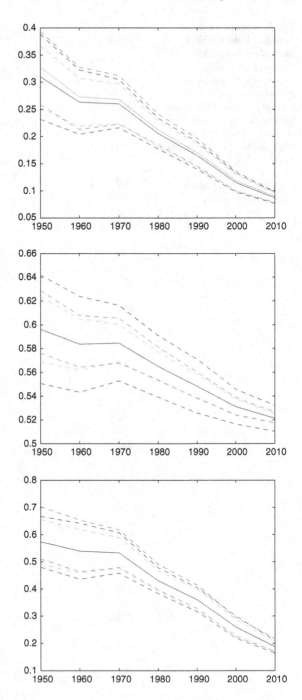

- Inequality in the Net State Domestic Product has risen slightly over the years. All indices show a sharp rise in inequality around 2004, which was the election year, and a sharp drop right afterwards.
- Inequality in Percentage of Population Below Poverty line has risen slightly over the years. There was a sharp rise in 1999–2000 and then a subsequent drop in 2004–2005. Since data is not available at a higher frequency, it is difficult to draw conclusions from here, as the changes might have happened at some intermediate time.
- Inequality in Infant mortality rate was a little on the rise until 2002, from when it has steadily declined. This can be associated with the National Health Policy of 2002.
- Inequality in literacy rate has steadily decreased over the years. In fact, the variance of the measure has also decreased. So this can be measured with very high level of precision.

6 Conclusions

In this paper, we have proposed point and interval estimators of the K-index. The estimator is shown to have asymptotic normal distribution. The estimator is compared to the Gini index and Zenga index through simulation and real data examples. The confidence intervals for the K-index give better coverage probabilities as well as average lengths as compared to the Gini index for all the distributions when the sample size is small. The variance of the estimator for K-index is also lower than those for Gini and Zenga indices. Thus for small sample sizes, it is better to estimate inequality using the K-index than the traditionally used Gini and Zenga indices.

Based on simulation studies, the bias and variance of the estimator are quite small even for small sample sizes. However, the skewness is quite large. When the underlying variable comes from a lognormal and Weibull distributions, the coverage probability of the confidence interval is quite high. However, for Dagum and Singh-Maddala distributions, the coverage probability of the 95 % confidence interval is around 70–80 %. These are the distributions for which the skewness of the measure is also high.

In the real data examples, we quantify the inequality between different states on India with respect to certain socio economic parameters and study the evolution of this inequality over time. It is seen that inequality in the Net State Domestic Product and percentage of population below poverty line have declined marginally over the years while inequality in infant mortality rate and literacy rate have steadily decreased.

Appendix A: Values of Inequality Indices Under Different Models

The theoretical values of the indices were calculated as

1. Lognormal distribution

$$F(y) = \Phi(\frac{lnz - \mu}{\sigma})$$

Gini	$G = 2\Phi(\frac{\sigma}{\sqrt{2}}) - 1$
Lorenz curve	$L(p) = \Phi(\Phi^{-1}(p) - \sigma)$
Zenga	$Z(p) = 1 - \frac{L(p)}{p}\frac{1-p}{1-L(p)}$

where Φ is the cumulative standard normal distribution.

2. Dagum distribution

$$F(y) = (1 + (\frac{x}{b})^{-a})^{-p}$$

$$G = \frac{\Gamma(p)\Gamma(2p + \frac{1}{a})}{\Gamma(2p)\Gamma(p + \frac{1}{a})} - 1$$

3. Singh-Maddala distribution

$$F(y) = 1 - (1 + (\frac{x}{b})^{a})^{-q}$$

$$G = \frac{\Gamma(q)\Gamma(2q - \frac{1}{a})}{\Gamma(2q)\Gamma(q - \frac{1}{a})}$$

4. Weibull distribution

$$F(y) = 1 - \exp(kx)^{\alpha}$$

$$G = 1 - (\frac{1}{2})^{1/\alpha}$$

Appendix B: Proof of Theorem 1

It has been shown, for e.g. in Beach and Davidson [2], that if $L_n(\theta)$ denotes $\int_0^\theta F_n^{-1}(u)du$ and $L(\theta)$ is the corresponding population quantity, then

$$\sqrt{n}(L_n - L) \Rightarrow N(0, v^2) \quad \text{where,} \quad v^2 = \theta\lambda^2 + \theta(1 - \theta)(\xi - \gamma)^2$$

$\hat{K}$ is the solution of $g_n(\theta) = 0$ where $g_n = \bar{Y}(\theta - 1) + L_n(\theta)$ is the sample version of $g(\theta) = \mu(\theta - 1) + L(\theta)$.

Hence by the theory of estimating equations,

$$\sqrt{n}(k_1 - K) \Rightarrow N(0, \sigma_1^2),$$

$$\text{where} \quad \sigma_1^2 = \frac{v^2}{(\partial g / \partial \theta)^2}.$$

The partial derivative in the denominator can be shown to equal $\xi + \mu$. This completes the proof.

References

1. E.A. Aly, M.O. Hervas, Nonparametric inference for Zenga's measure of income inequality. Metron **LVII**, 69–84 (1999)
2. C.M. Beach, R. Davidson, Distribution-free statistical inference with Lorenz curves and income shares. Rev. Econ. Stud. **50**, 723–735 (1983)
3. D. Chotikapanich, *Modelling Income Distributions and Lorenz Curves* (Springer, New York, 2008)
4. F.A. Cowell, *Measuring Inequality, Perspectives in Economic Analysis* (Oxford University Press, Oxford, 2011)
5. C. Dagum, A new model of personal income distribution: specification and estimation. Econ. Appl. **30**, 413–437 (1977)
6. R. Davidson, Reliable inference for the Gini index. J. Econ. **150**(1), 30–40 (2009)
7. G. Deltas, The small-sample bias of the Gini coefficient: results and implications for empirical research. Rev. Econ. Stat. **85**(1), 226–234 (2003)
8. B. Efron, R. Tibshirani, Improvements on cross-validation: the bootstrap method. J. Am. Stat. Assoc. **92**, 548–560 (1997)
9. J. Fei, G. Ranis, S. Kuo, Growth and the family distribution of income by factor components. Q. J. Econ. **92**, 17–53 (1979)
10. A. Ghosh, N. Chattopadhyay, B.K. Chakrabarti, Inequalities in societies, Academic Institutions and Science. Gini k-indices Physica **A410**, 30–34 (2014)
11. C. Gini, *Variabilita e Mutabilita* (Bologna University Press, Bologna, 1912)
12. G.J. Glasser, Variance formulas for the mean difference and coefficient of concentration. J. Am. Stat. Assoc. **57**, 54–648 (1962)
13. F. Greselin, L. Pasquazzi, R. Zitikis, Zenga's new index of economic inequality, its estimation, and an analysis of incomes in Italy. J. Probab. Stat. **2010**, 26 (2010). doi:10.1155/2010/718905
14. A. Jedrzejczak, Asymptotic properties of some estimators for income inequality measures-a simulation study, in *Proceedings 59th ISI World Statistics Congress, CPS203* (2013), pp. 5474–5479
15. C. Kleiber, S. Kotz, *Statistical Size Distributions in Economics and Actuarial Sciences* (Wiley, Hoboken, 2003)
16. M.O. Lorenz, Methods for measuring the concentration of wealth. J. Am. Stat. Assoc. **9**, 209–219 (1905)
17. J.B. McDonald, J. Sorensen, P.A. Turley, Skewness and Kurtosis properties of income distribution models. Rev. Income Wealth **59**, 360–374 (2013)

18. Y. Qin, J.N.K. Rao, C. Wu, Empirical likelihood confidence intervals for the Gini measure of income inequality. Econ. Model. **27**, 1429–1435 (2010)
19. A. Sen, Poverty - an ordinal approach to measurement. Econometrica **44**, 219–231 (1976)
20. S.K. Singh, G.S. Maddala, A function for size distribution of incomes. Econometrica **44**, 963–970 (1976)
21. S. Yitzhaki, Calculating jackknife variance estimators for parameters of the Gini method. J. Bus. Econ. Stat. **9**, 235–239 (1991)
22. A. Zeileis, Package ineq for R (Technical Report, CRAN, 2009)
23. M. Zenga, Concentration curves and concentration indices derived from them, in *Income and Wealth Distribution, Inequality and Poverty*, ed. by C. Dagum, M. Zenga (Springer, Berlin, 1990), pp. 94–110

A Dynamical View of Different Solution Paradigms in Two-Person Symmetric Games: Nash Versus Co-action Equilibria

V. Sasidevan and Sitabhra Sinha

Abstract The study of games and their equilibria is central to developing insights for understanding many socio-economic phenomena. Here we present a dynamical systems view of the equilibria of two-person, payoff-symmetric games. In particular, using this perspective, we discuss the differences between two solution concepts for such games—namely, those of Nash equilibrium and co-action equilibrium. For the Nash equilibrium, we show that the dynamical view can provide an equilibrium refinement, selecting one equilibrium among several possibilities, thereby solving the issue of multiple equilibria that appear in some games. We illustrate in detail this dynamical perspective by considering three well known 2-person games namely the Prisoner's Dilemma, game of Chicken and the Stag-Hunt. We find that in all of these cases, co-action equilibria tends to correspond to 'nicer' strategies than those corresponding to Nash equilibria.

1 Introduction

Games represent strategic interactions between entities generally referred to as agents. Here, the term "agents" could refer to a variety of entities, ranging from human beings or animals to computer programs or robots. In games, each agent receives a payoff depending upon the strategy choice made by all agents including herself. Thus, an agent who wants to optimize her payoff should consider not only the payoff structure of the game, but also the decision making processes of other agents. The choice of strategy by each agent in such an interaction leads to a collective outcome that may or may not be globally optimal. In this context, it is imperative to understand how two agents facing a game situation, who have to make a strategic decision, will go about doing it, since the strategic interaction between agents is

V. Sasidevan (✉) · S. Sinha
The Institute of Mathematical Sciences, CIT Campus, Taramani,
Chennai 600113, India
e-mail: sasidevan@imsc.res.in

S. Sinha
e-mail: sitabhra@imsc.res.in

© Springer International Publishing Switzerland 2015
F. Abergel et al. (eds.), *Econophysics and Data Driven Modelling of Market Dynamics*,
New Economic Windows, DOI 10.1007/978-3-319-08473-2_9

the basis of the collective behavior in a system comprising such agents. Financial markets, for example, may be viewed as the collective outcome of strategic interactions between a large number of people participating in it. Another example is that of evolution, where one may view evolution by natural selection as a result of the interaction between competing genes. Cooperation and conflict is at the heart of such systems and forms the subject matter of the study of games. In games, in general, each agent should have a behavior model of other agents so that she has a way to describe the decision making process of other agents. In this regard, standard game theory makes several assumptions about the agent's behavior. It assumes that agents are fully rational and would like to optimize their payoff and they are perfect in execution of their strategies (see for e.g. [1] for a detailed discussion). While the applicability of these assumptions in any particular situation is open to criticism, they form an important benchmark for optimal behavior. In fact, these assumptions form an important part of modern economic theory in which the participating agents are often assumed to be fully rational.

The simplest of games consists of the strategic interaction between two agents in a single play of the game. In fact, 2-person games like Prisoners Dilemma, Stag-Hunt etc., describe very general socio-economic scenarios, towards the analysis of which considerable effort has been devoted. A key concept in the study of games is that of an "equilibrium". It refers to a state of affairs where each agent has decided her strategy for the game at hand. How the agents pick their equilibrium strategy is given by a solution concept. A solution concept thus is a formal rule for predicting how a game will be played between agents and employs certain assumptions regarding agent's behavior. An important solution concept for non-cooperative games is that of Nash equilibrium. Informally, it is a state where after every agent has selected their 'Nash' strategies, none of the agents can improve their payoff by unilaterally deviating from it. It is to be noted that a game may have more than one Nash equilibrium.

In this article, we show that the equilibria of a game may be viewed as the "fixed-point" equilibria of a dynamical system. In particular, we present a dynamical view of the equilibria obtained by two different solution concepts, viz., Nash [2] and co-action [3], the latter being a concept that makes use of the symmetry between the agents for payoff-symmetric games. The vector flow diagrams on the strategy space that is generated using the dynamics approach makes the differences between the equilibria obtained in the two solution concepts visually apparent. For the Nash equilibrium, we argue that a dynamical perspective may be regarded as an equilibrium refinement selecting one equilibrium out of several possible ones, thus solving the multiplicity issue. We illustrate these points by considering three well known examples of 2-person games, namely the Prisoners Dilemma, Game of Chicken and Stag-Hunt.

2 A Dynamical Framework for Analysing 2-Person Games

Here we describe a dynamical perspective for analyzing games, focusing on 2-person single-stage games in which two agents interact only once. No communication is allowed between the agents. Furthermore, we consider the simple case where each agent has to choose one of two possible actions (say, Action 1 and Action 2) available to her. Each agent receives a payoff according to the pair of choices made by them, such that the game may be represented by a payoff matrix that specifies all possible outcomes (Fig. 1). We consider situations where the game is payoff symmetric, i.e., on exchanging the identities of the players (A, B), the payoff matrix remains unchanged. Note that most 2-person games that are studied in the literature fulfil the above criteria. Given the payoff matrix, an agent can have a mixed strategy, where she chooses Action 1 with some probability p and Action 2 with probability $(1 - p)$. If p is either 0 or 1, it is called a pure strategy. Given a game, represented by a matrix containing the numerical values of R, S, T and P (or a hierarchical relation among them), Nash equilibrium is defined as a state—i.e., a set of the choices made by all the agents—where no agent can increase her payoff by unilaterally deviating from the Nash state. A Nash equilibrium comprising pure strategies may be found by a search procedure, whereby each possible state is explicitly examined for the above criterion. Note that a given game can have more than one Nash equilibrium, possibly involving mixed strategies. In such cases, the choice of a particular equilibrium will have to involve additional refinement criteria, which is an important area of research in game theory [4].

We now illustrate a dynamical perspective on Nash equilibria by first defining payoff functions for all possible mixed strategies of the two agents. Assuming that agent A (B) chooses Action 1 with probability p_1 (p_2) and Action 2 with probability $1 - p_1$ ($1 - p_2$, respectively), the expected payoffs of the agents are

$$W_A = p_1 p_2 R + p_1(1 - p_2)S + (1 - p_1)p_2 T + (1 - p_1)(1 - p_2)P, \quad (1)$$
$$W_B = p_1 p_2 R + p_1(1 - p_2)T + (1 - p1)p_2 S + (1 - p1)(1 - p_2)P. \quad (2)$$

As the payoffs are continuous functions of p_1 and p_2, they can be represented as two-dimensional surfaces (Fig. 2) analogous to fitness landscapes in biology or energy landscapes in physics. However, unlike the latter, there are two distinct surfaces

	Agent B	
	Action 1	Action 2
Action 1	(R,R)	(S,T)
Action 2	(T,S)	(P,P)

(rows labelled Agent A: Action 1, Action 2)

Fig. 1 A generic representation of the payoff matrix for a 2-person symmetric game where each agent has two actions available to her

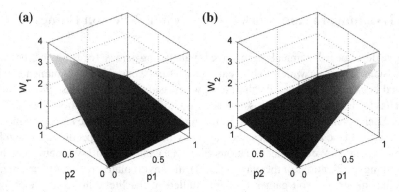

Fig. 2 The payoff functions W_A and W_B for two agents playing the game of Chicken, shown as functions of p_1 and p_2, i.e., the probability of each agent to choose Action 1. The payoffs (in terms of the terminology given in Fig. 1) are $T = 3.5$, $R = 1$, $S = 0.5$ and $P = 0$

for the two agents, and each of them would like to achieve the maximum of their respective payoff functions, a goal that may not be mutually compatible. By contrast, the evolution of the state of a physical system can be seen as a convergence process to a minimum of a single function, e.g., the free energy that describes the entire system.

Given the payoff function surfaces we can now proceed to find the strategy pairs (p_1^*, p_2^*) that correspond to a Nash equilibrium. Note that while the Nash solution is usually not defined in terms of a dynamical perspective, one can view (p_1^*, p_2^*) as an equilibrium point for flow dynamics in the p_1-p_2 plane, as described below. The initial condition for this dynamical system can be any arbitrary point in this plane. Each agent is then allowed to change its strategy infinitesimally (i.e., $p_1 \rightarrow p_1 + dp_1$, $p_2 \rightarrow p_2 + dp_2$) in order to improve their respective payoffs, taking into consideration that the other agent would also be doing the same. A sequence of such incremental changes, which will be manifested as a flow in the p_1-p_2 plane would eventually converge to an equilibrium point (p_1^*, p_2^*). Note that, while such a strategy would correspond to a stable equilibrium of the flow dynamics, there may also be unstable equilibria.

The dynamical equations governing the flow can be derived by considering the change in the payoffs (dW_A, dW_B) of the two agents as a result of the infinitesimal change in their strategies dp_1, dp_2:

$$\frac{\partial W_A}{\partial p_1} = p_2(R - T) + (1 - p_2)S, \tag{3}$$

$$\frac{\partial W_B}{\partial p_2} = p_1(R - T) + (1 - p_1)S.$$

Thus, on any point in the p_1-p_2 plane, the magnitude and direction of the flow can be obtained by a vector sum of the two components given by Eq. 3. The resulting flow diagram will describe the trajectory in strategy space starting from any arbitrary strategy pair (p_1, p_2). This will be illustrated with specific examples of 2-person games in the next section.

As mentioned earlier, Nash equilibrium is not the only possible solution of a payoff symmetric game. Recently, an alternative paradigm referred to as co-action equilibrium for solving such games has been introduced in the specific context of minority game [3]. Here we study this novel solution concept in the context of generic 2-person games with symmetric payoff from a dynamical perspective. The key notion of co-action equilibrium is: as the two agents are aware that they face an exactly symmetric situation, the choice made by agent A should be identical to the choice of agent B, assuming that they are equally rational (for a detailed discussion see Ref. [5]). Thus, in terms of the flow dynamics introduced above, in this solution concept, each agent will take into account in her calculation for revising her strategy that the other agent is not only using the same strategy (i.e., $p_1 = p_2$) but will also make exactly the same infinitesimal change, i.e., $dp_1 = dp_2$. Then the change in the payoffs of the two agents, as a result of changing p_1, p_2 (analogous to Eq. 3 for Nash equilibrium) is:

$$\frac{\partial W_A}{\partial p_1} = 2p_1 R + (1 - 2p_1)(T + S), \tag{4}$$

$$\frac{\partial W_B}{\partial p_2} = 2p_2 R + (1 - 2p_2)(T + S).$$

Note that the above equations hold not only when $p_1 = p_2 = p$ (so that the dynamics is confined to the diagonal line in the $p_1 - p_2$ plane), but also for situations where the two agents initially start with different strategies ($p_1 \neq p_2$), believing however that the other agent is using exactly the same strategy.

The co-action solution yields results that differ remarkably from those obtained using the concept of Nash equilibrium, some of which will be described in the next section in the context of specific 2-person games. An important distinction is that while there could be multiple Nash equilibria for a game, the corresponding co-action equilibrium is unique. The dynamical perspective allows us to also distinguish between Nash and co-action solutions for 2-person symmetric games in that a stable mixed strategy equilibrium is possible for the latter unlike in the former (Nash) where a mixed strategy equilibrium, if it exists, is always unstable.

Note that while the flow diagrams produced by the dynamical process presented here may resemble the trajectories generated by solving replicator equations [6], the two approaches are essentially distinct. In particular, the latter approach is based on the concept of evolutionary stable strategies, which is an equilibrium refinement of the Nash solution. Also, instead of being stages in the evolutionary progression of a population, the sequence of infinitesimal changes in strategies in the dynamical approach presented here, can be interpreted as steps in the deductive reasoning of the two agents, at the end of which they choose the strategy corresponding to the equilibrium they converge to. When our approach is applied to study the Nash solution of a game, it can also be viewed as an equilibrium refinement as, if there are multiple

Nash equilibria, it allows agents to choose a particular equilibrium depending on the arbitrarily chosen initial state. Thus, in an ensemble of many realizations of a game, the fraction of cases where agents will converge to a particular equilibrium is proportional to the size of its basin of attraction. An unstable equilibrium (if it exists) will lie on the separatrix that demarcates the basins of different stable equilibria.

3 Examples

We now illustrate the approach outlined in the previous section using three well-studied 2-person symmetric games, viz., Prisoner's Dilemma, Game of Chicken and Stag-Hunt, each of which can be connected to real-life scenarios involving interactions between a pair of agents who have to choose between two possible actions. Each of these games is defined in terms of a specific hierarchical relationship between the payoffs R, S, T and P (using the terminology of the payoff matrix in Fig. 1). Without loss of generality, we can set $P = 0$ and $R = 1$ (thereby fixing the origin and the scale for the payoffs), leaving only S and T as free parameters. In the following subsections, we discuss each of these games in turn, describing the meaning of the different choices available to the agents (viz. Action 1 and Action 2) in a particular game, and exploring the different equilibria obtained by using Nash and co-action solution concepts.

3.1 Prisoner's Dilemma

Prisoner's Dilemma (PD) [7] can be regarded as one of the most well-known games in the literature. It has evoked great interest among researchers from a multitude of disciplines ranging from social sciences and politics to biology and physics, from the 1950s onwards and continues to do so at present (a good place to read about historical developments in PD is the corresponding entry in the online Stanford Encyclopedia of Philosophy [8]). The game represents the strategic interaction between two players who have to choose between cooperation (Action 1) and defection (Action 2). The different payoffs are interpreted as follows: R is a "reward" for both players cooperating, P is a "punishment" for both players defecting, while, in the event that one agent defects while the other cooperates, T and S are the "temptation" received by former and the "sucker's payoff" of the latter. In PD, the hierarchical relation between the payoffs is $T > R > P > S$, which makes achieving mutual cooperation non-trivial as each player will benefit more by defecting (assuming that the other will cooperate).

It is easy to see that mutual defection is the only Nash equilibrium for PD. As Action 1 represents cooperation, p_1 (p_2) corresponds to the probability that agent A (B) will choose cooperation. As discussed in the previous section, we can associate a vector with each point in the (p_1, p_2) plane for the game which describes the flow from that point. Figure 3 shows the resulting flow diagrams obtained using the Nash

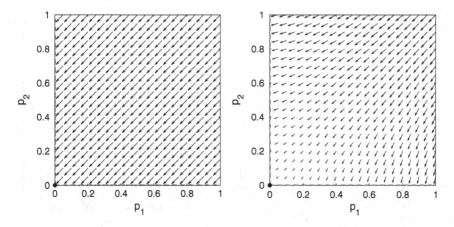

Fig. 3 Vector flow diagram representation of the Nash solution of the 2-person Prisoner's Dilemma game for temptation payoffs (*left*) $T = 1.5$ and (*right*) $T = 3.5$. The abscissae and ordinate correspond to the probabilities (p_1 and p_2) that players 1 and 2, respectively, choose to cooperate. The *broken lines* represent the best response (or reaction) correspondence of the players (*red* for player 1, *blue* for player 2). The intersection of the lines, represented by a *filled circle*, represent the single Nash equilibrium corresponding to both players defecting (i.e., $p_1 = 0$, $p_2 = 0$)

solution concept for two different values of the temptation payoff T (keeping S fixed at -0.5). In both cases, the system converges to the pure strategy $p_1 = p_2 = 0$ (mutual defection), which is the Nash equilibrium for PD.

By contrast, using the co-action solution concept, for low values of T we observe mutual cooperation (i.e., $p_1 = p_2 = 1$) as the stable equilibrium of the system (Fig. 4, left). For larger values of T, the stable equilibrium corresponds to a mixed strategy, $0 < p_1 = p_2 < 1$ (Fig. 4, right). Thus, as discussed in detail in Ref. [5], using the co-action concept for solving PD we can show that selfish agents trying to maximize their individual payoffs can also achieve the state of maximum collective benefit. This resolves a contentious aspect associated with the Nash solution of PD, where the agents end up worse off in trying to optimize their individual payoffs [9].

3.2 Chicken

The Game of Chicken (also referred to as Snowdrift) [2] is another well-studied 2-person game which is relevant in the context of social interactions [10] as well as evolutionary biology [11] (where it is also known as Hawk-Dove). The game represents a strategic interaction between two players, who are driving towards each other in a potential collision course, and have the choice between "chickening out", i.e., swerving away from the path of the other (Action 1) or continuing on the path (Action 2). Thus, the choices correspond to being docile or aggressive, respectively. If both players decide to swerve away, they receive the payoff R, while if one swerves

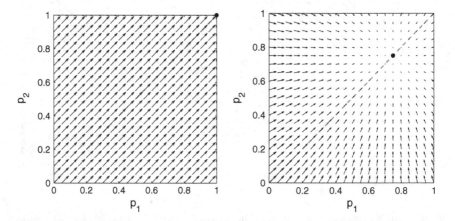

Fig. 4 Vector flow diagram representation of the co-action solution of the 2-person Prisoner's Dilemma game for temptation payoffs (*left*) $T = 1.5$ and (*right*) $T = 3.5$. The abscissae and ordinate correspond to the probabilities (p_1 and p_2) that players 1 and 2, respectively, choose to cooperate. The *broken line* represents the situation where the two agents have the same probability of cooperation. The *filled circles* represent the unique co-action equilibrium for each value of T corresponding to the players cooperating with equal probability ($= 1$ for $T = 1.5$ and $= 0.75$ for $T = 3.5$)

and the other continues on the path, the former loses face (getting the payoff S) and the latter wins (payoff T). However, the worst possible outcome corresponds to when both players continue on the path, eventually resulting in a collision which is associated with payoff P. The hierarchical relation between the payoffs in Chicken is $T > R > S > P$, which suggests that a player will benefit from being aggressive as long as the other is docile, but is better off being docile if it is sure that the other will play aggressively, as the cost of mutually aggressive behavior is high.

Analyzing this game using the dynamical perspective described earlier yields the flow diagram shown in Fig. 5 (obtained for two different values of T, with $S = 0.5$) on using the Nash solution concept. As can be seen, two of the multiple Nash equilibria are stable, corresponding to the pure strategies (i) $p_1 = 1$, $p_2 = 0$ and (ii) $p_1 = 0$, $p_2 = 1$ (i.e., when one player is aggressive, the other is docile). The remaining equilibrium is an unstable mixed strategy located on the $p_1 = p_2$ line (which defines the separatrix demarcating the basins of attraction of the two stable equilibria). With increasing T, the unstable equilibrium—which dynamically corresponds to a saddle point in the p_1–p_2 plane—moves closer to $p_1 = 0$, $p_2 = 0$ corresponding to mutual aggression.

Using the co-action solution concept gives rise to a qualitatively different solution, as seen in the flow diagrams in Fig. 6. When T is low, the system has a stable equilibrium at $p_1 = 1$, $p_2 = 1$, i.e., both agents choose docile behavior to avoid the potential damages associated with mutual aggression. For higher values of T the stable equilibrium is a mixed strategy $0 < p_1 = p_2 < 1$. As in PD, the co-action paradigm yields a single, stable solution of the game.

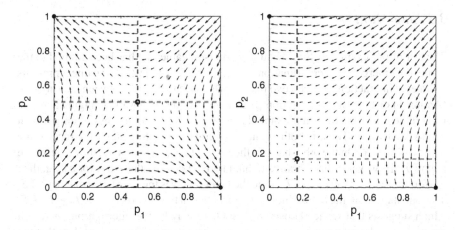

Fig. 5 Vector flow diagram representation of the Nash solution of the 2-person Chicken game for "temptation" payoffs (*left*) $T = 1.5$ and (*right*) $T = 3.5$. The abscissae and ordinate correspond to the probabilities (p_1 and p_2) that players 1 and 2, respectively, choose to be docile (i.e., non-aggressive). The *broken lines* represent the best response (or reaction) correspondence of the players (*red* for player 1, *blue* for player 2). The intersections of the lines, represented by unfilled and *filled circles*, represent the unstable and stable Nash equilibria respectively. The stable equilibria correspond to the pure strategy combination corresponding to one player being aggressive, the other being docile, while the unstable equilibrium in each case corresponds to a mixed strategy

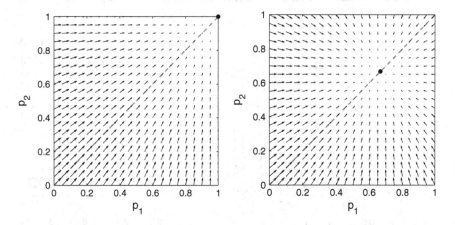

Fig. 6 Vector flow diagram representation of the co-action solution of the 2-person Chicken game for "temptation" payoffs (*left*) $T = 1.5$ and (*right*) $T = 3.5$. The abscissae and ordinate correspond to the probabilities (p_1 and p_2) that players 1 and 2, respectively, choose to be docile (i.e., non-aggressive). The *broken line* represents the situation where the two agents have the same probability of being docile. The *filled circles* represent the unique co-action equilibrium for each value of T corresponding to the players choosing to be docile with equal probability ($=1$ for $T = 1.5$ and $= 2/3$ for $T = 3.5$)

3.3 Stag-Hunt

The last of our examples, Stag-Hunt is a 2-person game that has been studied in the context of emergence of coordination in social interactions [12]. The game represents a strategic interaction between two players who have to choose between hunting stag (Action 1) or hunting hare (Action 2). A hare may be caught by a single agent but is worth less than a stag. On the other hand, hunting a stag is successful only if both agents hunt for it. Thus, if both agents cooperate by hunting stag they receive the highest payoff R. On the other hand, if they both choose to hunt hare, they receive the payoff P. However, if one chooses to hunt hare while the other goes for a stag, then the former receives the payoff T while the latter receives the worst possible payoff S. Thus, in Stag-Hunt, the hierarchical relation between the payoffs is $R > T \geq P > S$, which suggests that while choosing to hunt hare may be the safer option, there is a possibility of doing much better by choosing to hunt stag if one is confident that the other will also do the same.

The vector flow diagrams for Nash and co-action solution concepts in the Stag-Hunt are shown in Fig. 7 (obtained for $T = 0.5$ and $S = -0.5$). For Nash, as in the game of Chicken, there are three equilibria (Fig. 7, left), of which the pure strategies, corresponding to (i) $p_1 = 1$, $p_2 = 1$ and (ii) $p_1 = 0$, $p_2 = 0$ are stable (i.e., when both players hunt for stag or when both players hunt hare). The remaining equilibrium is an unstable mixed strategy located on the $p_1 = p_2$ line which again defines the separatrix demarcating the basins of attraction of the two stable equilibria.

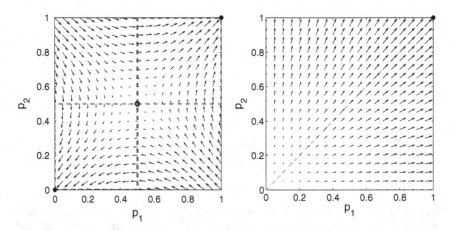

Fig. 7 Vector flow diagram representation of the (*left*) Nash and (*right*) co-action solutions of the 2-person Stag-Hunt game for $T = 0.5$ and $S = -0.5$. The abscissae and ordinate correspond to the probabilities (p_1 and p_2) that players 1 and 2, respectively, choose 'Stag' instead of 'Hare'. (*left*) The *broken lines* represent the best response (or reaction) correspondence of the players (*red* for player 1, *blue* for player 2). The *broken line* represents the situation where the two agents have the same probability of choosing 'stag'

The co-action solution for the games (Fig. 7, right) is a simple one in which both agents always choose hunting stag. i.e., $p_1 = p_2 = 1$. Thus, under the co-action concept, the players always converge to the best possible outcome. In this case, there is no mixed strategy equilibrium for any value of the parameters.

4 Conclusions

In this article we have shown that using a dynamical perspective allows us a visually appealing way to differentiate between two solution concepts, viz., Nash and co-action, for 2-person, symmetric games which lead to spectacularly different conclusions. To illustrate these differences in details we looked at three examples of such games in detail: Prisoners Dilemma, Chicken and Stag-Hunt. In all of these games, one action—in particular, Action 1 in the terminology used here—corresponds to the players being "nicer" to each other (e.g., cooperating in PD, etc.) compared to the other action. The vector flow diagrams generated by the approach presented here clearly show that co-action more often results in nicer strategies being converged at by the agents than in the case for Nash. Our results are intriguing in view of the experimental literature on 2-person games (see discussion in Ref. [9]), in particular PD, which seems to suggest that when these games are played between real human individuals they tend to be far nicer than suggested by the Nash solution.

Acknowledgments We thank Deepak Dhar for useful discussions and Shakti N Menon for a careful reading of the manuscript.

References

1. Shaun P. Hargreaves Heap, Yanis Varoufakis, *Game Theory: A Critical Text* (Routledge, Taylor and Francis, London, 2004)
2. M.J. Osborne, A. Rubenstein, *A Course in Game Theory* (MIT Press, Cambridge, 1994)
3. V. Sasidevan, D. Dhar, Strategy switches and co-action equilibria in a minority game. Phys. A **402**, 306–317 (2014)
4. J.C. Harsanyi, R. Selten, *A General Theory of Equilibrium Selection in Games* (MIT Press, Cambridge, 2003)
5. Sasidevan V, Sinha S, in preparation (2014)
6. J. Hofbauer, K. Sigmund, *Evolutionary Games and Population Dynamics* (Cambridge University Press, Cambridge, 1998)
7. A. Rapoport, *Prisoner's Dilemma: A Study in Conflict and Cooperation* (University of Michigan Press, Ann Arbor, 1965)
8. http://plato.stanford.edu/entries/prisoner-dilemma/
9. M.S. Morgan, *The World in the Model: How Economists Work and Think* (Cambridge University Press, Cambridge, 2009)
10. A. Rapoport, A.M. Chammah, The game of chicken. Am. Behav. Sci. **10**, 10–14 (1966)
11. J. Maynard Smith, G.R. Price, The logic of animal conflict. Nature **246**, 15–18 (1973)
12. B. Skyrms, *The Stag Hunt and the Evolution of Social Structure* (Cambridge University Press, Cambridge, 2003)

Spatio-Temporal Patterns of Development in India: Identifying Geographical Features of Growth and Urbanisation

S. Sridhar, Tara Thiagarajan and Sitabhra Sinha

Abstract Why do some places evolve into large sprawling metropolitan settlements over time, while other initially similar sites decay into obscurity? Identifying the factors underlying the phenomenon of urban growth has sparked the curiosity of scientists ever since Walter Christaller proposed the Central Place Theory in order to explain the observed number, sizes and locations of settlements in southern Germany. However, lack of availability of sufficient empirical data has hampered progress in developing a quantitative understanding of this process. In order to initiate a data-driven approach to answer questions on the growth of settlements, we have undertaken the analysis of a large database of economic, demographic and infrastructural factors associated with different sites of habitation in India. Here we present preliminary results of our analysis for a few of the most populous states of the Indian Union, viz., Maharasthra, Tamil Nadu and Uttar Pradesh. As rapid urbanization taking place in many parts of the country provides a window into the fast-changing rural-urban landscape, we have investigated the growth/decay of population centers in these states using information gleaned from government census reports. In particular, we show that combining demographic data with geographical information allows the identification of specific locations as being either growth "hot-spots" and decay "cold-spots". In addition, we compare the process of growth in different states (which have distinct trajectories for the evolution of the total population size) across multiple scales of settlement sizes. We also show that, for all the states considered here, the nature of the population distribution at different scales (of settlement sizes)

S. Sridhar (✉) · T. Thiagarajan
Scimergent Analytics and Education Pvt. Ltd.,
36, 2nd Main Road, Kasturba Nagar, Adyar, Chennai 600020, India
e-mail: dharmails@gmail.com

T. Thiagarajan
e-mail: tara@scimergent.com

S. Sinha
The Institute of Mathematical Sciences, CIT Campus,
Taramani, Chennai 600113, India
e-mail: sitabhra@imsc.res.in

© Springer International Publishing Switzerland 2015
F. Abergel et al. (eds.), *Econophysics and Data Driven Modelling of Market Dynamics*,
New Economic Windows, DOI 10.1007/978-3-319-08473-2_10

appear to change from a sharply bounded to a long-tailed one as one considers larger settlement size scales, implying that distinct population growth processes are at work in different scales.

1 Introduction

Understanding the growth and development of settlements, in particular, the processes leading to the transition from small agriculture-based villages to large sprawling cities with diverse economic activities, has interested academicians from different disciplines for over a century. Walter Christaller (1893–1969), a German economic geographer of the previous century, was one of the first scientists to construct theoretical models for explaining the spatial ordering of settlements [1]. According to Christaller, settlements serve as "central places" that provide services to the neighboring regions. In order to explain the observed spatial organization of settlements in southern Germany, he developed a model using various simplifying assumptions, such as, the availability of an isotropic unbounded flat surface, a homogeneous initial distribution of population, perfectly rational sellers of products who are trying to maximize their profit, etc. Under these conditions, Christaller deduced that in order to maximize efficiency without having any overlaps in the areas over which they provide services, settlements would tend to self-organize into a hexagonal lattice. He also showed that settlements of smaller sizes would be placed at locations equidistant from the positions of the neighboring larger settlements, which in turn would be located equidistant from even larger settlements and so on. Following Christaller there have been several theoretical attempts at explaining the emergence of spatial patterns and hierarchies in settlement sizes.

Apart from the spatial aspect, researchers have also investigated patterns in the distribution of settlement sizes, possibly the most famous being the power-law scaling of city sizes, referred to as Zipfs' Law [2]. Of possibly equal importance is the principle of proportionate growth that is widely considered to be valid for the evolution of city sizes, based on empirical observation of the evolution of settlements in industrialized nations [3, 4]. In comparison to the literature available on western industrialized nations, studies of the spatio-temporal aspects of demographic dynamics in developing countries are still relatively rare. However, identifying and understanding any regularities in the empirical data available about settlement growth in such countries has become important in recent times in view of the rapid urbanization occurring in much of the developing world. In the Indian context, following the economic liberalization of the 1990s, a large number of towns have seen a massive increase in population, and the cities have grown even bigger. This is at least partly driven by large-scale migrations across the country that is significantly altering pre-existing demographic profiles of different regions. Thus, understanding the spatio-temporal aspects of settlement growth dynamics is not just a problem of theoretical interest but has crucial implications for planning, policy and investment. India, in view of its size and diversity, serves as an ideal laboratory to study the phenomenon of growth and urbanization.

In this article we present a sampler of preliminary results of our analysis of the spatio-temporal patterns in settlement growth in India. For this purpose we have primarily used data obtained by the Government of India through its decennial census, as well as, annual data on other factors such as birth rate, death rate, natural growth rate, etc., that are based on sample surveys. In order to provide a geographic dimension to our study we have associated data with specific geographic locations, the spatial co-ordinates being obtained from the website of India Place Finder [5]. In the following sections, we discuss patterns seen in three different contexts, *viz.* temporal, spatio-temporal and settlements size scales.

2 Temporal Patterns

We first compare the population growth/decay rates in individual states due to natural processes of birth and death with the increase/decrease in population due to migration from/to other parts of the country. In order to make this comparison, we calculate state-wise z-scores of statistical significance for both natural growth rate and decadal population return (defined below). Natural growth rate for a given state is calculated as the difference between the overall birth and death rates. Figure 1 (top) shows the z-scores, obtained by subtracting the mean and dividing by the standard deviation calculated over the entire country, for the natural population growth rates of the various Indian states for the year 2006. Lighter the color of a state in the map, larger is the z-score of the natural growth rate. Thus, Bihar has a high positive growth rate, as against states like Tamil Nadu or Kerala which show a negative population growth.

Next, we quantify the overall change in population of a state—that results not only from births and deaths, but also from migration. For this we use the concept of population return for a state defined as, Population Return $= log_{10}(P_{11}/P_{01})$, where P_{11} and P_{01} are the populations for the years 2011 and 2001 respectively. The z-score of the population return quantifies the degree of statistical significance in the increase or decrease in the population of a state over a decade by subtracting from it the mean over the entire country and dividing by the standard deviation (calculated over all the states). The z-scores for the population return for the period 2001–2011 is shown in Fig. 1 (bottom). Thus large positive (negative) values of z-score correspond to a greater than average increase (decrease) in population. Thus, states like Uttar Pradesh and Bihar show a significant increase in population, while states like Kerala and Nagaland show a marked decline in population. Comparing the z-scores for the natural growth rate and the population return enables us to differentiate between the population dynamics of states like Tamil Nadu (TN) and Maharashtra (MH) from that of Uttar Pradesh (UP) and Bihar (BH). Both TN and MH show a positive z-score for decadal population return. However the natural growth rate of population in both these states are negative as indicated by the z-scores. This suggests that the increase in population observed over a decade is probably because of migration into these

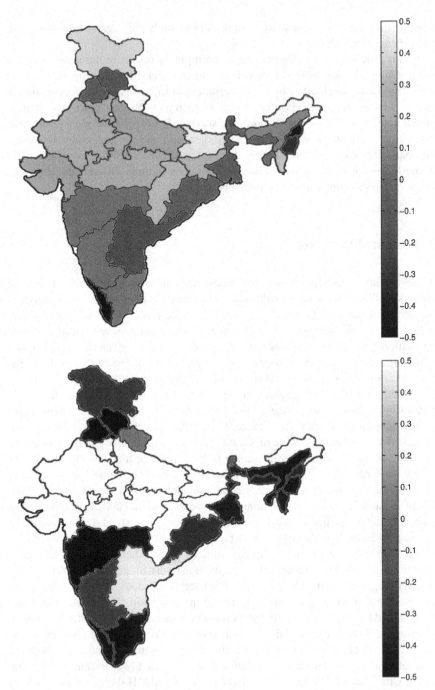

Fig. 1 Map of India showing the z-scores for the decadal return of population for the period (2001–2011) (*top*) and for the natural growth rate of the individual states calculated at the mid-point of this period, viz., 2006 (*bottom*)

states and not so much due to a natural increase in population. On the other hand, states like UP, Chattisgarh or Jharkhand that have large positive decadal changes in population also have very large growth rates, suggesting that natural growth rather than migration may be the primary reason for the increase in their population.

Figure 2 shows the scatter plot of the z-scores calculated for the population return and natural growth rate for 30 Indian states. We observe that z-scores for both the variables are correlated, with a linear correlation coefficient $r = 0.3414$ ($p < 0.05$).

3 Patterns in Settlement Size Scale

Next we look at the patterns in the distribution of settlement size in the different Indian states. An important theoretical idea associated with settlement size dynamics is *Gibrat's Law*, which states that the rate of growth of a variable (such as population) associated with a settlement is independent of the initial size of the settlement [3]. This implies that the variable in question (decadal population return in this case) will be log-normally distributed. While many papers have empirically highlighted the validity of this law for city size distributions at all size scales [6], others point out that the law holds only for settlement sizes determined over long time periods [7]. More recently, Sasaki et al. showed that while cumulative population distributions of all municipalities in Japan satisfy the Gibrat's law [8], when the different kinds of settlements namely, villages, towns and cities, are considered separately, deviations from log-normal behavior are observed—especially for towns. Sasaki et al. conclude that this deviation occurs because of the promotion condition, which is unique to each municipality, that changes the classification of a town to a city.

In this section we compare the distribution of population and the decadal growth for two demographically different states, UP and MH (highlighted in Figs. 1 and 2). While MH shows a negative natural growth rate, UP's population growth rate is one of the highest in the country. For both these states, we plot the cumulative distribution for the settlement size (Fig. 3 (top)) and the probability distribution for the population return (Fig. 3 (bottom)) over different settlement size scales. We categorize the 2001 population into four groups with the bins corresponding to $P_{01} <= 100$, $101 \leq P_{01} \leq 1,000$, $1001 \leq P_{01} 10,000$ and $P_{01} > 10,000$. For both UP (Fig. 3 top(left)) and (MH Fig. 3 top(right)), we observe that there is a change in the nature of the cumulative distribution for settlements having populations greater than 1,000. For values less than 1,000, the distribution is bounded and follows the Gibrat's law, with the population following a log-normal behavior. Settlements with population greater than 1,000 show a long-tailed behavior. This dependence of the nature of a distribution on the scale of settlement sizes that is being considered is also observed in the probability distribution of the decadal population return.

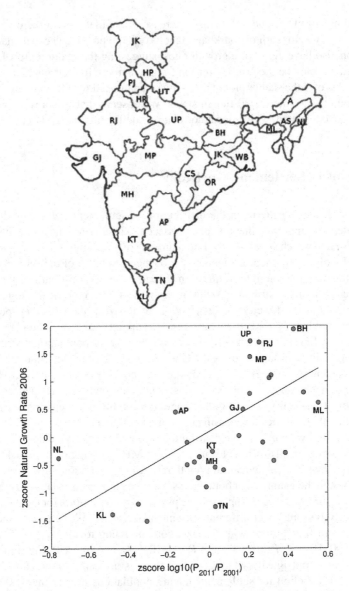

Fig. 2 (*top*) Map showing the different states of the Indian Union, the abbreviated names corresponding to *NL* Nagaland, *KL* Kerala, *TN* Tamil Nadu, *KT* Karnataka, *MH* Maharashtra, *ML* Meghalaya, *AP* Andhra Pradesh, *GJ* Gujarat, *UP* Uttar Pradesh, *RJ* Rajasthan, *MP* Madhya Pradesh, *BH* Bihar, *HR* Haryana, *CS* Chattisgarh, *AS* Assam, *AR* Arunachal Pradesh, *JK* Jammu Kashmir, *PJ* Punjab, *HP* Himachal Pradesh, *UT* Uttarakhand. (*bottom*) Scatter plot of the *z*-scores for decadal return for population growth plotted against the *z*-scores for the natural growth rates. The best fit linear regression is shown. The abbreviations refer to the identity of the biggest states

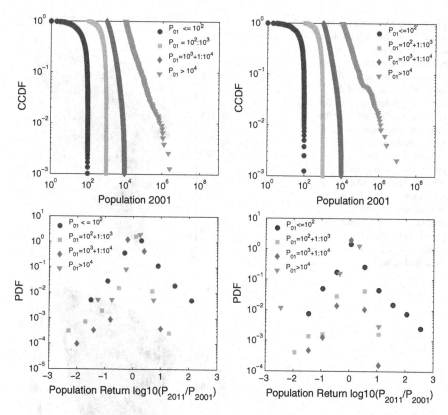

Fig. 3 (*top*) Cumulative distribution of the population for the year 2001 and (*bottom*) distribution for the decadal population return for the period 2001–2011 for the states of Uttar Pradesh (*left*) and Maharashtra (*right*). Both CCDF and PDF are calculated for four different population size bins: P_{01} less than 100 individuals (*circle*), P_{01} between 100 and 1,000 (*square*), P_{01} between 1,000 and 10,000 (*diamond*) and P_{01} more than 10,000 people (*downward triangle*)

4 Spatiotemporal Patterns

While in Figs. 1 and 2 we studied the gross state level processes of population growth over a decade, we next incorporate a spatial aspect to the problem. Using the settlement coordinates (i.e., latitude and longitude) in conjunction with the decadal population return, we identify the regions with very significant population flux. We consider the state of Tamil Nadu (TN) in order to study the spatial profile of the population flux. In Fig. 4 (top), we first identify the spatial distribution of the settlements. Settlements are binned into six categories based on their population with the marker size increasing with the size of the settlement. Next we include the temporal aspect by using the interpolated z-score for the decadal population return calculated for every settlement in the state (Fig. 4 (center)). The bottom panel in Fig. 4, is a magnification of the spatial profile of z-scores for the district of (*Tiruchi*), falling roughly within

Fig. 4 (*top*) Settlements in Tamil Nadu shown with the node size scaling with the population. (*center*) Pseudocolor images of the interpolated zscores of the decadal population return for the period 2001–2011 for the different settlements in Tamil Nadu. (*bottom*) Magnified view of the interpolated zscores for the district of Tiruchi. The individual settlements in the district are marked with the node size scaling with the settlement population. The district headquarters Tiruchi, is indicated using a *square*

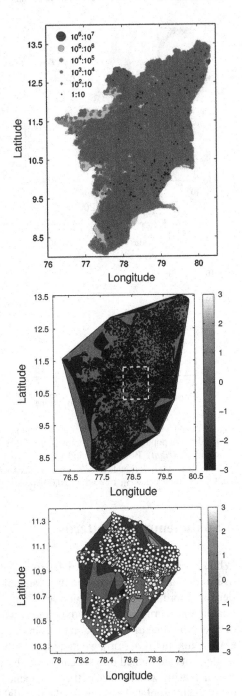

the square with the broken lines in Fig. 4 (center). The different settlements in the district are identified using circles with the radius increasing with population. The district headquarters, Tiruchi, is marked using a square. The spatial z-score picture in both the figures, allow us to visualize spatiotemporal changes in the population. Population growth "hot spots" (light colored regions) are settlements that have witnessed significant increase in size, possibly through inward migration. On the other hand, the dark regions correspond to population decline "cold spots". Settlements in the cold spots have witnessed significant drop in population, that may be a result of outward migration to other regions. Thus, by using demographic and geographical data in conjunction, we can identify spatiotemporal distribution of settlements.

5 Conclusions

We have described a few preliminary observations in the patterns of population dynamics in India based on empirical data. These patterns can be classified into three types, viz., temporal, settlement size and spatiotemporal. On the temporal front, we have compared the growth rates of different states with their population return. This enables the determination of temporal patterns of migration in the different states of India. At the settlement size level, we compare the population distributions at different scales. Unlike as in the paper of Sasaki et al., here we have not grouped settlements based on any pre-existing classification. Instead we have used the absolute population numbers themselves to classify settlements. We report deviations from the log-normal behavior for settlements with populations greater than 1,000. Note that this number is much below the population threshold of 5,000 used by the Government of India to classify census towns. Finally, we describe a novel way to incorporate both demographic and geographic data in order to capture the spatiotemporal changes in populations over a long period of time. Our results can serve as an invitation for researchers to pursue the study of the spatiotemporal pattern dynamics underlying the rapidly changing demographic profile of India.

Acknowledgments Authors will like to thank the team at Scimergent Analytics who helped organize the census and latitude-longitude data. We would also like to thank Dr T. Jesan and Dr Arnab Chatterjee for helpful discussions.

References

1. W. Christaller, *Central Places in Southern Germany*(Prentice-Hall, 1966)
2. G. Zipf, *Human Behaviour and the Principle of Least Effort* (Addison-Wesley, Cambridge, 1949)
3. R. Gibrat, Une loi des rparations conomiques: leffet proportionnel. Bull. Stat. Gen. Fr. **19**, 469 (1930)
4. M. Kalecki, On the Gibrat distribution. Econom.: J. Econom. Soc. **13**, 161–170 (1945)

5. http://india.csis.u-tokyo.ac.jp/
6. J. Eeckhout, Gibrat's law for (All) cities. Am. Econ. Rev. **94**, 1429–1451 (2004)
7. R. Gonzlez-Val, L. Lanaspa, F. Sanz, In Gibrat's law for cities revisted (2010), http://www.eefs. eu/conf/athens/Papers/503.pdf
8. Y. Sasaki, H. Kuninaka, N. Kobayashi, M. Matsushita, Characteristics of population distributions in municipalities. J. Phys. Soc. Jpn. **76**, 074801-6 (2007)

Part III
Reviews

Physicists' Approaches to a Few Economic Problems

Anirban Chakraborti, Yoshi Fujiwara, Asim Ghosh,
Jun-ichi Inoue and Sitabhra Sinha

Abstract We review some of the recent approaches and advances made by physicists in some selected problems in economics, given that this interdisciplinary field popularly called "Econophysics" is now two decades old. These approaches, mainly originating from statistical physics, have not been free of drawbacks and criticisms, but we intend to discuss these advancements and highlight some of the promising aspects of this research. We hope the readers will be able to judge the positive impact that these efforts have created; perhaps even improve the methods and the results or remove the shortcomings, and eventually strengthen the field with their inputs.

A. Chakraborti (✉)
School of Computational and Integrative Sciences, Jawaharlal Nehru University,
New Delhi 110067, India
e-mail: anirban@jnu.ac.in

Y. Fujiwara
Graduate School of Simulation Studies, University of Hyogo,
Kobe 650-0047, Japan
e-mail: yoshi.fujiwara@gmail.com

A. Ghosh
Condensed Matter Physics Division, Saha Institute of Nuclear Physics,
1/AF Bidhannagar, Kolkata 700064, India
e-mail: asim.ghosh@saha.ac.in

J. Inoue
Graduate School of Information Science and Technology, Hokkaido University,
N14-W9, Kita-Ku, Sapporo 060-0814, Japan
e-mail: j_inoue@complex.ist.hokudai.ac.jp

S. Sinha
The Institute of Mathematical Sciences, CIT Campus,
Taramani, Chennai 600113, India
e-mail: sitabhra@imsc.res.in

© Springer International Publishing Switzerland 2015   237
F. Abergel et al. (eds.), *Econophysics and Data Driven Modelling of Market Dynamics*,
New Economic Windows, DOI 10.1007/978-3-319-08473-2_11

1 Background and Motivation

The econophysicists, among many other things, have advocated that one should rely primarily on the empirical observations in order to construct models and validate them. Thus, a major part of the efforts in econophysics have been the study of empirical data and financial time series analyses. Often, the empirics have guided the theoreticians in designing more realistic and practical models. It has also been found that very simple yet elegant models or mechanisms are able to reproduce much of the features of the observed data. In many cases, these (idealized) models could serve as a test-bed for many complex properties, and the models could be further improved to fit more realistic situations. Recently, due to the advent of very powerful and cheap computation, many multi-agent models could be simulated and tested, without having to wait for a long time to validate the predictions of the models. Interestingly, very simple models (with a very few parameters and minimal assumptions) inspired from statistical physics, have been easily adapted in recent occasions, to gain deeper understanding and insights of many complex economic problems.

In this article, we have chosen to review certain representative efforts or approaches of the econophysicists; it is certainly not exhaustive by any means. Neither do we claim that the efforts are the only ones—correct and free from error; nor do we say that they should replace the tools and techniques of mainstream economics. We only modestly suggest that these interdisciplinary approaches may also prove to be very effective, and possibly could compliment and strengthen the existing ones. Our goal is to only highlight the positive aspects and important outcomes of these research efforts, in order to arouse the interests of open minded economists and social scientists, as well as the physicists.

There exists a number of books [1–3], monographs [4–7], edited volumes [8–16], reviews [17–19] and a huge number of journal articles in this field. There has been a surge of research activities in this interdisciplinary field, and the references can only be too many to be possibly included here. Thus, we try to point out in this article, some general books, important reviews and key references, which further contain more details and references of original research in this field. We encourage the readers to go through the referred literature carefully, and then contribute to this new interdisciplinary field so that this field may develop further.

In the following sections, we will present some distinct and disjoint topics, each with sufficiently short introductions or motivations, a few important results and summaries. These topics do not comprise the whole field of econophysics, and give only partial (and perhaps biased) glimpses of the research conducted over the years by the authors and their collaborators. The order of discussions of the topics do not reflect in any manner either their chronology or their importance.

2 Income and Wealth Distributions: Kinetic Exchange Models

2.1 Introduction

The subject of statistical mechanics provides a theoretical framework for relating the microscopic properties of individual atoms and molecules to the macroscopic or "bulk" properties of materials that we observe in our everyday life. It can be applied to various systems with an inherently *stochastic* nature in the fields of physics, chemistry, biology, and even economics and sociology. In fact, the application of statistical physics ideas and tools in modelling economics and sociology, have led to the interdisciplinary fields of "Econophysics"[1] and "Sociophysics".[2]

A natural question that may rise in one's mind is: "How can such a physical theory like statistical mechanics, which deals with particles, be applied to an economic system, composed of (human) agents?" Well, that is what we set to describe next, while dealing with the problem of income and wealth distributions in the society. Since 2000s, the physicists have come up with some very elegant and simple kinetic exchange models in recent times to the problem of economic inequality, based on a simple philosophy—the economy can be described in terms of simple observables as in the case of the physical systems. To elucidate this belief: A single molecule of gas does not have a temperature (T), or a pressure (p). It is simply a point-like particle that moves at a particular speed, depending on how much energy it has, along with many other particles and are altogether governed by the statistical law of Maxwell-Boltzmann distribution of molecular speeds. When there are of the order of 10^{23} or so molecules in an isolated and sealed box of volume V, their collective behaviour can be captured by the ideal gas equation of state: $pV = RT$, where R is the gas constant; and even though each individual particle of the gas is moving at *random*, one can predict with extraordinary accuracy, for example, how many of them will hit the walls of the box at any one time. Similarly, an individual person is neither an economy, nor has any of the characteristics of the entire economy. However, a million such persons acting individually builds up the economy, and may be described by some rules that perhaps allow an economy to be predicted, just as the ideal gas equation of state mathematically describes pressure and temperature, and predicts the aggregate behaviour of gaseous particles. Also, the standard economic theory would like to consider that the activities of individual agents are driven by the *utility maximization principle*. The alternate picture proposed by physicists is that the agents can be simply viewed as gaseous particles exchanging "money", in the place of energy, and trades as money (energy) conserving two-body scatterings, as in

---

[1] The term was coined by the American physicist, H. Eugene Stanley, in a conference on statistical physics in Kolkata (erstwhile Calcutta) in 1995, and first appeared in its proceedings published in the Journal Physica A (1996).

[2] The term was first used by the French physicist, S. Galam, and appeared in an article published in 1982.

the *entropy maximization* based kinetic theory of gases [20]. This qualitative analogy between the two maximization principles seems to be quite old—both economists and natural scientists had already noted it earlier in many contexts, but this equivalence has gained firmer ground only recently.

It would be indeed difficult to find in the annals of time, any society or country where income or wealth was equally (or fairly) distributed amongst its people. Socio-economic inequality has not been just limited to the modern times; it has been a persistent phenomenon and a constant source of vexation to most. Inequality has been one of the most studied subjects in economics, and the economists and philosophers have spent much time on the normative aspects of this issue [21–24], or the direct and indirect effects of inequality on the society, or the effects of inequality on the growth of the economy [25–28], or even the econo-political scenario [29–32]. There are several related non-trivial questions that are still challenging: *How are income and wealth distributed? What are the forms of the distributions? Are they universal, or do they depend upon specific conditions of a country?*

Vilfredo Pareto and many great personalities in the past have been intrigued by such questions. More than a century ago, Pareto made thorough studies in Europe and found that wealth distribution possesses a power law tail for the richer section of the society [33], known now as the *Pareto law*. Roger Gibrat worked on the similar aspects of distribution and he proposed a "law of proportionate effect" [34] to explain his observations. Many years later, Champernowne also considered this problem and proposed a probabilistic theory to justify Pareto's claim [35, 36]. Following numerous studies, today we realize that the distributions of income and wealth indeed possess some globally stable and robust features (see, e.g., [17] for a detailed review). In general, it is observed that the bulk of the distribution of both income and wealth seems to fit both the *log-normal* and the *Gamma* distributions, reasonably well. Economists usually prefer the log-normal distribution [37, 38], whereas statisticians [39] and more recently, physicists [17, 40, 41] tend to use the alternate fitting forms such as the Gamma distribution (for the probability density) or Gibbs/exponential distribution (for the cumulative distribution). There is considerably more consensus on the upper end of the distribution, that is the tail of the distribution—described by a power law, as was first proposed by Pareto.

Mathematically, we can define the distribution of income $P(x)$ as follows: $P(x)dx$ is the probability in the "equilibrium" or "steady state" of the system,[3] a randomly chosen person (out of a myriad number of people) would be found to have income between x and $x + dx$. Empirical data analysis, especially with recent income data, have been extensively reviewed, e.g., in the chapter of book by Chakrabarti et al. [7], and it may be mentioned that compared to the empirical work done on income distribution, relatively fewer studies have looked at the distribution of wealth.[4] The lack of an easily available data source for measuring wealth, analogous to income

---

[3] We often use for simplicity the terms "equilibrium" or "steady state" interchangeably, though for systems that are "non-ergodic", strictly speaking one should only use the term "steady state".

[4] This consists of the *net* value of assets (financial holdings and/or tangible items) owned at a given instant.

tax returns for measuring income, compels researchers to resort to indirect methods. Interestingly, the detailed empirical analyses of the income distribution indicate

$$P(x) \sim x^n \exp(-x/T), \quad \text{for} \quad x < x_c, \tag{1}$$

and

$$P(x) \sim x^{-\alpha-1}, \quad \text{for} \quad x \geq x_c, \tag{2}$$

where n and α are two numerical exponents, and the parameter T denotes a scaling factor. The latter exponent α is now known as the *Pareto exponent* and its value ranges between one and three (see, e.g., [42, 43]). A historical account of Pareto's data and that from recent sources can be found in [44]. The crossover point x_c is extracted from the numerical fittings of the initial Gamma distribution form to the eventual power law tail. As mentioned earlier, economists often fit the region below x_c to a log-normal form: $\log P(x) = \text{const} - (\log x)^2$. The statisticians and physicists rather think that the Gamma distribution form fits better with the data (see [17, 39, 45]). Figure 1 shows schematically the features of the cumulative income or wealth distribution. These observed regularities in the income distribution indicate sort of a "natural" law of economics.

In the limited empirical studies of wealth distribution using indirect methods, one again observes the general feature: A power law behavior for the wealthiest 5–10% of the population, and exponential or log-normal distribution for the rest

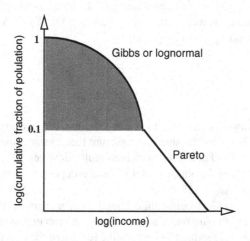

Fig. 1 When one plots the cumulative wealth (income) distribution against the wealth (income), almost 90–95 % of the population fits the Gibbs distribution, or often fitted also to log-normal form (Gibrat law)—indicated by the shaded region in the distribution; for the rest (very rich) 5–10 % of the population in any country, the number density falls off with their wealth (income) much slowly, following a power law (Pareto law). It is found that about 40–60 % of the total wealth of any economy is possessed by 5–10 % of the people in the Pareto tail. Taken from [7]

of the population. It is noteworthy that the Pareto exponent as measured from the wealth distribution is always found to be lower than that for the income distribution.[5] Interestingly, instead of focussing on the income of individuals when one shifts attention to the income of companies or firms, one still observes the power law tail. A study of the income distribution of Japanese firms ([42]; see also [6]) concluded that it follows a power law (with exponent value near unity, which is also often referred to as the Zipf's law). Similar observation has been reported for the income distribution of companies in the USA [47].

Having analysed enormous amount of data available today, and observing such strikingly robust features of the distribution $P(x)$, in income or wealth, the important questions arise—*if inequality is universal (as some of its gross features, studied by Pareto, Gibrat and others, indicate), then what is the reason for such universality? Is it plausible that this only reflects a basic natural law, with simple physical explanation?* Many econophysicists actually believe so. According to them, the regular patterns observed in the income (and wealth) distribution are indeed indicative of a natural law for the statistical properties of a many-body dynamical system representing the entire set of economic interactions in a society, analogous to those previously derived for gases and liquids. Hence 2,000 onwards, they proposed many simple models inspired by the statistical physics of gases. The class of kinetic exchange models [48–53] are simple microeconomic models with a large number of "agents" and the "asset" transfer equations among the agents due to "trading" in such an economy, closely resemble the process of "energy" transfer due to "collisions" among "particles" like those in a thermodynamic system of ideal gas. In these models, the system is assumed to be made up of N agents with assets $\{x_i \geq 0\}$ $(i = 1, 2, \ldots, N)$. At every trade, an agent j exchanges a part Δx with another agent k chosen randomly. The total asset $X = \sum_i x_i$ is constant, as well as the average asset $\langle x \rangle = X/N$. After the exchange the new values x'_j and x'_k are $(x'_j, x'_k \geq 0)$

$$x'_j = x_j - \Delta x, \tag{3}$$
$$x'_k = x_k + \Delta x.$$

The form of the function $\Delta x = \Delta x(x_j, x_k)$ defines the underlying dynamics of the model. Figure 2, shows the schematic picture that captures the essence of these models. Very similar kind of models with both multiplicative and additive exchanges were proposed earlier by Ispolatov et al. [54], and independently by social scientists, Bennati [55–57] and Angle [58–61].

As first proposed by Gibbs more than a hundred years ago, the steady state distribution for a system with pure random asset exchange is an exponential one (see e.g., [17, 41, 53]). The novel feature of Gamma-like feature of the distribution $P(x)$, was brought forth with the introduction of "saving propensity" [48] and more importantly such a random exchange model with uniform saving propensity for all agents, was

---

[5] This is believed to be consistent with the general observation that, in market economies, wealth is much more unequally distributed than income [46].

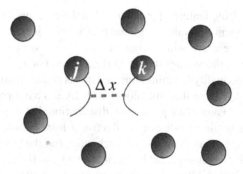

Fig. 2 The kinetic exchange models prescribe a microscopic interaction between two units analogously to a kinetic model of gas in which, during an elastic collision, two generic particles j and k exchange an energy amount Δx, as in (3). Taken from [7]

subsequently shown to be equivalent to a commodity clearing market where each agent maximizes his/her own "utility" [62]. A further modification of the model, with non-uniform (or distributed) saving propensities for the agents, produces [50] a power law for the upper or tail end of the distribution of money, as has been found empirically. These are explained in the following sub-section.

2.2 Model with Uniform Savings

In any trading, savings comes naturally [46]. A saving propensity factor λ was introduced in the random exchange model [48], where one trader i at time t saves a fraction λ of its money $x_i(t)$ and trades randomly with the another trader j:

$$x_i(t+1) = \lambda x_i(t) + \varepsilon_{ij}\left[(1-\lambda)(x_i(t)+x_j(t))\right], \tag{4}$$

$$x_j(t+1) = \lambda x_j(t) + (1-\varepsilon_{ij})\left[(1-\lambda)(x_i(t)+x_j(t))\right], \tag{5}$$

where

$$\Delta x = (1-\lambda)[\varepsilon_{ij}\{x_i(t)+x_j(t)\} - x_i(t)], \tag{6}$$

where ε_{ij} being a random fraction. This randomness reflects the *stochastic* nature of the trading. By definition, λ is a proper fraction, i.e., $0 \le \lambda \le 1$.

Interestingly, in this model, the market (non-interacting at $\lambda = 0$ and 1) becomes 'interacting' for any other non-vanishing λ: For fixed λ (uniform for all agents), as can be seen in Fig. 3, the steady state distribution $P(x)$ of money is rapidly decaying on both sides with the most-probable money per agent shifting away from $x = 0$ (for $\lambda = 0$) to X/N as $\lambda \to 1$; the inequality in the distribution is a function of λ.

Here, the *self-organizing* feature of the market,[6] induced simply by the *self-interest* of saving by each agent without any global perspective, is quite significant as the fraction of paupers decrease with saving fraction λ and most people end up with some finite fraction of the average money in the market. For $\lambda \to 1$, the economy is ideally 'socialist', inequality is a minimum, and this is achieved purely with people's self-interest of saving. Note that this model with fixed saving propensity ($\lambda \neq 0$) does not give yet the Pareto-like power-law distribution, but the Markovian nature of the scattering or trading processes is effectively lost. Indirectly through λ, the agents get to know (start interacting with) each other and the system co-operatively self-organizes towards a most-probable distribution ($x_p \neq 0$) (see Fig. 3).

Based on numerical results and heuristic arguments, it was conjectured that the distribution is a Gamma distribution [64]:

$$P(x) = \frac{n^n}{\Gamma(n)} x^{n-1} \exp(-nx) \tag{7}$$

where $\Gamma(n)$ is the Gamma function whose argument n is related to the savings factor λ as:

$$n = 1 + \frac{3\lambda}{1 - \lambda}. \tag{8}$$

This result and its *ergodic* behaviour was numerically studied in [65]. However, later studies [66, 67] analyzed the moments, and found that moments up to the third order agree with those obtained from the form of the Eq. (7), and discrepancies start from fourth order onwards. Hence, the actual form of the distribution for this model remains an open question.

2.3 Model with Distributed Savings

In a real society or economy, the amount of saving varies from person to person, which implies that λ may be a very inhomogeneous parameter. To reproduce this situation, one proposed that the saving factor λ is widely distributed (non-uniform) within the population [49–51]. The evolution of money in such a trading can be written as:

$$x_i(t+1) = \lambda_i x_i(t) + \varepsilon_{ij} \left[(1 - \lambda_i) x_i(t) + (1 - \lambda_j) x_j(t) \right], \tag{9}$$

$$x_j(t+1) = \lambda_j x_j(t) + (1 - \varepsilon_{ij}) \left[(1 - \lambda_i) x_i(t) + (1 - \lambda_j) x_j(t) \right]. \tag{10}$$

---

[6] Self-organization also occurs in other market models when there is restriction in the commodity market [63].

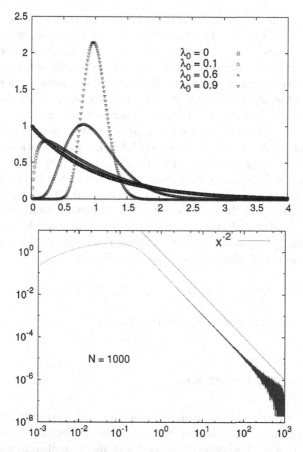

Fig. 3 Steady state money distribution $P(x)$ *versus* x for: (*Top*) The model with uniform savings. The data shown are for different values of λ: 0, 0.1, 0.6, 0.9 for a system size $N = 100$. All data sets shown are for average money per agent $X/N = 1$. Taken from [41]. (*Bottom*) The distributed λ model with $0 \le \lambda < 1$ for a system of $N = 1{,}000$ agents. The x^{-2} is a guide to the observed power-law, with $1 + \alpha = 2$. Again, the average money per agent $X/N = 1$. Taken from [41]

The trading rules are similar as before, except that

$$\Delta x = \varepsilon_{ij}(1 - \lambda_j)x_j(t) - (1 - \lambda_i)(1 - \varepsilon_{ij})x_i(t), \tag{11}$$

where λ_i and λ_j are the saving propensities of traders i and j. In this model, the traders have fixed (over time) saving propensities, distributed independently, randomly and uniformly, within an interval 0 to 1. A particular agent i saves a random fraction λ_i ($0 \le \lambda_i < 1$) and this λ_i value is *quenched* for each agent, i.e., λ_i does not change with time t.

For this model too, we start with an arbitrary initial (uniform or random) distribution of money among the agents, and the market evolves with the trading. At each time step, two traders are randomly selected and the money exchange among them occurs, following the above mentioned scheme. One checks for the steady state, by looking at the stability of the money distribution in successive Monte Carlo steps t (one Monte Carlo time step is defined as N pairwise exchanges). Eventually, after a typical *relaxation time* the money distribution becomes *stationary*. Note that this relaxation time is dependent on system size N and the distribution of λ (e.g., $\sim 10^6$ for $N = 1,000$ and uniformly distributed λ). After this, one takes the average of the money distributions over $\sim 10^3$ time steps. Finally, one takes the configurational average over $\sim 10^5$ realizations of the λ distribution to get the money distribution $P(x)$. Interestingly, this is a *non-ergodic* model [65], but it gives rise to a strict power-law decay, and the decay fits to Pareto law (Eq. 2) with $\alpha = 1.01 \pm 0.02$ (Fig. 3). One may also note, for finite size N of the market, the distribution has a narrow initial growth up to a most-probable value x_p after which it falls off with a power-law tail for several decades. This Pareto law (with $\alpha \simeq 1$) covers almost the entire range in asset x of the distribution $P(x)$ in the limit $N \to \infty$. This power law is very robust in the sense that apart from the uniform λ distribution used in these simulations in Fig. 3, this decay can also be reproduced for a distribution

$$\rho(\lambda) \sim |\lambda_0 - \lambda|^\alpha, \quad \lambda_0 \neq 1, \quad 0 < \lambda < 1, \tag{12}$$

of quenched λ values among the agents, for all $\alpha > 0$.

2.4 Summary and Discussions

In summary, viewing the economy as a "thermodynamic" system [53, 68–70], one can identify the income/wealth distribution with the distribution of energy among the particles in a gas. This general framework can also be applied to studying other socio-economic problems like firm-dynamics and opinion formation [71, 72]. Several attempts by social scientists (see e.g., [55–61]) also provide impetus to this interdisciplinary approach. Also, the class of kinetic exchange models [48–51, 53] have provided a simple mechanism for understanding the unequal accumulation of assets. While being simple from the perspective of economics, they have the benefit of capturing a key factor—*savings*—in socio-economic interactions, that results in very different societies converging to similar forms of unequal distribution. Interestingly, the economic inequality is a natural outcome of this framework of stochastic kinetics of trading processes in the market, *independent* of any exogenous factors. Thus, the kinetic exchange models demonstrate how inequality may arise naturally. They also indicate how its effects may be partially reduced by modifying the saving habits.

Several analytical aspects of this class of models have been studied (see e.g., [5, 54, 67, 73, 74]). It is noteworthy that presently this is the only known class of models which, starting from microeconomics of utility maximization and solving for the resultant dynamical equations in the line of statistical physics, can reproduce well the major empirical features of income and wealth distributions in real economies.

These developments have received critical comments (see e.g., [75–77] for criticism and [44] for rebuttal). In view of the failure of main stream economic schools to anticipate or correctly analyse the recent economic crisis, there have been some recent interests by the main stream economists to revert to such physically motivated models of the market dynamics and their solutions (see, e.g., [78]).

3 Market Mechanism: Agent-Based Models

3.1 Introduction

In this section, we will discuss some games on agent-based model [3, 79, 80] which may be considered as toy models of the market mechanism. One of the most famous games proposed related to this issue, renowned also for its simplicity, is the El Farol bar problem [81]. Brian Arthur introduced in 1994 the game to illustrate the idea of 'inductive reasoning'. In Santa Fe town, there was a bar named El Farol Bar, where every Thursday a musical programme was organized. People went to the bar for entertainment. But when the bar was too crowded then the bar was not enjoyable. Based on this observation, he proposed the repetitive game model where it was suggested the people in the bar would be enjoying only if less than, say, 60 % of capacity was occupied. Assuming that all agents were not interacting with each other, and taking their decisions parallelly, he modeled the problem considering only previous attendance history. He defined a strategy space based on previous history and argued that the attendances of the bar can be around 60 % of the total number people. The solution therefore is completely based on inductive reasoning, i.e., people learned strategies from previous history and corrected from past errors.

Later a variant of the El Farol bar problem was introduced by Challet and Zhang, was named as Minority Game [18, 80, 82–85]. In the Minority Game problem, two restaurants are considered (to allow for symmetric choices). The agents on the less crowded side will be the winners (payoff one for each) and agents in the more crowded side will be loser (payoff zero for each). People again learn from the past history and change their strategies to minimize their error/loss. The steady state fluctuation associated with the population is important and need to be minimized for efficient 'reasoning' or learning by the players. Many methods of collective learning have been proposed and studied for this game to reduce fluctuation in this problem and also the convergence time.

In the next part of this section we will discuss another variation of the (two choices) El Farol Bar problem, called Kolkata Paise Restaurant problem [86–88]. In this problem there are many choices (of restaurants) and many agents. It is again a repetitive game. Every restaurant can serve only one agent each day and the price of a meal is same for all restaurants. The agents here also do not interact with each other and everyone wants to choose an unique restaurant each day. If more than one agent arrive at a restaurant then one of them will be randomly chosen by the restaurant and will be served; the rest will not get any meal for that day. The utilization fraction is defined as the ratio of the average number of agents getting dinner in a day to the total number of the restaurants. The main aim of this problem is finding out a strategy which will give a maximum utilization taking the smaller time to converge to the solution.

3.2 El Farol Bar Problem

Before the El Farol Bar problem was introduced, the economists had mainly modeled the problems based on deductive rationality. Although it was useful to generate a theoretical problem, but it was observed that deductive rationality breaks down under some complications [81]. It was pointed out that inductive reasoning was needed to model the problems in a more realistic way. It was argued that in our society, we may make our decisions based on inductive reasoning and not based on deductive rationality. The inductive reasoning gives good sense as an intellectual process and it is very easy to model.

Historically, the El Farol Bar problem was introduced as follows: In Santa Fe, New Mexico, there was a bar and every Thursday night, a finite number of people went to the bar. But as the capacity of the bar was finite, it was no fun to go there if it was too crowded and people therefore decided to go or not depending on their personal past experience. Inspired by this fact, Brian Arthur defined his model and it can be described as follows: Suppose total 100 agents are available to go to the bar and capacity of the bar is only 60. Therefore, if any one finds that the attendance of the bar exceeds the capacity value, in that case, staying at home would be a better choice than going to the bar. Similarly, if the attendance of the bar does not exceed the capacity value, in that case, going to the bar would be a better choice than staying at home. Also the agents do not interact with each other while taking their decisions, but previous attendance history of the bar is available to everyone. Depending upon the previous history, every agent will choose either to go to the bar or to stay at home. By doing computer simulation, Brian Arthur surprisingly found that the mean attendance converges to 60 [81].

3.3 Minority Game

As mentioned already, a symmetric choice variant of the El Farol Bar problem was introduced by Challet and Zhang [80, 82–85] which is known as the Minority game problem. In their model, they considered two restaurants ($n = 2$) and N ($= 2M + 1$; M integer) agents. The agents are going to one of the two restaurants each day. Every day, one restaurant will always be more crowded than the other. The agents in the less crowded restaurant are said to be in minority side (or winner side) and will receive a positive payoff. At the same time, the other restaurant (say majority side or looser

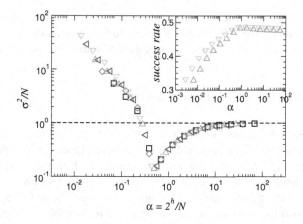

Fig. 4 *Square* of fluctuation versus $\alpha = 2^h/N$ for different number of agents $N = 101, 201, 301, 501, 701$ ($\square$, $\lozenge$, $\triangle$, $\triangleleft$, $\triangledown$, respectively). *Inset* Variation of mean success rate of the agents' with α. From [84]

side) is more crowded and every agent in that restaurant will get a payoff 0. In this problem, if all the agents choose any one of two restaurants randomly every day then the population distribution of each restaurant will be Gaussian and the peak of the distribution will be around M with a mean-square deviation σ (or fluctuation) which is the order of $\sqrt{N}$. The fluctuation in the problem actually is a measure of the loss of resources in this game. Therefore, the fluctuation in the problem should be optimized and researchers developed algorithm to adjust weights in the strategy space depending upon the past history of his/her success or failure. Now if anyone use only h days as a memory size for taking decision on the next day then total number of possible strategies will be 2^h which is a fast-growing function of h. They observed that the fluctuation in the problem could be reduced to a certain minimum level by increasing the memory size of the agents. The fluctuation is numerically found to have a minimum value (about $1/10$ of $\sqrt{N}$, the random case value) between short memory and large memory size of agents. A Monte Carlo simulation of the variability of fluctuation in memory size is shown in Fig. 4.

3.3.1 Market Toy Model

El Farol bar problem or Minority Game (two choices game) can be considered as a kind of very simple "market toy model". The markets are interpreted as a flexible competitive system in which minority group plays an important role. Game of two choices can be mapped to the market model by identifying the two choices as 'buy' or 'sell' options for any trader. At each time step, agents can buy or sell for trading. The price is determined by a simple supply-demand rule. When there are more buyers than sellers, the price is high, and similarly when there are more sellers than buyers, the price is low. If the price is high, sellers are winner, when the price is low, buyers win the game. Therefore, the minority group always wins, irrespective of buyers or sellers. Many other complex systems are operated by this minority rule, like traffic on roads, packet traffic in networks, and ecologies of animal food searching [80, 84].

3.4 Kolkata Paise Restaurant Problem

The Kolkata Paise Restaurant Problem was introduced by Chakrabarti et al. in 2009 [86] to accommodate many choices by the agents or players. In this problem, the number of choices (restaurants) and number of agents, both are large. Again here, the agents do not interact with each other for making their decisions and they take their decisions parallelly. Specifically one considers n restaurants and N agents where $n \sim \mathcal{O}(N)$. Also the previous history of the game is available to everyone. Every agent can choose only one restaurant in any day. In this game, price of a meal is same for all the restaurants and each restaurant can serve only one agent each day. So, if more than one customer arrives in a day in a restaurant, one of them will be randomly chosen and will be served and so the rest of the agents will not get food for that day.

3.4.1 Strategies of the Game

In KPR problem, the utilization fraction is defined as the ratio of average number of agents getting food every day to the total number of restaurants. In this part, we will discuss different strategies and their corresponding utilization fraction. The efficient strategy is such that it will give the maximum utilization fraction within a finite time limit.

Random Choice

First let us talk about random choice strategy [86]. In this case, an agent will choose any restaurant at random. So, the agents do not use their memory related to previous attendances of the restaurants for making their choices. Also the agents are not discussing with each other to make their choices and all decisions are taken parallelly each day. In this case every agent can choose only one restaurant for a day. We know that every restaurant can serve only one agent for a day so it is not guaranteed that every agent will get food every day. Next part, we will calculate how many agents on an average, get served every day.

Suppose there are N agents and n restaurants and in this case all agents choose any restaurant with probability $p = 1/n$. Now the probability of restaurants chosen by m agents for a day is given by

$$\Delta(m) = \binom{N}{n} p^m (1-p)^{N-m}; \quad p = 1/n$$

$$= \frac{(N/m)^m}{m!} exp(-N/n) \quad \text{for} \quad N \to \infty, \quad n \to \infty. \tag{13}$$

Therefore, the probability of the restaurants not chosen by any agent can be written as $\Delta(0) = exp(-N/n)$. Now we can write an average fraction of restaurants visited by at least one agent as

$$\bar{f} = 1 - \Delta(0) \sim 0.63 \quad \text{for} \quad N = n. \tag{14}$$

Therefore, in the random choice case we get about 63 % utilization and obviously the convergence time to reach this utilization fraction value is zero.

Rank Dependent Choice

Although the price of a meal in all the restaurants is the same, we can assume that all restaurants have different rank depending upon the service quality, food quality etc. To make a model, we assume that the kth restaurant has rank of k and probability to go to that restaurant is proportional to k. Again all agents do not use previous history for this strategy. If all agents follow this strategy then it was observed numerically that utilization fraction is about 57 % [88]. Therefore, this strategy is less efficient compared to the random choice case. Again no time is needed to reach this steady value means same as the random choice case.

Crowd Avoiding Cases

(A) Fully Crowd Avoiding Case: In this case, all the agents use their memory of previous day attendances history for taking their decisions. If all the agents decide to choose only the previous day's vacant restaurants randomly and completely avoiding other restaurants then it was observed numerically that the average utilization fraction becomes about 46 % [88]. Therefore, this strategy is much less efficient compared to the previously discussed strategies.

(B) Stochastic Crowd Avoiding Case: In this case, the agents do not avoid the previously crowded restaurants completely. Suppose $n_k(t-1)$ is the number of agents arriving or choosing kth restaurant on $(t-1)$-th day, then next day (t) these agents will visit the same restaurant with probability $p = 1/[n_k(t-1)]$ and will go to any other restaurants with uniform probability $p\prime = (1-p)/(N-1)$. If all the agents are playing with this strategy then utilization fraction becomes about 80 % [88] which is much larger than that for the previously described strategies. But the time to reach the steady state value of the efficiency here is of the order of $\log N$ which is much larger than the other strategies (but still much smaller compared to N).

An approximate estimate for above utilization can be made as follows: Suppose in the steady state a_i is the fraction of restaurants visited by i agents and we assume $a_i = 0$ for $i \geq 2$; we approximate that number of restaurants where three or more customers arrive on any day are negligible. If we take the equal number of agents and restaurants (i.e. $N = n$) then we can write the equations as

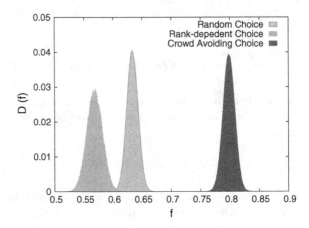

Fig. 5 Probability distribution of utilization for different strategies; using Monte Carlo simulation for $N = n = 10^5$ are shown. From [89]

$$a_0 + 2a_2 = 1$$
$$a_0 + a_1 + a_2 = 1.$$

According to this strategy, the agent who visited any restaurant alone on a given day, will surely go the same restaurant next day($p = 1$). But if any restaurant is visited by two agents then these two agents will go to the same restaurant with probability $p = 1/2$ the next day. In this process every time ($a_2/4 - a_2a_2/4$) fraction of restaurants will be vacant from the restaurants previously visited by two agents. Similarly a_0a_2 fraction of restaurants will be visited by one agent from restaurants previously visited by none. Therefore one can write

$$a_0 - a_0a_2 + a_2/4 - a_2a_2/4 = a_0.$$

If the above three equations are solved, we get $a_0 = 0.2$, $a_1 = 0.6$ and $a_2 = 0.2$. So the average fraction of restaurants visited by at least one agent is $a_1 + a_2 = 0.8$, which is very close to the results of the Monte Carlo simulation as shown in Fig. 5.

3.4.2 KPR and Link with Minority Game

Another important use of the stochastic crowd avoiding strategy is in the Minority Game. Dhar et al. [90] showed that the fluctuation of Minority Game (MG) can be reduced to zero within very short time scale by applying a variation of the stochastic crowd avoiding strategy. The strategy was used as follows: Suppose at any instant of time (day) population of the two restaurants are $M + \Delta + 1$ and $M - \Delta$ in MG problem. Next day majority side agents will change the choice with probability $p_+ = \Delta/(M + \Delta + 1)$ and minority side people will not change their previous choice ($p_- = 0$). Therefore the agents use previous day information only and it can be shown that Δ will become zero order within $\log\log(2M + 1)$ time [90].

But in this stochastic strategy there was a problem. When Δ becomes 0 the flipping process will stop which means that majority people ($M + 1$ in number) will remain in the same state for the rest of the game (i.e., there is no dynamical movement of the agents over time). Therefore, though the solution gives zero fluctuation, the situation is very unjust solution for the majority group. To overcome this situation, the probability was later modified as—$p_+ = (g\Delta)/(M + g\Delta + 1)$ (flipping probability for majority side agents; g any real positive parameter) and $p_- = 0$ (flipping probability for minority side agents) [91]. It was shown that for $g < 2$ the dynamic stops after some time but for $g > 2$ it will never stop, thought $\Delta(g) \sim (g-2)$ (which is associated with fluctuation) can be made arbitrarily small.

3.5 Summary and Discussions

We have discussed different games and their efficiencies. First, we considered the El Farol Bar problem which was a game, based on inductive reasoning. In this game, many agents went to a restaurant in a city. The agents would be happy if they found that crowd of the restaurant did not exceed a certain threshold value. In the El Farol Bar problem, the agents would not discuss for making their decisions and one would go to the Bar if he/she was expecting a less crowded situation. The game was actually represented as a toy model of 'inductive learning' strategy. It was also shown that models based on the inductive learning were more effective than the deductive rationality for collective or social decision making contexts.

Later, we have discussed here a variant of El Farol Bar problem which was called Minority Game. In this game there were two restaurants and many agents. All the agents choose one of the these restaurants every day without discussing with each other and the agents in less crowded restaurant would receive payoff and the agents in the crowded restaurant lose. Both the restaurants are similar regarding the price of the meal but the agents in crowded side are loser for that day. In this problem the fluctuation associated with attendances of the agents in the restaurants is an important quantity which can be identified as similar to the volatility in financial markets. We saw that the strategy based on memory size could reduce the fluctuation up to about $1/10$ of the random process fluctuation.

In the next part we have discussed the Kolkata Paise Restaurant (KPR) problem. In this problem, there are N agents and n ($\sim \mathcal{O}(N)$) restaurants. Here also the agents are non-interacting and they use the past performance records only for making their decisions. On the other side, the price of a meal for each restaurant is assumed to be the same so that no budget restriction or specification can dictate the choice and any restaurant could serve one agent each day and agents choose only one restaurant in any day. Therefore, if more than one agent had arrived at any restaurant in a day then one of the them would be randomly picked up and served. And the rest of them would not get food for that day. We discussed different strategies of the game and their efficiency (by measuring utilization). We observed that the most efficient was the stochastic crowd avoiding strategy where the utilization fraction attain a value about 80 % within a time, bounded by $\log N$.

In last part, we have discussed a variant of the stochastic crowd avoiding strategy (developed for KPR problem) was applied in the Minority Game problem. Using this strategy, the fluctuation associated with the Minority Game reduced to the order of zero by taking few time steps (magnitude of convergence time in order of $\log \log N$). But it was also observed that the dynamics of the game would be stopped after that. As discussed, this problem can be avoided by taking some noise trader in the game, although the fluctuation remains non zero, though very small compared to the random choice case.

4 Economic Success and Failures: Analyses and Modelling

4.1 Introduction

Economic phenomena rarely give rise to equitable distributions where everyone has the same share of the market. In fact, the same emergence of inequality that marks wealth and income distributions (see Sect. 2 of this paper), also characterizes other aspects of economic life, such as the respective fates of an ensemble of products (or for that matter, services or ideas) that maybe of more or less similar quality. This applies even to larger economic structures such as organizations and companies which drive economic growth in most societies; Ormerod has pointed out in the book *Why Most Things Fail* [92] that of the successful companies that existed in the past, only a handful have managed to survive to the present. In fact, the relative success of firms appear to have a similar long-tailed distribution described by a power law function that characterizes the distribution of personal income or wealth [93]. It is thus of some importance to identify features that characterize the process of economic success and failure.

What decides whether a newly introduced entity in the market will eventually succeed in face of severe competition from several other competitors is often not so much a result of intrinsic differences between them but rather a result of a social collective consensus that emerges stochastically from the interactions between the choice behavior of individual agents. The choices in many such decision problems often constitute a discrete set, sometimes even having a binary nature, for instance, whether one should cooperate with others or defect by exploiting (free-loading off) the cooperators. These problems resemble the formulation of spin models used by statistical physicists to analyze in detail cooperative phenomena [94, 95]. It is therefore not surprising that the phenomenon of how and why certain entities become successful through competition has been a topic of interest among econophysicists.

To understand how people choose from a set of options, mainstream economics assumes that each individual decides on a specific alternative that maximizes his/her utility function. Moreover, except possibly for the formulation of the utility function, the behavior of other agents are usually not considered as directly affecting the decision to select a particular choice. It has however become apparent over the past few decades, e.g., as exemplified by Schelling's analysis of the reasons for housing segregation along racial lines [96], that in many cases the apparent freedom of choice of an agent may well be illusory. Indeed, the decision taken by an individual is affected by those taken by his/her peers or rather, the actions of neighboring elements in the corresponding social network [97, 98]. A physics-based approach which stresses on the role of interactions (and hence the social environment) in taking a decision contrasts with the conventional economic approach of utility maximization by individual rational agents. This can be crucial for explaining why a very popular product can suddenly emerge even though it may difficult to distinguish it from its competitors.

The response of mainstream economic theory to this may well be that it suggests the existence of an unobservable property that should be included as a term in the utility function which differentiates the popular entity from its competitors. However, as this assertion cannot be verified empirically, we cannot comment on its scientific validity. By contrast, an interactions-based mechanism may suggest that although a specific choice did not have any intrinsic advantage over others, stochastic fluctuations may have resulted in a relatively larger number of individuals opting for it initially. This may generate a small advantage in favor of the choice being adopted by others. For example, more people buying a product may make it more economical to produce (economy of scale) or the use of a particular product by some may make it more likely to be adopted by their acquaintances (network externalities). Eventually, through a process of self-reinforcing or positive feedback via interactions, an inexorable momentum is created in favor of the entity that builds into an enormous advantage in comparison to its competitors (see e.g. Ref. [99] for an application of this idea into high-technology markets). While the idea of such feedback or externalities has been discussed in economics from quite early on (see the article by Paul Krugman [100] and responses to it by others, including Kenneth Arrow [101]), the quest for analytically tractable linear economic models among mainstream practitioners has meant that a nonlinear interactions-based perspective for analyzing economic phenomena has become popular only recently with the advent of econophysics. The study of economic popularity is one such area that has benefited from this incursion of physics ideas into economics.

4.2 The Economic Fate of Movies: A Case Study

Movie popularity provides an arena for analyzing the phenomena of economic success or failure arising through interactions among agents—not least because of the availability of large quantities of publicly available digital data. In fact, there have been recent attempts to use temporal patterns in the digital data, e.g., the increase in the number of blog posts on specific movies [102] or in rising activity in the

Wikipedia entries for soon-to-be or newly released movies [103], to provide early prediction for the success of a movie. However, we shall here focus on the data about the box-office gross receipts of a movie during its initial run at theaters. Note that, unlike the popularity of several other types of products (e.g., to measure the popularity of a car, we look at how many people are driving it), in the case of a movie it is not completely obvious how to identify a unique observable that will be efficient at capturing all the different dimensions of its popularity. For example, one can consider the average of the ratings given by different film critics in various media, votes received in movie-related forums online or the total number of DVDs bought or rented. For example, we can take the case of popular movies decided by votes of registered users of the Internet Movie Database (IMDb) (http://www.imdb.com), one of the largest movie-related online sites. As voters can give a score between 1 and 10 to a movie, with 1 corresponding to "awful" and 10 to excellent, the rating of a movie can be decided by taking the average over all votes. Unfortunately, there are obvious limitations in using such a score for accurately measuring the popularity of movies. In particular, different scores may be only reflecting the amount of information about the movies available with voters. Thus, the older, so-called "classic" movies may be judged by a completely different yardstick compared to recently released films in view of the differences in the voters knowledge about them. Possibly more important from a economic agent's point of view is that as it does not cost the user anything to vote for a movie in the online forums, the vital element of competition for viewers that governs which product/idea will eventually become popular is missing from this measure. Therefore, focusing on the box-office gross earnings of movies after they are newly released in theaters is a reasonable measure of their relative popularity, as the potential viewers have a roughly similar kind of information available about the competing items. Moreover, such "voting with one's wallet" is arguably a more honest indicator of individual preference for movies.

An important property to note about the distribution of movie income is that it deviates significantly from a Gaussian form with a much more extended tail. In other words, there are many more highly popular movies than one would expect from a normal distribution. This immediately suggests that the process of emergence of popularity may not be explained simply as the outcome of many agents independently making binary (namely 'yes' or 'no') decisions to adopt a particular choice, such as going to see a particular movie. As this process can be mapped to a random walk, we expect it to result in a Gaussian distribution that, however, is not observed empirically. Previous studies of movie income distribution [104–106] have looked at limited datasets and found some evidence for a power-law fit. A more rigorous analysis using data on a much larger number of movies released across theaters in the USA was performed in Ref. [107]. While the tail of the distribution for both the opening gross and the total gross for movies may appear to follow an approximate power law $P(I) \sim I^{-\alpha}$ with an exponent $\alpha \simeq 3$ [107], an even better fit is achieved with a log-normal form [108],

$$P(x) = \frac{1}{x\sigma\sqrt{2\pi}} e^{-(\ln x - \mu)^2/2\sigma^2}, \qquad (15)$$

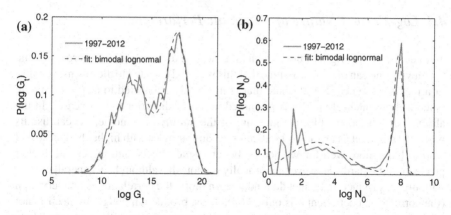

Fig. 6 Distribution of the logarithms of **a** the total gross income of a movie, G_t and **b** the number of theaters in which it opened, N_o, for all movies released in USA during 1997–2012. Fit with bimodal log-normal distributions shows that the empirical data can be well described by this theoretical form

where μ and σ are parameters of the distribution, being the mean and standard deviation of the variable is natural logarithm (see Fig. 6a). The lognormal form has also been seen in the income distribution of movies released in India and Japan [108]. It is of interest to note that a strikingly similar feature has been observed for the popularity of scientific papers, as measured by the number of their citations, where initially a power law was reported for the probability distribution with exponent three but was later found to be better described by a log-normal form [109, 110].

Instead of focusing only on the tail (which corresponds to the top grossing movies), if the entire income distribution is considered, we notice another important property: a bimodal nature. There are two clearly delineated peaks, which correspond to a large number of movies having either a very low income or a very high income, with relatively few movies that perform moderately at the box office. The existence of this bimodality can often mask the nature of the distribution, especially when one is working with a small dataset. For example, De Vany and Walls, based on their analysis of the gross for only about 300 movies, stated that log-normality could be rejected for their sample [111]. However, they had clearly assumed that the underlying distribution can be fitted using a single unimodal form. This assumption was evidently incorrect as evident from the histogram of their data. A more detailed and comprehensive analysis with a much larger dataset shows that the distribution of the total (as well as the opening) gross is in fact a superposition of two different log-normal distributions [108].

4.3 Log-Normal Nature of Economic Performance

To understand the origin of the bimodal log-normal distribution of the gross income for movies one can of course assume that this is directly related to the intrinsic quality of a movie or some other attribute that is intimately connected to a specific movie (such as how intensely a film is promoted in the media prior to its release). In the absence of any other objective measure of the *quality* of a movie, we can use its production budget as an indirect indicator because movies with higher budget would tend to have more well-known actors, better visual effects and, in general, higher production standards. However, empirically we note that although, in general, movies with higher production budget do tend to earn more, the correlation is not very high (the correlation coefficient r is only 0.63). Thus, production budget by itself is not enough to guarantee economic success. Another possibility is that the immediate success of a movie after its release is dependent on how well the movie-going public have been made aware of the film by pre-release advertising through various public media. Ideally, an objective measure for this could be the advertising budget of the movie. However, as this information is mostly unavailable, one can use instead data about the number of theaters that a movie is initially released at. As opening a movie at each theater requires organizing publicity for it among the neighboring population and wider release also implies more intense mass-media campaigns, we expect the advertising cost to roughly scale with the number of opening theaters. Unfortunately, the correlation between this quantity and per theater movie income is essentially non-existent. In this context, one may note that De Vany and Walls have looked at the distribution of movie earnings and profit as a function of a variety of variables, such as genre, ratings, presence of stars, etc., and have not found any of these to be significant determinants in movie performance [106].

In fact, the bimodal log-normal nature appears as a result of two independent factors, one responsible for the log-normal form of the component distributions and the other for the bimodal nature of the overall distribution. First, turning to the log-normal form, we observe that it may arise from the nature of the distribution of gross income of a movie normalized by the number of theaters in which it is being shown. The income per theater gives us a more detailed view of the popularity of a movie, compared to its gross aggregated over all theaters. It allows us to distinguish between the performance of two movies that draw similar numbers of viewers, even though one may be shown at a much smaller number of theaters than the other. This implies that the former is actually attracting relatively larger audiences compared to the other at each theater and hence is more popular locally. Thus, the less popular movie is generating the same income simply on account of it being shown in many more theaters, even though fewer people in each locality served by the cinemas may be going to see it. The appearance of the log-normal distribution may not be surprising in itself, as it is expected to occur in any linear multiplicative stochastic process. The decision to see a movie (or not) can be considered to be the result of a sequence of independent choices, each of which have certain probabilities. Thus, the final probability that an individual will go to the theater to watch a movie is a product of each of these constituent probabilities, which implies that it will follow a log-normal distribution. It is worth noting here that the log-normal distribution also appears in other areas where the popularity of different entities arises as a result of collective decisions, e.g. in the context of proportional elections [112], citations of scientific papers [110, 113] and visibility of news stories posted by users on an online website [114].

4.4 Bimodality of Success and Failure

Turning now to the bimodality in the income distribution, this appears to be related to an observed bimodality in the distribution of the number of theaters in which a motion picture is released (see Fig. 6b). Thus, most movies are shown either at a handful of theaters, typically a hundred or less (these are usually the independent or foreign movies), or at a very large number of cinema halls, numbering a few thousand (as is the case with the products of major Hollywood studios). Unsurprisingly, this also decides the overall popularity of the movies to an extent, as the potential audience of a film running in less than 100 theaters is always going to be much smaller than what we expect for blockbuster films. In most cases, the former will be much smaller than the critical size required for generating a positive word-of-mouth effect spreading through mutual acquaintances, which will gradually cause more and more people to become interested in seeing the film. There are occasional instances where such a movie does manage to make the transition successfully, when a major distribution house, noticing an opportunity, steps in to market the film nationwide to a much larger audience and a ësleeper hiti is created. An example is the movie *My Big Fat Greek Wedding*, which opened in only 108 theaters in 2002 but went on to become the fifth highest grossing movie for that year, running for 47 weeks and at its peak being shown in more than 2,000 theaters simultaneously.

Bimodality has also been observed in other popularity-related contexts, such as in the electoral dynamics of US Congressional elections, where over time the margin between the victorious and defeated candidates has been growing larger [115]. For instance, the proportion of votes won by the Democratic Party candidate in the federal elections has changed from about half of all votes cast to one of two possibilities: either about 35n40 % (in which case the candidate lost) or about 60n65 % (when the candidate won). This can be explained using a theoretical framework for describing how collective decisions arise from individual binary choice behavior [116, 117]. Here, individual agents take ëyesi or ënoi decisions on issues based on information about the decisions taken by their neighbors and are also influenced by their own previous decisions (adaptation) as well as how accurately their neighborhood had reflected the majority choice of the overall society in the past (learning). Introducing these effects in the evolution of preferences for the agents lead to the emergence of two-phase behavior marked by transition from a unimodal behavior to a bimodal distribution of the fraction of agents favoring a particular choice, as the parameter controlling the learning or global feedback is increased [116]. In the context of the movie income data, we can identify these choice dynamics as a model for the decision process by which theater owners and movie distributors agree to release a particular movie in a specific theater. The procedure is likely to be significantly influenced by the previous experience of the theater and the distributor, as both learn from previous successes and failures of movies released/exhibited by them in the past, in accordance with the assumptions of the model. Once released in a theater, its success will be decided by the linear multiplicative stochastic process outlined earlier and will follow a log-normal distribution. Therefore, the total or opening gross distribution for movies may be considered to be a combination of the lognormal distribution of income per theater and the bimodal distribution of the number of theaters in which a movie is shown.

4.5 Power Law Decay of Income with Time

To go beyond the simple blockbusternsleeper distinction and have a detailed view of the time evolution of movie performance, one has to consider the trend followed by the daily or weekly income of a movie over time. This shows an exponential decay with a characteristic rate, a feature seen not only for almost all other blockbusters, but for bombs as well (the rate is different for different movies). The only difference between blockbusters and bombs is in their initial, or opening, gross. However, sleepers may behave differently, showing an initial increase in their weekly gross and reaching the peak in the gross income several weeks after release. For example, in the case of *My Big Fat Greek Wedding* (referred earlier) the peak occurred 20 weeks after its initial opening. It was then followed by exponential decay of the weekly gross until the movie was withdrawn from circulation.

Instead of looking at the income aggregated over all theaters, if we consider the weekly gross income per theater, a surprising universality is observed. As previously mentioned, the income per theater gives us additional information about the movieis popularity because a movie that is being shown in a large number of theaters may have a bigger income simply on account of higher accessibility for the potential audience. Unlike the overall gross that decays exponentially with time, the gross per theater of a movie shows a power-law decay in time measured in terms of the number of weeks from its release, W: $g_W \sim W^{-\beta}$, with exponent $\beta \sim 1$ [108]. Thus, the local popularity of a movie at a certain point in time appears to be inversely proportional to the duration that has elapsed from its initial release. This shares a striking similarity with the time evolution of popularity for scientific papers in terms of citations as the citation probability to a paper published t years ago decays approximately as $1/t$ [110]. In a very different context, namely, the decay over time in the popularity of a website (as measured by the rate of download of papers from the site) and that of individual web pages in an online news and entertainment portal (as measured by the number of visits to the page), power laws have also been reported but with different exponents [118, 119]. More recently, the relaxation dynamics of popularity with a power-law decay have been observed for other products, such as book sales from Amazon.com [120] and the daily views of videos posted on YouTube [121], where the exponents appear to cluster around multiple distinct classes.

4.6 The Stylized Facts of "Popularity"

Thus, we observe that the complex process of economic success can be understood, at least in the case of movies, in terms of three robust features that (using the terminology of economics) we can term as the *stylized facts of popularity*: (i) log-normal distribution of the success of individual agents (theaters), (ii) the bimodal distribution of the number of agents taking part in a particular round (the theaters in which a movie is shown) and (iii) power-law decay with time of the economic performance of agents (gross income per theater). Some of these features have been seen in areas outside economics in which popularity dynamics play a role, such as citations of scientific papers or political elections. This suggests that it is possible that the above three properties apply more generally to the processes by which a few entities emerge to become a successful product or idea. Possibly a unifying framework may be provided by understanding successful or popular entities as those which have repeatedly survived a sequential failure process [108].

5 Inter-firms and Banks-firms Network Structures: Empirical Studies

5.1 Introduction

Credit-debt relationships among economic agents comprise as a whole large-scale networks of the economic system at nation-wide and global scales. There are different layers in such networks even at the core of real-economic and financial systems. One layer is an arena of real economy, namely supplier-customer links among firms as nodes. The firms activities are financed from financial institutions as well as directly from financial markets. The layer of supplier-customer network is thus linked to another layer of financial network between firms and banks. Furthermore, the banks are also creditors and debtors of themselves comprising another layer of inter-banks network.

As a financial system, the inter-bank network resides at the core, which is connected with firms, via banks-firms network, at a periphery of the system; the periphery is a large network of supplier and customer for the engine of real economy. These networks are actually further linked to financial markets, but one may depict the basic picture in a way given in Fig. 7.

Systemic risk is a network effect caused by failures or financial deterioration of debtors and creditors through the credit-debt links to other nodes even in a remote part of the networks (see [122] and references therein). The systemic risk often has

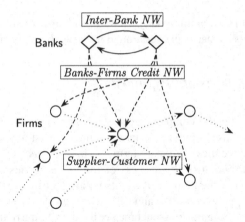

Fig. 7 Inter-bank, banks-firms credit, and supplier-customer networks schematically depicted. Financial institutions or banks (*squares*) and firms (*circles*) are creditors and debtors in the links of inter-bank credit (*lines*), lending-borrowing between banks-firms (*dashed lines*), and supplier-customer links among firms (*dotted lines*)

considerable consequences at a nation-wide scale, and sometimes to a world-wide extent, as we experience today in repeated financial crises.

While the understanding of inter-bank network at the core of financial system is crucial (see [123–128] and references therein for a surge of research focusing on inter-bank networks in different countries), no less important is the propagation of risk from the core of banks to the periphery of firms, vice versa, as well as the propagation of risk among firms. Unfortunately, empirical study based on real-data of banks-firms network or supplier-customer network at a large scale is still lacking. Only recently, there are literature in economics including the studies on US trade network among sectors using input-output (IO) data [129], propagation of sectoral shocks through the IO network [130], US inter-sectoral trade [131], for example; see also reviews [122, 132] and conference reports [133, 134] for collaborative works between physics and economics.

This section reviews recent empirical studies in Japan on banks-firms lending-borrowing credit network including all financial institutions and listed firms for decades [135, 136], and on supplier-customer network covering a million firms and all bankruptcies in a year [137]. We present new materials here in addition to the description on the unique and exhaustive properties of the Japanese data, but mainly focused is to review recent availability of large-scale networks at nation-wide scale, which can potentially open a new empirical and theoretical studies.

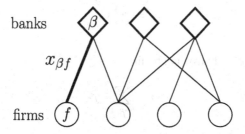

Fig. 8 Credit network as a bipartite graph. An edge connecting between bank β and firm f is associated with an amount of credit $x_{\beta f}$ as a weight

5.2 Banks-firms Credit Network

5.2.1 Data of Lending-Borrowing Between Banks and Firms

The dataset is based on a survey of firms quoted in the Japanese stock-exchange markets (Tokyo, Osaka, Nagoya, Fukuoka and Sapporo, in the order of market size). The data were compiled from the firms' financial statements and survey by Nikkei. They include the information about each firm's borrowing obtained from financial institutions such as the amounts of borrowing from 1980 to 2012 including the years of Japanese financial crisis in the late 90s.

For financial institutions, we select commercial banks as a set of leading suppliers of credit. The set comprises long-term, city, regional (primary and secondary), trust banks, insurance companies and other institutions including credit associations. During the examined period, more than 200 commercial banks existed, while the number of listed firms is more than 1,600.

5.2.2 Bipartite Network Structure

Annual snapshot of the lending-borrowing network can be regarded as a bipartite graph. Nodes are either banks or firms.[7] Banks and firms are denoted by Greek letters β ($\beta = 1, \ldots, n$) and Latin letters f ($f = 1, \ldots, m$) respectively. n is the number of banks, and m is that of firms. An edge between a bank β and a firm f is defined to be present if there is a credit relationship between them. In addition, a positive number $x_{\beta f}$ is associated with the edge, which is defined to be the amount of the credit. We can depict the network as shown in Fig. 8.

$x_{\beta f}$ is the amount of lending by bank β to firm i, which precisely equals to the amount of borrowing by firm i from bank β. The total amount of lending by bank β is

---

[7] Note that banks are not included in the side of firms, even if they are borrowing from other banks. Because our dataset includes banks' borrowing only partially, the interbank credit is not considered here, though it is no less important than the bank-firm credit studied here.

$$x_\beta = \sum_f x_{\beta f},$$ (16)

and the total amount of borrowing by firm f is

$$x_f = \sum_\beta x_{\beta f}.$$ (17)

The distributions for the amount of credit, x_β, x_f, and the number of borrowers and lenders, denoted by k_β, k_f respectively, have long-tails. They are shown, for the data of credit relationships in the year 2011, in Fig. 9a–d. There is a significant correlation between w_β and k_β in a natural way, and also for w_f and k_f, as shown in Fig. 9e, f respectively. In particular, from the Fig. 9e, we can observe an empirical relation of $k_\beta \propto w_\beta^a$, where $a \approx 0.67 \pm 0.04$ (least-square fit; error 95 % level). This implies the relation of $w_\beta/k_\beta \propto k_\mu^{0.49 \pm 0.07}$ meaning that the average loan is larger for the larger degree k_μ, or roughly speaking, for the larger banks.

Important properties of the large-scale structure of banks-firms network can be summarized as follows:

- Long-term and city banks are lenders to a number of listed firms. Calculation of Herfindahl index, defined by the sum of squares of shares in the lending amount, shows that they typically lend to 100 firms in the set of 1,600 firms.
- Regional banks have much narrower scope of lending, typically a tenth of long-term and city banks; the lending patterns are closely related to geographical regions.
- From a similarity measure defined by lending patterns of banks, one can construct a minimum-spanning tree (MST), for example (see also [138]). The resulting MST reveals a community structure between banks and firms, the modules of which are related to geographical locations and historical developments of the financial institutions.

See also [135].

5.2.3 Distress Propagation on the Firms-Banks Network

We note that a same value $x_{\beta f}$ has different meanings as a weight to the bank β and to the firm f. For example, even if 90 % of the total lending of the bank β goes to the firm f, it may be the case that f depends on β by only 20 % for all the loans from banks. It is therefore natural to define $w_{f \to \beta} = x_{\beta f}/x_\beta$, which represents the relative weight of lending by bank β to firm f, or dependency of β on f. Similarly, we define $w_{\beta \to f} = x_{\beta f}/x_f$, which represents the relative amount of borrowing by firm f from bank β, or dependency of f on β. See Fig. 10.

Suppose that bank β has a state of financial deterioration. Then it may shrink the amount of its supplied credit, increase interest-rate, shorten the due time of payment by firms, and so forth. This would eventually influence firm f to an extent that can

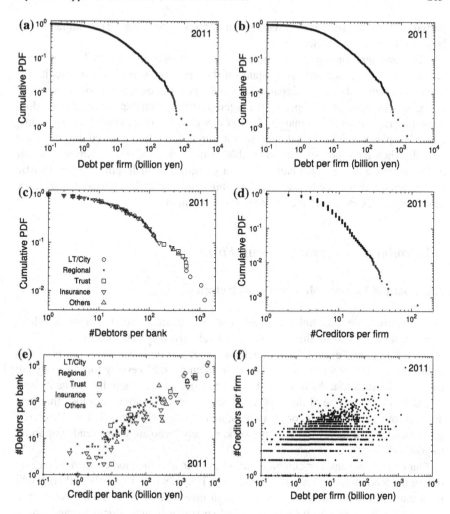

Fig. 9 **a** Cumulative distribution for banks' lending, x_β. **b** For firms' borrowing, x_f. **c** For the number of banks' lending relationships, k_β. **d** For the number of firms' borrowing relationships, k_f. **e** Scatter plot for banks' x_β and k_β. **f** Scatter plot for firms' w_f and k_f. All the plots are for the data in the year 2011. Rank correlations (Kendall's τ) for **e** and **f** are $\tau = 0.814(15.0\sigma)$ and $\tau = 0.384(23.5\sigma)$ respectively (σ calculated under the null hypothesis of statistical independence)

be quantified by $w_{\beta \to f}$, because it represents the dependency of firm f on bank β for the source of financing. See Fig. 10a.

Similarly for the reverse direction of influence, from firms to banks. Firm f with some level of financial deterioration may delay its repayment, have defaults, even fail into bankruptcy, and so forth. Then the lending banks will not be able to fully enjoy profits in expected amounts due to the delay, may possibly have bad loans partially, if not totally, for the credit given to bankrupted firms. This influence can be quantified by $w_{f \to \beta}$. See Fig. 10b.

This consideration can lead one to a methodology of evaluating the level of financial distress that potentially propagate on the network of banks-firms credit. We invented a method based on eigen-vectors and eigen-values structure of the matrices of weights in [136]. By comparing the eigen-structure with that obtained in random bipartite graphs, we found that the largest few (non-trivial) eigenvalues are significant. We performed historical analysis for our datasets, and showed that there are periods when the eigen-structure is stable or unstable, and that a particular set of banks, mostly a few regional banks, have large values of the fragility scores. Drastic change occurs in the late 80s during the bubble and also at the epochs of financially unstable periods including the financial crisis in Japan.

5.3 Production Network Among Firms

5.3.1 Data of Large-Scale Production Network

Let us say that a directed link is present as $A \to B$ in the production network, where firm A is a supplier to another firm B, or equivalently, B is a customer of A. While it is difficult to record every transaction among suppliers and customers, it is pointless to have a record that a firm buys a pencil from another. Necessary for our study are data of links such that the relation $A \to B$ is crucial for the activity of one or both A and B. If at least one of the firms at either end of a link nominates the other firm as most important suppliers or customers, then the link should be listed. This has a good analogy to a survey of social network, namely "who are your friends important to yourself?"

Our dataset for supplier-customer links has been accumulated on such an idea by one of the leading credit research agencies in Tokyo, which regularly gathers credit information on most of active firms through investigation of financial statements, corporate documents and by hearing-based survey at branch offices located across the nation.

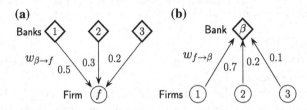

Fig. 10 Dependency between banks and firms. **a** Dependency of firm f on banks β. **b** Dependency of bank β on firms f

A typical number of active firms in Japan is roughly estimated to be two million.[8] We employ a snapshot of production networks compiled in September 2006. In the data, the number of firms is roughly a million, and the number of directional links is more than four million.

5.3.2 Network Structure

The entire network can be represented as a directed graph. To understand the global connectivity, the following graph-theoretical method is useful as was performed in the study of the hyperlink structure of the world-wide web [139].

NW   The whole network.

GWCC   Giant weakly connected component: the largest connected component when viewed as an undirected graph. An undirected path exists for an arbitrary pair of firms in the component.

DC   Disconnected components: other connected components than GWCC.

GSCC   Giant strongly connected component: the largest connected component when viewed as a directed graph. A directed path exists for an arbitrary pair of firms in the component.

IN   The firms from which one can reach the GSCC by a directed path.

OUT   The firms that are reachable from the GSCC by a directed path.

TE   "Tendtrils"; the rest of GWCC. Note that TEs may not look like tendrils.

It follows from the definitions that

$$NW = GWCC + DC \qquad (18)$$

$$GWCC = GSCC + IN + OUT + TE \qquad (19)$$

For the benefit of readers, a small example is given in Fig. 11.

The result for the numbers of firms is given as follows Table 1:

The shortest-path lengths (distances) from the GSCC and firms in the IN and OUT are given respectively, in Table 1 below:

Any two firms in the network are mutually reachable within eight directed links as seen from the above table. The inspection of industrial sectors of firms in these components shows that the IN has a relative excess of agriculture, forestry, real estate, and a relative shortage of government-related firms; and that the OUT has a relative excess of medical/health care/welfare, food establishments, and a relative shortage of manufacturing. The majority of manufacturing firms is present in the GSCC.

Important properties of the large-scale structure of production network can be summarized as follows:

---

[8] The National Tax Agency Annual Statistics Report. Other major sources are Establishment and Enterprise Census by the Ministry of Internal Affairs and Communications, and the Ministry of Justice's records on the entry and exit of firms, which are said to have under—or over-estimation problems due to the counting of non-active firms and so forth.

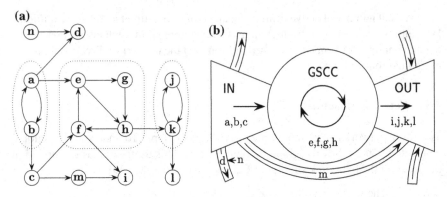

Fig. 11 A simple example of bowtie structure for a graph. **a** A directed graph. **b** "Bowtie structure"

Table 1 XXX

Component	#firms	Note
NW	1,019,854	
GWCC	1,009,597	99 % × NW
DC	10,257	Component-size ≤ 4
GSCC	462,563	46 % × GWCC
IN	182,018	18 % × GWCC
OUT	324,569	32 % × GWCC
TE	40,447	4 % × GWCC
Total	1,009,597	Equal to GWCC

Distance from GSCC to IN	
Distance	#firms
1	175,855
2	5,949
3	206
4	8
Total	182,018

Distance from GSCC to OUT	
Distance	#firms
1	308,572
2	15,441
3	536
4	20
Total	324,569

- The distributions for the numbers of suppliers/customers (in/out degrees) have long-tails. Denoting the degree by k, the cumulative distributions obey the power-law:

$$P_>(k) \sim k^{-\mu} \tag{20}$$

where the exponents $\mu \sim 1.3$.

- There exists a significant positive correlation between the degree and firm-size. Large firms are basically big suppliers/customers, while small and medium enterprises (SME) have relatively smaller numbers of links.

- There exists a weak negative correlation between the degrees at the ends of each link. This means in typical cases of manufacturing sectors that large firms have a number of SMEs as suppliers.

- Transitivity or clustering coefficients, the probability of finding triangles, is small compared with what is expected by random graphs preserving degree-distributions (see [140] for a lucid introduction).

- There exists a hierarchical modular or community structure. For example, manufacturing sectors have communities including electronics, heavy industry, foods, automobiles, construction material, pulp/paper and apparel. The electronics can be recursively divided into sub-communities, which are groups of industrial organization having historical developments and the so-called *keiretsu*, and/or are located in divided geographical sectors. Examples include sub-communities of Hitachi, Fujitsu, NEC; Panasonic, Sharp; Canon, Epson, Nikon, etc.

- Furthermore, these communities can be found to be quasi-cliques in a corresponding bipartite graph as follows. A supplier-customer link $u \rightarrow v$ for a set of nodes V ($u, v \in V$) can be considered as an edge in a bipartite graph that has exactly two copies of V as V_1 and V_2 ($u \in V_1$ and $v \in V_2$). Large and competing firms quite often share a set of suppliers to some extent, depending on the industrial sectors, geographical locations and so on. For example, Honda (v_1), Nissan (v_2) and Toyota (v_3) possibly have a number of suppliers u_i of mechanical parts, electronic devices, chassis and assembling machines, etc., in common. Then the links form a clique or a quasi-clique in the bipartite graph, where most possible links from u_i to $v_1, v_2, v_3, \ldots$ are present. This forms a portion in the entire graph with a higher density than other portions, which is basically the community structure in the production network.

We refer the readers to the reference [137] and tables and figures therein.

5.3.3 Chain of Bankruptcies

Supplier-customer link is a credit relation [141]. Whenever one delivers goods to others without an immediate exchange of money or goods of full value, credit is extended. Frequently, suppliers provide credit to their customers, who supply credit to their customers and so forth. Also customers can provide credit to their suppliers so as to have them produce an abundance of intermediate goods beforehand. In either case, once a firm goes into financial insolvency state, its creditors will possibly lose the scheduled payment, or goods to be delivered that have been necessary for production. The influence propagates from the bankrupted customer to its upstream in the former cases, and similarly from the bankrupted supplier to its downstream in the latter cases. Thus a creditor has its balance-sheet deteriorated in accumulation, and may eventually go into bankruptcy. This is an example of a *chain of bankruptcy*.

A bankruptcy chain does not occur only along the supplier-customer links. Ownership relation among firms is another typical possibility for such creditor-debtor relationship. It is, however, also frequently observed in our dataset that supplier-customer links are also present between holding and held companies, and sibling and related firms. We assume that most relevant paths along which the chain of bankruptcy occurs are the creditor-debtor links of the production network.

Corresponding to the snapshot of the network taken in September 2006, we employ an exhaustive list of all the bankruptcies for exactly one-year period from October. The number of bankruptcies amounts to roughly 0.13 million, daily mean being 30, and includes a few bankruptcies of listed firms. Nearly half of the bankrupted firms, precisely $N_b \equiv 6264$, were present on the network at the beginning and went into bankruptcy during the period. One can define the probability of bankruptcy by

$$p = N_b/N \approx 0.620\% \tag{21}$$

Note that the probability has inverse of time in its physical dimension. A year was chosen for the time-scale so that it should be longer than the time-scale for financial activities of firms, typically weeks and months, and be shorter than that for the change of network itself.

By using these data, we examined the size distribution for chains of bankruptcies, or avalanche-size distribution. We used a method to evaluate the frequencies of accidental chain in randomized networks, and found that the actual avalanche has a heavy tail distribution in its size. Combining with the large-scale properties and heterogeneity in modular structures, we claim that the effect to a number of creditors, non-trivially large due to the heavy tail in the degree distribution, is considerable in the real economy of the nation [137].

5.4 Summary

We briefly review recent empirical studies on financial networks based on large-scale datasets at nation-wide scale in Japan, banks-firms credit network and production network of suppliers and customers. These datasets provide a quite unique opportunity to investigate the structure and dynamics of networks as well as propagation of financial distress on the them. Because the networks are an arena with different levels of economic agents and relationships among them, on which economic activities take place with possible propagation of financial fragility and distress, it is crucial to understand them based on empirical study. We believe that the current and future collaboration with economists, physicists, computer scientists and practitioners in central banks all over the world would be of great value potentially leading to new ways to monitor and control financial crises that we experience more and more frequently today in the complex connected systems of economy.

6 Financial Time-Series Analyses: Wiener Processes and Beyond

6.1 Introduction

Nowadays, many people accept the conjecture that there exist probabilistic nature behind almost all of events around us. In economic science, Bachelier [142] dealt with the time-series in financial markets using random walk concept which was five years before Einstein's seminal paper on Brownian motion to determine the Avogadro number. Bachelier attempted to describe the up-down movement of the price changing by means of the Chapman-Kolmogorov equation and found that now-called *Wiener process* is a solution of the equation (see e.g. [143] for the details).

To see the Wiener process, let us define X_t as a price of commodity at time t. Then, we assume that the price updates according to the following rule:

$$X_{t+1} = X_t + Y_t \tag{22}$$

where Y_t is an additive white Gaussian noise satisfying $\langle Y_t \rangle = 0$ and $\langle Y_t Y_s \rangle = \sigma^2 \delta_{t,s}$ under the definition: $\langle \cdots \rangle = \int_{-\infty}^{\infty} (\cdots)(dY_t/\sqrt{2\pi}\sigma)\, e^{-Y_t^2/2\sigma^2}$. Repeating the above recursion relation (22), we obtain the price at time N as a cumulative return as $X_N \equiv \sum_{t=1}^{N} Y_t$, where we set $X_1 = 0$ for simplicity. It is easy for us to show that X_N also obeys a Gaussian with mean zero and the variance $N\sigma^2$. The model described by (22) to generate a time-series is referred to as Wiener process. It should be noted that even if Y_t does not follow a Gaussian, X_N can posses a Gaussian density with mean zero and the variance $N\sigma^2$ if the variance of each independent component Y_t in X_N is finite, namely, $\langle Y_t Y_s \rangle = \sigma^2 \delta_{t,s} < \infty$ in the limit of $N \to \infty$. In other words, the density of the stochastic variable $Z_N \equiv X_N/\sqrt{N}\sigma$ follows

$$\lim_{N \to \infty} P(Z_N) = \mathcal{N}(0, 1). \tag{23}$$

This fact is known as *central limit theorem*.

6.2 Empirical Evidence and Stable Distributions

However, several extensive empirical data analysis (see e.g. [1, 144]) revealed that real financial time series does not follow the Wiener process. This means that the return in financial markets is not generated by a Gaussian distribution and especially the empirical distribution possess a heavy tail. In Fig. 12, we show the probability distribution of the Standard & Poor's (S&P) 500 index observed at various time intervals Δt (namely, $X_{\Delta t} = \sum_{l=1}^{\Delta t} Y_l$ in terms of (22)) [144]. Obviously, these plots do not look like Gaussian distributions and they have much large kurtosis and exhibit

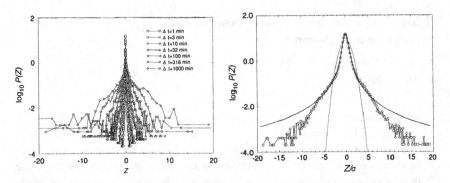

Fig. 12 The *left panel* shows the probability distribution of the Standard & Poor's (S&P) 500 index observed at various time intervals Δt. From the *right panel*, we find that the empirical distribution of the *left panel* is well-described by a Levy stable distribution with $\alpha = 1.4$ and $\gamma = 0.00375$. The *broken line* is a Gaussian with the same mean 0.0508 as in the empirical distribution of S&P 500 index with $\Delta t = 1$ (The *both panels* are taken from the reference [144])

heavy tails. From the central limit theorem, this fact means that the return of the data Y_t does not have a finite variance. Hence, we cannot describe the data in terms of the Wiener process.

In order to describe more generalized stochastic process including the Wiener process, we here introduce the so-called *stable process*. Let us consider that independent stochastic variables $Y_1, Y_2, \ldots, Y_N$ obey the identical distribution $P(y)$. Then, the Fourier transform of the distribution of variable $X (= Y_1 + \cdots + Y_N)$, say, $P_N(X)$ is given by $\phi_N(q) = \{\phi(q)\}^N$, where we defined

$$\phi(q) = \int_{-\infty}^{\infty} P(y)e^{-iqy}dy. \tag{24}$$

The inverse transform of $\phi_N(q)$ is immediately written as

$$P_N(X) = \frac{1}{2\pi} \int_{-\infty}^{\infty} \phi_N(q)\,e^{iqX}dq. \tag{25}$$

One can conclude that $P(y)$ is *stable* if $P(y)$ possesses the same function as the $P_N(X)$ does. Especially, for a specific choice $\phi(q) = e^{-\gamma \Delta t |q|^\alpha}$ where Δt is a scaling factor due to the sampling interval, the $P(y)$ leads to

$$P(y : \Delta t) = \frac{1}{\pi} \int_{0}^{\infty} dq\, e^{-\gamma \Delta t |q|^\alpha} \cos(qy) \equiv P_L(y : \Delta t) \tag{26}$$

The $P_L(y : 1)$ is refereed to as *Levy distribution*. We should keep in mind that $P_L(y : 1)$ is identical to a Gaussian when we set $\alpha = 2$, and Lorentzian for $\alpha = 1$. At the tail regime, that is, for $|y| \gg 1$, we have the power-law behavior as $P_L(y : 1) \sim y^{-(\alpha+1)}$.

As shown in Fig. 12 (left), the shape of return distribution is dependent on the sampling intervals Δt. Obviously, if one chooses a large Δt, it is a very rare event to obtain the large $|Y_t|$ and we need huge data points to confirm the shape of the distribution. To avoid this difficulty, we rescale the variables according to Mantegna and Stanley [144], namely, $y_s = y/(\Delta)^{1/\alpha}$, $P_L(y_s : 1) = P_L(y : \Delta t)/(\Delta t)^{-1/\alpha}$. All empirical data having various sampling intervals Δt collapse on the $\Delta t = 1$ distribution by accompanying with the above rescaling with $\alpha = 1.4$ and it is well-described by a Levy distribution $P_L(y) \equiv P_L(y_s : 1)$ as shown in Fig. 12 (right).

6.3 Time-Dependent Volatility and the Prediction Models

In the Wiener process, the standard deviation (the volatility in the context of finance) σ is independent of time. However, as empirical analysis for financial time series has revealed, the volatility itself is dependent on time and usually exhibits several distinct behavior, namely, it possess a long memory [1, 2]. The long memory is observed through the power-law behavior of the auto-correlation function with respect to the volatility, that is,

$$\overline{\sigma_l \sigma_{l+t}} \equiv \lim_{L \to \infty} \frac{1}{L} \sum_{l=1}^{L} \sigma_l \sigma_{l+t} \sim t^{-\beta} \tag{27}$$

where we defined $\sigma_t^2 \equiv (1/T) \sum_{l=t-1-T}^{t} Y_l^2 - \{(1/T) \sum_{l=t-1-T}^{t} Y_l\}^2$ which is evaluated in the time window with width T assuming the stationarity of Y_t. This is one of the remarkable features of the volatility in comparison with the fact that the auto-correlation function of return Y_t decays exponentially as $\overline{Y_l Y_{l+t}} \sim e^{-\beta t}$.

For the time-series having the time-dependent volatility, several models to predict the behavior have been proposed. One of the most famous models is refereed to as *ARCH (AutoRegressive Conditional Heteroskedasticity) model* [145] and the simplest version of the model, the so-called ARCH (1) model is given by

$$X_{t+1} = X_t + Y_t, \quad P(Y_t) = \mathcal{N}(0, \sigma_t), \quad \sigma_{t+1}^2 = \alpha_0 + \alpha_1 X_t^2 \tag{28}$$

where we should keep in mind that Y_t obeys a Gaussian, however the volatility is not constant but is updated by (28).

The ARCH model is easily extended to the GARCH (Generalized ARCH) model [146]. The update of the volatility in the simplest GARCH (1, 1) is described by $\sigma_{t+1}^2 = \alpha_0 + \alpha_1 X_t^2 + \beta_1 \sigma_t^2$ instead of (28) (see Fig. 13 for its typical behaviour).

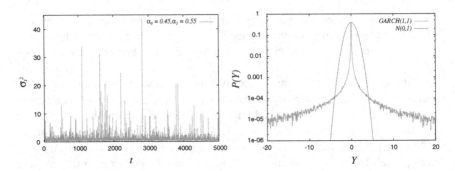

Fig. 13 The time-dependence of the volatility σ_t^2 for setting $\alpha_0 = 0.45$, $\alpha_1 = 0.55$ (*left*). The *right panel* shows the distribution of Y_t for the GARCH(1, 1) process with $(\alpha_0, \alpha_1, \beta_1) = (0.4, 0.3, 0.3)$

6.4 Duration Between Price Changes: First-Passage Process

In the previous subsections, our argument was restricted to the stochastic variables of the price changes (returns) and most of them concern a key-word: *Fat tails* of the distributions or deviation from a Gaussian. However, also the distribution of time intervals can deliver useful information on the markets and it is worth while to investigate these properties extensively [147–153] and if possible, to apply the gained knowledge to financial engineering (Fig. 13).

In fact, the *Sony bank rate* is one of the suitable examples. The Sony bank rate is the rate for individual customers of the Sony bank [154] in their on-line foreign exchange trading service via the internet. If the USD/JPY market rate changes by greater or equal to 0.1 yen, the Sony bank USD/JPY exchange rate is updated to the market rate. In this sense, the Sony bank rate can be regarded as a first-passage processes [155–161]. In Fig. 14, we show the mechanism of generating the Sony bank rate from the market rate (this process is sometimes referred to as a *first exit process* [162]). As shown in the figure, the time difference between two consecutive points in the Sony bank rate becomes longer than the time intervals of the market rates. We also should notice that the first passage time fluctuates even if the underlying stochastic process possesses a constant duration.

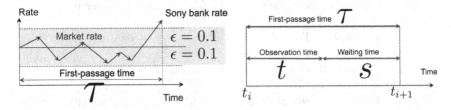

Fig. 14 An illustration of generating the filtered rate by the rate window with width 2ε from the market rate. If the market rate changes by a quantity greater or equal to 0.1 yen, the Sony bank USD/JPY exchange rate is updated to the market rate. The *right panel* shows the relation of the points τ, t and s in time axis. The first-passage time τ is given by $\tau = t_{i+1} - t_i$. The observation time is measured from the point t_i

To qualify the system in terms of the duration, in following, let us suppose that the difference between two consecutive points of the Sony bank rate change, namely, the first-passage time τ follows the distribution with probability density function $P_W(\tau)$ [163, 164]. Then, the customers observe the rate at time t $(0 \leq t \leq \tau)$ that should be measured from the point at which the rate previously changed. In Fig. 14 (right), we show the relation among these variables τ ($=$ first-passage time), t ($=$ observation time) and s ($=$ waiting time) in the time axis. The waiting time for the customers is naturally defined as $s \equiv \tau - t$. Then, we should notice that the distribution $\Omega(s)$ can be written in terms of the first-passage time distribution (with density $P_W(\tau)$) and the observation time distribution (with density $P_O(t)$) of the customers as a convolution $\Omega(s) \propto \int_0^\infty d\tau \int_0^\tau dt\, Q(s|\tau, t) P_O(t) P_W(\tau)$. In this equation, the conditional probability density $Q(s|\tau, t)$ that the waiting time takes the value s provided that the observation time and the first-passage time were given as t and τ, respectively, is given by $Q(s|\tau, t) = \delta(s - \tau + t)$, where $\delta(\cdot)$ is Dirac's delta function. Taking into account the normalization constant of $\Omega(s)$, we have

$$\Omega(s) = \frac{\int_0^\infty d\tau P_W(\tau) \int_0^\tau dt\, \delta(s - \tau + t) P_O(t)}{\int_0^\infty ds \int_0^\infty d\tau P_W(\tau) \int_0^\tau dt\, \delta(s - \tau + t) P_O(t)} \tag{29}$$

where t denotes the observation time for the customers. The result of the renewal-reward theorem: $w = \langle s \rangle = E(\tau^2)/2E(\tau)$ (see for example [165, 166]) is recovered by inserting a uniformly distributed observation time distribution $P_O(t) = 1$ into the above expression. Indeed, we have

$$w = \langle s \rangle = \int_0^\infty ds\, s \Omega(s) = \frac{\int_0^\infty ds\, s \int_s^\infty d\tau P_W(\tau)}{\int_0^\infty ds \int_s^\infty d\tau P_W(\tau)} = \frac{E(\tau^2)}{2E(\tau)} \tag{30}$$

where we defined the nth moment of the first-passage time $E(\tau^n)$ by $E(\tau^n) = \int_0^\infty ds\, s^n P_W(s)$. For a Weibull distribution: $P_{m,a}(t) = (mt^{m-1}/a) \exp(-t^m/a)$ which is required from the empirical evidence for the Sony bank rate [167–169], we have

$$\Omega(s) = \frac{m\, e^{-s^m/a}}{a^{1/m}\, \Gamma\left(\frac{1}{m}\right)}, \quad w = a^{1/m} \frac{\Gamma\left(\frac{2}{m}\right)}{\Gamma\left(\frac{1}{m}\right)} \tag{31}$$

We show the distribution $\Omega(s)$ with $a = 1$ and $m = 0.59, 1$ and 2 in the left panel of Fig. 15.

Here we encounter the situation which is known as *inspection paradox* [164]. For the Weibull distribution, the paradox occurs for $m < m_c = 1$. Namely, for this regime, we have $\langle s \rangle > \langle \tau \rangle$ (see Fig. 15). In general, it means that the average of durations (first-passage times) is shorter than the average waiting time. This fact is quite counter-intuitive because the customer checks the rate at a time between arbitrary consecutive rate changes. This fact is intuitively understood as follows.

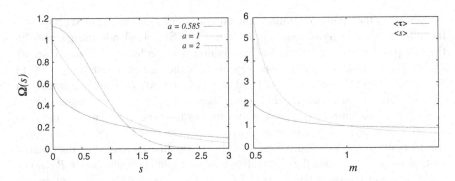

Fig. 15 The distribution of waiting time for a Weibull distributing $\Omega(s)$ with $a = 1$ and $m = 0.59$, 1 and 2. The *right panel* shows average duration $\langle\tau\rangle$ and average waiting time $\langle s\rangle$ as a function of m for a Weibull duration distribution with $a = 1$. The inspection paradox occurs for $m < m_c = 1$

When the parameter m is smaller than m_c, the bias of the duration is larger than that of the exponential distribution. As a result, the chance for customers to check the rate within large intervals between consecutive price changes is more frequent than the chance they check the rate within shorter intervals. Then, the average waiting time can become longer than the average duration.

6.5 Microscopic Reconstruction of Prices

From statistical physics point of view, the price change should be explained from the decision making of huge amount of teaders. Here we show the model proposed by Kaizoji [170] as such an attempt. Recently, we modified the Ising model approach by taking into account the cross-correlations in stocks [171].

As we saw, the return, which is defined as the difference between prices at successive two time steps t and $t + 1$ is given by (22). To reconstruct the return Y_t from the microscopic view point, we assume that each trader ($i = 1, \ldots, N$) buys or sells unit-volume at each time step t and write the total volumes of buying and selling are explicitly given by $\psi_+^{(t)}$ and $\psi_-^{(t)}$, respectively. Then, the return Y_t is naturally defined by means of $\psi_\pm^{(t)}$ as $Y_t = \lambda(\psi_+^{(t)} - \psi_-^{(t)})$, where λ is a positive constant. Namely, when the volume of buyers is greater than that of sellers, $\psi_+^{(t)} > \psi_-^{(t)}$, the return becomes positive $Y_t > 0$ (the price increases from (22)).

We should notice that the making decision of each trader ($i = 1, \ldots, N$) is obtained simply by an *Ising spin*:

$$S_i^{(t)} = \begin{cases} +1 \ (\text{buy}) \\ -1 \ (\text{sell}) \end{cases} \tag{32}$$

The return is also simplified as $Y_t = \lambda(\psi_+^{(t)} - \psi_-^{(t)}) = \lambda \sum_{i=1}^N S_i^{(t)} \equiv m_t$ where we set $\lambda = N^{-1}$ to make the return:

$$m_t = \frac{1}{N} \sum_{i=1}^{N} S_i^{(t)} \tag{33}$$

satisfying $|m_t^{(t)}| \leq 1$. Thus, m_k corresponds to the so-called *magnetization* in statistical physics, and the update rule of the price is written in terms of the magnetization m_t.
We next introduce the energy function.

$$E_t(S) = -\frac{J_t}{N} \sum_{ij} S_i S_j - h_t^{(k)} \sum_i \sigma_\tau^{(t)} S_i \tag{34}$$

where the above first term induces human collective behaviour, namely, each agent inclines to take the same decision as the others to decrease the total energy. The effect of this first term on the minimization of total energy might be recognized as the so-called *Keynes's beauty contest*. It means that traders tend to be influenced by the others' decision makings, in particular, at the crisis. The second term appearing in the right hand side of (34) represents the cross-correlation between the decision of trader and market information $\sigma_\tau^{(t)}$. Here we choose the 'trends':

$$\sigma_\tau^{(t)} = \frac{(p_t - p_{t-\tau})}{\tau} \tag{35}$$

for such information. It should be noticed that the state vectors of the agents: $S = (S_1, \ldots, S_N)$ are determined so as to minimize the energy function (34) from the above argument. For most of the cases, the solution should be unique. However, in realistic financial markets, the decisions by agents should be much more 'diverse'. Thus, here we consider statistical ensemble of traders S and define the distribution of the ensemble by $P(S)$. Then, we look for the suitable distribution which maximizes the so-called Shannon's entropy $H = -\sum_S P(S) \log P(S)$ under two distinct constraints $\sum_S P(S) = 1$, $\sum_S P(S)E(S) = E$. After some algebra, we immediately obtain the solution as *Gibbs-Boltzmann distribution*:

$$P(S) = \frac{\exp[-\beta E(S)]}{\sum_S \exp[-\beta E(S)]} \tag{36}$$

where β stands for the inverse-temperature given by $\beta = 1/T$. The equation of state at the equilibrium is obtained by $m = \sum_S (1/N) \sum_i S_i P(S)$, however, in financial markets, it might be assumed that the system is not at the equilibrium. To include the non-equilibrium property, we consider that the system described macroscopically by the following update rule which is based on the equation of state for m as

$$m_t = \tanh\left(J_t m_{t-1} + h_t \sigma_\tau^{(t)}\right). \tag{37}$$

It should be noted that the magnetization at the equilibrium is obtained by setting $m_t = m_{t-1} = m$ in the limit of $t \to \infty$.

In order to use the update rule (22) with $Y_t = m_t$ and (37), the information about parameters J_t, h_t appearing in the right hand side of (37) is needed. Hence, we should infer these parameters from the past data set in the financial market. In machine learning framework, the parameters are determined by the gradient descent

$$J_{t+1} = J_t - \eta \frac{\partial \mathscr{E}}{\partial J_t}, \quad h_{t+1} = h_t - \eta \frac{\partial \mathscr{E}}{\partial h_t} \tag{38}$$

for the cost function:

$$\mathscr{E}(J_t, h_t) \equiv \frac{1}{2} \sum_{l=1}^{t} \left[\overline{\Delta Z_l} - \tanh \left\{ J_t \overline{\Delta Z_{l-1}} + h_t \sigma_\tau^{(t)} \right\} \right] \tag{39}$$

where η is a learning coefficient and we defined $\overline{\Delta Z_l} \equiv (1/M) \sum_{i=l-M+1}^{l}$ $(Z_{i+1} - Z_i)$ for real (empirical) value of the price Z_t. Namely, the cost function (39) is an error measurement to evaluate how good the update rule of the return (37) is satisfied for the empirically observed return $\overline{\Delta Z_l}$ during the past t-time steps. Learning equations for the cost function (39) coupled with (37) and (22) determine the price at the next step. We show the result of the prediction in Fig. 16. After crush in Fig. 16, the parameters J, h converges to $J \to 1$ and $h \to 0$ which are corresponding to the solution $m = \tanh(Jm)$ as a critical point of the second order phase transition.

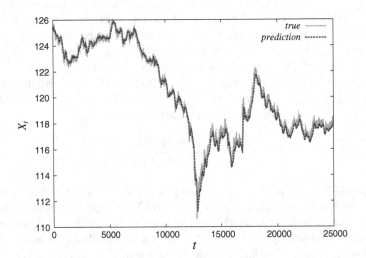

Fig. 16 The result of the prediction. The empirical data (true time-series) is picked-up from EUR/JPY exchange rate from 27th April 2010 to 13th May 2010. We set $\tau = M = 100$ and $\eta = 0.01$

6.6 Summary

We briefly showed several examples of analysis for the fluctuation in space and time of financial data sets. We found that the modelling of on-line trading system by means of renewal process and its first-passage process is practically useful and one can evaluate several macroscopic quantities such as 'waiting time' for the customers. We also introduced an agent-based modelling of financial markets. We have confirmed that Ising models, which have been used in the literature of statistical physics, are actually effective even in the context of finance. To explain the 'Stylized facts' from the agent-based microscopic viewpoint might be addressed as an important future direction.

7 Outlook

In this article, we have presented a few simple models and mechanisms based on statistical mechanics, and shown how they can be adapted to studying socio-economic systems. The knowledge from kinetic exchange models or other models of self-organised criticality (not discussed here) in the physical sciences, may help in understanding how the structurally similar connections between micro units might lead to similar or emergent collective behaviour in the social systems. We have discussed some agent based models of market mechanisms, which can in fact be studied with a more general perspective of resource allocation problems [172]. We have also discussed in this article that links in economic networks can be interpreted e.g., as relationships between creditors and debtors, or dependency among heterogeneous economic agents as nodes including financial institutions and firms. Once a certain set of agents has deterioration in their financial states, it can potentially propagate as a systemic risk to a considerable scale. It is crucial to understand the large-scale structure of economic networks at nation-wide or globe-wide scales, but such a study has been considered a formidable task so far. We have reviewed the study focusing on banks-firms lending-borrowing credit network and on production network of supplier and customers at a large-scale that covers essentially all domestic banks and firms. There have also been a few other phenomenological studies (which we could not discuss here due to space constraints) of particular segments in the interbank market in the Econophysics literature. We have also presented some robust features in the complex process of economic "success", such as in the case of movies, that (using the terminology of economics) can be termed as the stylized facts of "popularity". Since some of these features have also been observed in other areas outside economics in which popularity dynamics play a role, it is plausible that these properties apply more generally to the processes by which a few entities emerge to become a successful product or idea. We have also briefly showed several analyses for the fluctuation in space and time of financial data sets, and then studied agent-based models inspired from the Ising model, in the context of finance. It is imperative to now go beyond the first step of analogies, and relatively simple mechanical models, e.g., to understand the globally well-knit, strongly correlated or interdependent socio-economic systems and their resulting complex emergent dynamic behaviour.

In this article, we have also tried to portray the fact that Econophysics is not just about finance, by choosing many topics outside the realm of finance. Many of the early contributions in Econophysics were very much related to finance, and the phrases "finance", "financial markets", or "speculation" appeared in the titles of the first few books. However, it has constantly and successfully expanded over time from just financial markets to other branches of economics. And probably, in this respect we have seen only the tip of the iceberg!

Acknowledgments The authors would like to thank all their collaborators and students, whose works have been presented here. AG would like to thank B.K. Chakrabarti and S. Biswas for careful reading of the manuscript and useful comments. YF is supported in part by the European Community Seventh Framework Programme (FP7/2007–2013) under Socio-economic Sciences and Humanities, Grant agreement No. 255987 (FOC-II), by the Program for Promoting Methodological Innovation in Humanities and Social Sciences by Cross-Disciplinary Fusing of the Japan Society for the Promotion of Science, by Grant-in-Aid from the Zengin Foundation for Studies on Economics and Finance, and also by Grant-in-Aid for Scientific Research (B) No. 22300080 of the Ministry of Education, Science, Sports and Culture, Japan. JI thanks Enrico Scalas, Naoya Sazuka and Giacomo Livan for critical discussions and useful comments. He also acknowledges the financial support by Grant-in-Aid for Scientific Research (C) of Japan Society for the Promotion of Science, No. 22500195 and No. 25330278.

References

1. R.N. Mantegna, H.E. Stanley, *An Introduction to Econophysics: Correlations and Complexity in Finance* (Cambridge University Press, Cambridge, 2000)
2. J.-P. Bouchaud, M. Potters, *Theory of Financial Risk and Derivative Pricing* (Cambridge University Press, Cambridge, 2000)
3. S. Sinha, A. Chatterjee, A. Chakraborti, B.K. Chakrabarti, *Econophysics: An Introduction* (Wiley, Berlin, 2011)
4. B.M. Roehner, *Patterns of Speculation: A Study in Observational Econophysics* (Cambridge University Press, Cambridge, 2005)
5. U. Garibaldi, E. Scalas, *Finitary Probabilistic Methods in Econophysics* (Cambridge University Press, Cambridge, 2010)
6. H. Aoyama, Y. Fujiwara, Y. Ikeda, H. Iyetomi, W. Souma, *Econophysics and Companies: Statistical Life and Death in Complex Business Networks* (Cambridge University Press, New York, 2011)
7. B.K. Chakrabarti, A. Chakraborti, S.R. Chakravarty, A. Chatterjee, *Econophysics of Income and Wealth Distributions* (Cambridge University Press, Cambridge, 2013)
8. B. K. Chakrabarti, A. Chakraborti, A. Chatterjee (eds.), *Econophysics and Sociophysics: Trends and Perspectives* (Wiley, Weinheim 2006)
9. F. Abergel, H. Aoyama, B.K. Chakrabarti, A. Chakraborti, A. Ghosh (eds.), *Econophysics of Agent-Based Models* (Springer International Publishing, Switzerland, 2014)
10. F. Abergel, B.K. Chakrabarti, A. Chakraborti, A. Ghosh (eds.), *Econophysics of Systemic Risk and Network Dynamics* (Springer, Milan, 2012)
11. F. Abergel, B.K. Chakrabarti, A. Chakraborti, M. Mitra (eds.), *Econophysics of Order-Driven Markets* (Springer, Milan, 2011)

12. L. Pareschi, G. Naldi, G. Toscani (eds.), *Mathematical Modeling of Collective Behavior in Socio-Economic and Life Sciences* (Birkhauser, Berlin, 2010)

13. B. Basu, B.K. Chakrabarti, S.R. Chakravarty, K. Gangopadhyay (eds.), *Econophysics and Economics of Games, Social Choices and Quantitative Techniques* (Springer, Milan, 2010)

14. A. Chatterjee, B.K. Chakrabarti (eds.), *Econophysics of Markets and Business Networks* (Springer, Milan, 2007)

15. A. Chatterjee, B.K. Chakrabarti (eds.), *Econophysics of Stock and Other Markets* (Springer, Milan, 2006)

16. A. Chatterjee, S. Yarlagadda, B.K. Chakrabarti (eds.), *Econophysics of Wealth Distributions* (Springer, Milan, 2005)

17. V. Yakovenko, J. Barkley Rosser Jr, Colloquium: statistical mechanics of money, wealth and income. Rev. Mod. Phys. **81**, 1703 (2009)

18. A. Chakraborti, I. Muni Toke, M. Patriarca, F. Abergel, Econophysics: I. Empirical facts. Quant. Financ. **11:7**, 991 (2011)

19. A. Chakraborti, I. Muni Toke, M. Patriarca, F. Abergel, Econophysics: II. Agent-based models. Quant. Financ. **11:7**, 1013 (2011)

20. A. Chakraborti, M. Patriarca, Variational principle for the pareto power law. Phys. Rev. Lett. **103**, 228701 (2009)

21. A. Sen, *Poverty and Famines* (Oxford University Press, New Delhi, 1999)

22. M. Foucault, *Le Pouvoir psychiatrique:Cours au College de France 1973–74* (Gallimard, Paris, 2003)

23. R. Scruton, *Thinkers of the New Left* (Claridge Press, Longman, 1988)

24. J. Rawls, *A Theory of Justice* (Harvard University Press, Cambridge, 1971)

25. P. Aghion, E. Caroli, C. Garca-Pealosa, Inequality and economic growth: the perspective of the new growth theories. J. Econ. Lit. **37**, 1615 (1999)

26. R. Barro, *Inequality, Growth, and Investment*, Working Paper 7038, National Bureau of Economic Research (1999)

27. R. Benabou, Human capital, inequality, and growth: a local perspective. Eur. Econ. Rev. **38**, 817 (1994)

28. K. Forbes, A reassessment of the relationship between inequality and growth. Am. Econ. Rev. **90**, 869 (2000)

29. A. Alesina, D. Rodrik, Distribution, political conflict, and economic growth: a simple theory and some empirical evidence, in *Political Economy, Growth and Business Cycles*, ed. by A. Cukierman, Z. Hercowitz, L. Leiderman (MIT Press, Cambridge, 1992)

30. R. Benabou, Unequal societies: income distribution and the social contract. Am. Econ. Rev. **90**, 96 (2000)

31. A. Alesina, R. Perotti, Income distribution, political instability, and investment. Eur. Econ. Rev. **40**, 1203 (1996)

32. J. Blau, P. Blau, The cost of inequality: metropolitan structure and violent crime. Am. Soc. Rev. **47**, 114 (1982)

33. V. Pareto, *Cours dfeconomie politique* (Rouge, Lausanne, 1897)

34. R. Gibrat, *Les Ingalites conomiques* (Libraire du Recueil Sirey, Paris, 1931)

35. D. Champernowne, A model of income distribution. Econ. J. **63**, 318 (1953)

36. D. Champernowne, F. Cowell, *Economic Inequality and Income Distribution* (Cambridge University Press, Cambridge, 1998)

37. E. Montroll, M. Shlesinger, On 1/f noise and other distributions with long tails. Proc. Natl. Acad. Sci. **79**, 3380 (1982)

38. C. Gini, Measurement of inequality of incomes. Econ. J. **31**, 124 (1921)

39. R. Hogg, J. Mckean, A. Craig, *Introduction to Mathematical Statistics* (Pearson Education, Delhi, 2007)

40. A. Chatterjee, S. Yarlagadda, B.K. Chakrabarti (eds.), *Econophysics of Wealth Distributions, New Economic Windows Series* (Springer, Milan, 2005)

41. A. Chatterjee, B.K. Chakrabarti, Kinetic exchange models for income and wealth distributions. Eur. Phys. J. B **60**, 135 (2007)

42. H. Aoyama, Y. Nagahara, M. Okazaki, W. Souma, H. Takayasu, M. Takayasu, Pareto's law for income of individuals and debt of Bankrupt companies. Fractals **8**, 293 (2000)

43. S. Sinha, Evidence for power-law tail of the wealth distribution in India. Phys. A **359**, 555 (2006)

44. P. Richmond, S. Hutzler, R. Coelho, P. Repetowicz, A review of empirical studies and models of income distributions in society, in *Econophysics and Sociophysics: Trends and Perspectives*, ed. by B.K. Chakrabarti, A. Chakraborti, A. Chatterjee (Wiley, Weinheim, 2006)

45. A. Salem, T. Mount, A convenient descriptive model of income distribution: the gamma density. Econometrica **42**, 1115 (1974)

46. P.A. Samuelson, *Economics* (McGraw Hill, Auckland, 1998)

47. R. Axtell, Zipf distribution of U.S. firm sizes. Science **293**, 1818 (2001)

48. A. Chakraborti, B.K. Chakrabarti, Statistical mechanics of money: how saving propensity affects its distribution. Eur. Phys. J. B **17**, 167 (2000)

49. A. Chatterjee, B.K. Chakrabarti, S. Manna, Money in gas-like markets: Gibbs and Pareto laws. Phys. Scr. **T106**, 36 (2003)

50. A. Chatterjee, B.K. Chakrabarti, S. Manna, Pareto law in a kinetic model of market with random saving propensity. Phys. A **335**, 155 (2004)

51. B.K. Chakrabarti, A. Chatterjee, Ideal gas-like distributions in economics: effects of savings propensity, in *Applications of Econophysics*, ed. by H. Takayasu (Springer, Tokyo, 2004)

52. M. Patriarca, A. Chakraborti, Kinetic exchange models: from molecular physics to social science. Am. J. Phys. **81**, 618 (2013)

53. A. Drăgulescu, V. Yakovenko, Statistical mechanics of money. Eur. Phys. J. B **17**, 723 (2000)

54. S. Ispolatov, P.L. Krapivsky, S. Redner, Wealth distributions in asset exchange models. Eur. Phys. J. B **2**, 267 (1998)

55. E. Bennati, *La simulazione statistica nell'analisi della distribuzione del reddito: modelli realistici e metodo di Monte Carlo* (ETS Editrice, Pisa, 1988)

56. E. Bennati, Un metodo di simulazione statistica nell'analisi della distribuzione del reddito. Riv. Internazionale di Sci. Econ. e Commerciali **35**, 735 (1988)

57. E. Bennati, *Il metodo Monte Carlo nell'analisi economica*, Rassegna di lavori dell'ISCO X, 31 (1993)

58. J. Angle, The surplus theory of social stratification and the size distribution of personal wealth, in *Proceedings of the American Social Statistical Association, Social Statistics Section*, (Alexandria, VA 1983), p. 395

59. J. Angle, The surplus theory of social stratification and the size distribution of personal wealth. Soc. Forces **65**, 293 (1986)

60. J. Angle, The statistical signature of pervasive competition on wage and salary incomes. J. Math. Soc. **26**, 217 (2002)

61. J. Angle, The inequality process as a wealth maximizing process. Phys. A **367**, 388 (2006)

62. A. Chakraborti, B.K. Chakrabarti, Microeconomics of the ideal gas like market models. Phys. A **388**, 4151 (2009)

63. A. Chakraborti, S. Pradhan, B.K. Chakrabarti, A self-organising model of market with single commodity. Physica A **297**, 253–259 (2001)

64. M. Patriarca, A. Chakraborti, K. Kaski, Statistical model with a standard gamma distribution. Phys. Rev. E **70**, 016104 (2004)

65. K. Bhattacharya, G. Mukherjee, S.S. Manna, Detailed simulation results for some wealth distribution models in econophysics, in *Econophysics of Wealth Distributions*, ed. by A. Chatterjee, S. Yarlagadda, B.K. Chakrabarti (Springer, 2005), pp. 111–119

66. P. Repetowicz, S. Hutzler, P. Richmond, Dynamics of money and income distributions. Phys. A **356**, 641 (2005)

67. M. Lallouache, A. Jedidi, A. Chakraborti, *Wealth distribution: To be or not to be a Gamma?*, Science and Culture, vol 76 (Kolkata, India, 2010), p. 478

68. B.K. Chakrabarti, S. Marjit, Self-organisation and complexity in simple model systems: game of life and economics. Indian J. Phys. B **69**, 681 (1995)

69. B. Hayes, Follow the money. Am. Sci. **90**, 400 (2002)

70. M. Patriarca, E. Heinsalu, A. Chakraborti, Basic kinetic wealth-exchange models: common features and open problems. Eur. Phys. J. B **73**, 145 (2010)
71. M. Patriarca, A. Chakraborti, Kinetic exchange models: From molecular physics to social science. Am. J. Phys. **81**, 618 (2013)
72. S. Goswami, A. Chakraborti, Kinetic exchange models in economics and sociology, in *Nonlinear Maps and their Applications: Selected Contributions from the NOMA 2013 International Workshop*, ed. by R. Lopez-Ruiz, D. Fournier-Prunaret, Y. Nishio, C. Gracio (2015). arXiv:1408.1365v1 [physics.sos-ph]
73. B. Düring, D. Matthes, G. Toscani, Kinetic equations modelling wealth redistribution: a comparison of approaches. Phys. Rev. E **78**, 056103 (2008)
74. G. Toscani, C. Brugna, Wealth redistribution in Boltzmann-like models of conservative economies, in *Econophysics and Economics of Games, Social Choices and Quantitative Techniques*. New Economic Windows, ed. by B. Basu, S. Chakravarty, B.K. Chakrabarti, K. Gangopadhyay (Springer, Milan, 2010), p. 71
75. T. Lux, Emergent statistical wealth distributions in simple monetary exchange models: a critical review, in *Econophysics of Wealth Distributions*. New Economic Windows, ed. by A. Chatterjee, S. Yarlagadda, B.K. Chakrabarti (Springer, Milan, 2005), p. 51
76. J. Hogan, There's only one rule for the rich. New Sci. 21 (2005)
77. M. Gallegati, S. Keen, T. Lux, P. Ormerod, Worrying trends in econophysics. Phys. A **370**, 1 (2006)
78. T. Lux, F. Westerhoff, Economics crisis. Nat. Phys. **5**, 2 (2009)
79. R.N. Mantegna, H.E. Stanley, *Introduction to Econophysics: Correlations and Complexity in Finance* (Cambridge University Press, Cambridge, 1999)
80. D. Challet, M. Marsili, Y.C. Zhang, *Minority Games* (Oxford University Press, Oxford, 2005)
81. W. Brian Arthur, Inductive reasoning and bounded rationality. Am. Econ. Rev. (Papers and Proc.) **84**, 406 (1994)
82. D. Challet, Y.C. Zhang, Emergence of cooperation and organization in an evolutionary game. Phys. A **246**, 407 (1997)
83. D. Challet, Y.C. Zhang, On the minority game: analytical and numerical studies. Phys. A **256**, 514 (1998)
84. E. Moro, The minority game: an introductory guide, in *Advances in Condensed Matter and Statistical Mechanics*, ed. by E. Korutcheva, R. Cuerno (Nova Science Publishers, New York, 2004). arXiv:cond-mat/0402651v1 [cond-mat.dis-nn]
85. W. Kets, The minority game: an economics perspective, preprint: arXiv:0706.4432v1 [q-fin.GN] (2007)
86. A.S. Chakrabarti, B.K. Chakrabarti, A. Chatterjee, M. Mitra, The Kolkata paise restaurant problem and resource utilization. Phys. A **388**, 2420 (2009)
87. A. Ghosh, A.S. Chakrabarti, B.K. Chakrabarti, Kolkata Paise Restaurant problem in some uniform learning strategy limits, in *Econophysics and Economis of Games, Social Choices and Quantitative Techniques*. New Economic Windows, ed. by B. Basu, B.K. Chakrabarti, S.R. Chakravarty, K. Gangopadhyay (Springer, Milan, 2010), p. 3
88. A. Ghosh, A. Chatterjee, M. Mitra, B.K. Chakrabarti, Statistics of the Kolkata Paise Restaurant problem. New J. Phys. **12**, 075033 (2010)
89. A. Ghosh, S. Biswas, A. Chatterjee, A.S. Chakrabarti, T. Naskar, M. Mitra, B.K. Chakrabarti, Econophysics of systemic risk and network dynamics, in *Kolkata Paise Restaurant Problem: An Introduction*, New Economic Windows, ed. by F. Abergel, B.K. Chakrabarti, A. Chakraborti, A. Ghosh (Springer, Milan, 2013), p. 173
90. D. Dhar, V. Sasidevan, B.K. Chakrabarti, Emergent cooperation amongst competing agents in minority games. Phys. A **390**, 3477 (2011)
91. S. Biswas, A. Ghosh, A. Chatterjee, T. Naskar, B.K. Chakrabarti, Continuous transition of social efficiencies in the stochastic-strategy minority game. Phys. Rev. E **85**, 031104 (2012)
92. P. Ormerod, *Why Most Things Fail* (Faber & Faber, London, 2005)
93. R.L. Axtell, Zipf distribution of U.S. firm sizes. Science **293**, 1818 (2001)

94. S.N. Durlauf, How can statistical mechanics contribute to social science? Proc. Natl. Acad. Sci. USA **96**, 10582 (1999)

95. S.V. Vikram, S. Sinha, Emergence of universal scaling in financial markets from mean-field dynamics. Phys. Rev. E **83**, 016101 (2011)

96. T.C. Schelling, *Micromotives and Macrobehavior* (Norton, New York, 1978)

97. S. Wasserman, K. Faust, *Social Network Analysis: Methods and Applications* (Cambridge University Press, Cambridge, 1994)

98. F. Vega-Redondo, *Complex Social Networks* (Cambridge University Press, Cambridge, 2007)

99. W.B. Arthur, Competing technologies increasing returns, and lock-in by historical events. Econ. J. **99**, 116 (1989)

100. http://www.slate.com/articles/business/the_dismal_science/1998/01/the_legend_of_arthur. html

101. http://www.slate.com/articles/briefing/articles/1998/01/krugmans_life_of_brian.html

102. A. Ishii, H. Arakaki, N. Matsuda, S. Umemura, T. Urushidani, N. Yamagata, N. Yoshida, The 'hit' phenomenon: a mathematical model of human dynamics interactions as a stochastic process. New J. Phys. **14**, 063018 (2012)

103. M. Mestyan, T. Yasseri, J. Kertesz, Early prediction of movie box office success based on Wikipedia activity big data (2012), arXiv:1211.0970

104. A. De Vany, W.D. Walls, Uncertainty in the movie industry: does star power reduce the terror of the box office? J. Cult. Econ. **23**, 285 (1999)

105. D. Sornette, D. Zajdenweber, Economic returns of research: the Pareto law and its implications. Eur. Phys. J. B **8**, 653 (1999)

106. A. De Vany, *Hollywood Economics* (Routledge, London, 2003)

107. S. Sinha, S. Raghavendra, Hollywood blockbusters and long-tailed distributions: an empirical study of the popularity of movies. Eur. Phys. J. B **42**, 293 (2004)

108. R.K. Pan, S. Sinha, The statistical laws of popularity: universal properties of the box office dynamics of motion pictures. New J. Phys. **12**, 115004 (2010)

109. S. Redner, How popular is your paper? An empirical study of the citation distribution. Eur. Phys. J. B **4**, 131 (1998)

110. S. Redner, Citation statistics from more than a century of physical review. Phys. Today **58**, 49 (2004)

111. A. De Vany, W.D. Walls, Bose-Einstein dynamics and adaptive contracting in the motion picture industry. Econ. J. **106**, 1493 (1996)

112. S. Fortunato, C. Castellano, Scaling and universality in proportional elections. Phys. Rev. Lett. **99**, 138701 (2007)

113. F. Radicchi, S. Fortunato, C. Castellano, Universality of citation distributions: toward an objective measure of scientific impact. Proc. Natl. Acad. Sci. USA **105**, 17268 (2008)

114. F. Wu, B.A. Huberman, Novelty and collective attention. Proc. Natl. Acad. Sci. USA **104**, 17599 (2007)

115. D. Mayhew, Congressional elections: the case of the vanishing marginals. Polity **6**, 295 (1974)

116. S. Sinha, R.K. Pan, *Econophysics and Sociophysics: Trends and Perspectives* (Wiley, Weinheim, 2006), p. 417

117. S. Sinha, S. Raghavendra, *Advances in Artificial Economics: The Economy as a Complex Dynamic System* (Springer, Berlin, 2006), p. 177

118. A. Johansen, D. Sornette, Renormalization group analysis of the 2000–2002 anti-bubble in the US S&P 500 index: explanation of the hierarchy of 5 crashes and prediction. Phys. A **276**, 338 (2000)

119. Z. Dezso, E. Almaas, A. Lukacs, B. Racz, I. Szakadat, A.-L. Barabasi, Dynamics of information access on the web. Phys. Rev. E **73**, 066132 (2006)

120. D. Sornette, F. Deschatres, T. Gilbert, Y. Ageon, Stochastic gain in population dynamics. Phys. Rev. Lett. **93**, 28701 (2004)

121. R. Crane, D. Sornette, Robust dynamic classes revealed by measuring the response function of a social system. Proc. Natl. Acad. Sci. USA **105**, 15649 (2008)

122. F. Schweitzer, G. Fagiolo, F. Sornette, F. Vega-Redondo, A. Vespignani, D.R. White, Economic networks: the new challenges. Science **325**, 422 (2009)
123. S. Battiston, M. Puliga, R. Kaushik, P. Tasca, G. Caldarelli, DebtRank: too central to fail? financial networks, the FED and systemic risk, Sci. Rep. **2**, (Nature Publishing Group, 2012)
124. M. Boss, H. Elsinger, M. Summer, S. Thurner, Network topology of the interbank market. Quant. Financ. **4**, 677 (2004)
125. R. Cont, A. Moussa, E.B. Santos, in *Network Structure and Systemic Risk in Banking Systems*, ed. by e Edson Bastos (2010)
126. K. Imakubo, Y. Soejima, Network of fund transaction in call money market. Monet. Econ. Stud. **26**, (Bank of Japan, 2008)
127. G. Iori, G. De Masi, O. Precup, G. Gabbi, G. Caldarelli, A Network analysis of the Italian overnight money market, To appear in J. Econ. Dyn. Control (2007)
128. K. Soramäki, M.L. Bech, J. Arnold, R.J. Glass, W.E. Beyeler, The topology of interbank payment flows. Phys. A **379**, 317 (2007)
129. D. Acemoglu, A. Ozdaglar, A. Tahbaz-Salehi, The network origins of aggregate fluctuations. Econometrica **80**, 1977 (2012)
130. A. Foerster, G.S. Pierre-Daniel, M.W. Watson, Sectoral versus aggregate shocks: a structural analysis of industrial production. J. Polit. Econ. **119**, 1 (2011)
131. V.M. Carvalho, *Aggregate fluctuations and the network structure of intersectoral trade*, Working Paper (2008)
132. G. Caldarelli, *Scale-Free Networks* (Oxford University Press, Oxford, 2007)
133. F. Abergel, B.K. Chakrabarti, A. Chakraborti, A. Ghosh (eds.), *Econophysics of Systemic Risk and Network Dynamics* (Springer, Italia, 2012)
134. *Latsis Symposium: Economics on the Move, Trends and Challenges from the Natural Sciences* (ETH Zurich, 2012), http://web.sg.ethz.ch/Latsis_2012/main/main.html;. Accessed Jan. 2013
135. G. De Masi, Y. Fujiwara, M. Gallegati, B. Greenwald, J.E. Stiglitz, An analysis of the Japanese credit network. Evolut. Inst. Econ. Rev. **7**, 209 (2011)
136. Y. Fujiwara, H. Aoyama, Y. Ikeda, H. Iyetomi, W. Souma, Structure and temporal change of the credit network between banks and large firms in Japan. Econ. E-J. **3** (2009) (2009-7) http://www.economics-ejournal.org/economics/journalarticles/2009-7
137. Y. Fujiwara, H. Aoyama, Large-scale structure of a nation-wide production network. Eur. Phys. J. B **77**, 565 (2010)
138. R. Mantegna, Hierarchical structure in financial markets. Eur. Phys. J. B **11**, 193 (1999)
139. A. Broder, R. Kumar, F. Maghoul, P. Raghavan, S. Rajagopalan, R. Stata, A. Tomkins, J. Wiener, Graph structure in the web. Comput. Netw. **33**, 309 (2000)
140. M.E.J. Newman, The structure and function of complex networks. SIAM Rev. **45**, 167 (2003)
141. J.E. Stiglitz, B. Greenwald, *Towards a New Paradigm in Monetary Economics* (Cambridge University Press, Cambridge, 2003)
142. L. Bachelier, Theorie de la speculation. Annales Scientifique de l'Ecole Normale Superieure **III–17**, 21 (1900)
143. J. Voit, *The Statistical Mechanics of Financial Markets* (Springer, 2001)
144. R.N. Mantegna, H.E. Stanley, Scaling behaviour in the dynamics of an economic index. Nature **376**, 46 (1995)
145. R.F. Engle, Autoregressive conditional heteroskedasticity with estimates of variance of united kingdom inflation. Econometrica **50**, 987 (1982)
146. T. Ballerslev, Generalized autoregressive conditional heteroskedasticity. Econometrics **31**, 307 (1986)
147. R.F. Engle, J.R. Russel, Autoregressive conditional duration: a new model for irregularly spaced transaction data. Econometrica **66**, 1127 (1998)
148. F. Mainardi, M. Raberto, R. Gorenflo, E. Scalas, Fractional calculus and continuous-time finance II: the waiting-time distribution. Physica A **287**, 468 (2000)
149. M. Raberto, E. Scalas, F. Mainardi, Waiting-times and returns in high-frequency financial data: an empirical study. Phys. A **314**, 749 (2002)

150. E. Scalas, R. Gorenflo, H. Luckock, F. Mainardi, M. Mantelli, M. Raberto, Anomalous waiting times in high-frequency financial data. Quant. Financ. **4**, 695 (2004)
151. T. Kaizoji, M. Kaizoji, Power law for the calm-time interval of price changes. Phys. A **336**, 563 (2004)
152. E. Scalas, The application of continuous-time random walks in finance and economics. Phys. A **362**, 225 (2006)
153. R. Gorenflo, F. Mainardi, *The Asymptotic Universality of the Mittag-Leffler Waiting Time Law in Continuous Random Walks*, Lecture Note at WE-Heraeus-Seminar on Physikzentrum Bad-Honnef (Germany, 2006), pp. 12–16
154. http://moneykit.net
155. S. Redner, *A Guide to First-Passage Processes* (Cambridge University Press, Cambridge, 2001)
156. N.G. van Kappen, *Stochastic Processes in Physics and Chemistry* (North Holland, Amsterdam, 1992)
157. C.W. Gardiner, *Handbook of Stochastic Methods for Physics Chemistry and Natural Sciences* (Springer, Berlin, 1983)
158. H. Risken, *The Fokker-Plank Equation: Methods of Solution and Applications* (Springer, Berlin, 1984)
159. I. Simonsen, M.H. Jensen, A. Johansen, Optimal investment horizons. Eur. Phys. J. B **27**, 583 (2002)
160. S. Kurihara, T. Mizuno, H. Takayasu, M. Takayasu, in *The Application of Econophysics*, ed. by H. Takayasu (Springer, 2003), p. 169
161. J. Inoue, N. Sazuka, Crossover between Lvy and Gaussian regimes in first-passage processes. Phys. Rev. E **76**, 021111 (2007)
162. M. Montero, J. Masoliver, Mean exit time and survival probability within the CTRW formalism. Eur. J. Phys. B **57**, 181 (2007)
163. N. Sazuka, J. Inoue, E. Scalas, The distribution of first-passage times and durations in FOREX and future markets. Phys. A **388**, 2839 (2009)
164. J. Inoue, N. Sazuka, Queueing theoretical analysis of foreign currency exchange rates. Quant. Financ. **10**, 121 (2010)
165. H.C. Tijms, *A First Course in Stochastic Models* (Wiley, Chichester, 2003)
166. S. Oishi, *Queueing Theory* (CORONA PUBLISHING CO. LTD (in Japanese), Japan, 2003)
167. N. Sazuka, Analysis of binarized high frequency financial data. Eur. Phys. J. B **50**, 129 (2006)
168. N. Sazuka, On the gap between an empirical distribution and an exponential distribution of waiting times for price changes in a financial market. Phys. A **376**, 500 (2007)
169. N. Sazuka, J. Inoue, Fluctuations in time intervals of financial data from the view point of the Gini index. Phys. A **383**, 49 (2007)
170. T. Kaizoji, Speculative bubbles and crashes in stock markets: an interacting-agent model of speculative activity. Phys. A **287**, 493 (2000)
171. T. Ibuki, S. Higano, S. Suzuki, J. Inoue, A. Chakraborti, Statistical inference of co-movements of stocks during a financial crisis. J. Phys. Conf. Ser. **473**, 012008 (2013)
172. A. Chakraborti, D. Challet, A. Chatterjee, M. Marsili, Y.-C. Zhang, B.K. Chakrabarti, Statistical mechanics of competitive resource allocation using agent-based models. Phys. Rep. **1**, 1 (2013) (in press, 2014). available at arXiv:1305.2121[bibphysics.soc-ph]

Socio-Economic Inequalities: A Statistical Physics Perspective

Arnab Chatterjee

Abstract Socio-economic inequalities are manifested in different strata of our daily life. We discuss various aspects, beginning with the evolutionary and historical origins, and discussing the major issues from the social and economic point of view. The subject has attracted scholars from across various disciplines, including physicists, who bring in a unique perspective to the field. The major attempts to analyze the results, address the causes, and understand the origins using statistical tools and statistical physics concepts are discussed.

1 Introduction

If you were in Dharmatola in Kolkata, Manhattan in New York, Ikebukuro in Tokyo, Chandni Chowk in Delhi, Normalm in Stockholm, Soho in London, Andheri in Mumbai, Kallio in Helsinki, Hutong in Beijing, Montparnasse in Paris or San Salvario in Turin, you can observe different degrees of inequality in socio-economic conditions, from standard of living, sanitation, traffic, pollution, buying power or consumer behavior. The study of socio-economic inequalities has been important in identifying the causes to help minimize them by intervention of the state through laws and regulations.

Humans are social beings. Our social interactions are simple at times, but often more complex. Social interactions in many forms produce spontaneous variations manifested as inequalities while at times these inequalities result out of continued complex interactions among the constituent human units. With the availability of a large body of empirical data for a variety of measures from human social interactions, it is becoming increasingly possible to uncover the patterns of socio-economic inequalities. Additionally, the data can give us a clear picture of the origin of these variations as well as the dynamics of relevant quantities that produce these conditions, leading to their theoretical understanding. This brief review aims at giving a unique

A. Chatterjee (✉)
Condensed Matter Physics Division, Saha Institute of Nuclear Physics, 1/AF Bidhannagar, Kolkata 700064, India
e-mail: arnabchat@gmail.com

© Springer International Publishing Switzerland 2015
F. Abergel et al. (eds.), *Econophysics and Data Driven Modelling of Market Dynamics*,
New Economic Windows, DOI 10.1007/978-3-319-08473-2_12

perspective to the analysis of complex social dynamics using the statistical physics framework for computational social science [1]. With tools of statistical physics as a core, the knowledge and techniques from various disciplines like statistics, applied mathematics, information theory and computer science are incorporated for a better understanding of the nature and origin of socio-economic inequalities that shape the humankind.

Sociology and Economics are disciplines in their own right, with a huge body of modern literature developed independently of physical sciences. However, in their infancy, these disciplines were not much distinct from physical science. It is interesting to note that the development of statistical physics was also influenced by social statisticians recording the 'state' of a person by recording the various measures of his/her social conditions. Of course, one of the most well known facts in the history of modern economics is that many of its ideas have been largely influenced by physical sciences, with their logical basis and technicalities having close resemblance with statistical physics. A classic example is that of Jan Tinbergen, who with his colleague Ragner Frisch, was the first Nobel laureate in Economics (Nobel Memorial Prize in Economic Sciences in 1969) "for having developed and applied dynamic models for the analysis of economic processes". Tinbergen studied mathematics and physics at the University of Leiden under Paul Ehrenfest, who was one of the key figures in the development of statistical physics. During his years at Leiden, Tinbergen had numerous discussions with Kamerlingh Onnes, Hendrik Lorentz, Pieter Zeeman, and Albert Einstein, all Nobel laureates who left profound contributions to statistical physics. Tinbergen and many other icons of modern economics shaped their ideas with heavy influences from physical science, most of which were already developed in the literature of statistical physics.

From the other side, physicists seemed to have been thinking about socio-economic systems through the eyes of physical principles, and had often been inspired by those problems to lay foundations to the fields of mathematics and physics which became further relevant in broader contexts. Aristotle (384–322 BCE) is known to have laid the first foundations of physics in his *Physica*, where he even argued that barter is the form of exchange which is used in most societies. Huygens (1629–1695), known for his contribution to optics, had laid the foundations of probability theory while working on insurance and annuities. Astronomer Halley (1656–1742) was studying mortality and life expectancy besides studying celestial bodies. Quetelet (1796–1874) was the among the first to study the average man (*l'homme moyen*), and describe different facets of human social life through statistics, laying the foundations of what he termed as 'social physics', while his contemporary Comte (1798–1857) provided solid foundations to the subject [2], also calling it 'sociology'. Poincaré's student Bachelier [3] studied the dynamics of stock market, was the first to use Brownian motion to study stock price fluctuations, and predated Einstein's work on Brownian motion. Other notable discussions include Majorana's article on the value of statistical laws in physics and in social science [4]. Kadanoff's work on urban dynamics [5] was an interesting study on the labor and management aspects of cities. In subsequent years, physicists became more interested in social phenomena, asking interesting questions and proposing models to explain interesting aspects of social life (See e.g., Montroll and Badger [6] as well as Ref. [7]), which gave alternative views to the traditional approaches developed in mainstream social sciences.

Social scientists and economists have dealt with numerous interesting issues of the human society, uncovering behaviors which seem to be universal. However, it is often pointed out that certain patterns observed a couple of decades ago may have changed. This happens due to the essential nature of human interaction which has changed with the advent of technologies. This makes socio-economic systems distinct compared to physical systems—here we rarely find established 'laws' [2]. This naturally calls for a change in the theoretical analysis and modelling, making the field an extremely challenging one. Specifically, there has been wide discussions about addressing socio-economic phenomena in the light of *complexity science* [8] and embracing ideas from various disciplines [9] to understand socio-economic phenomena and a few big initiatives are already taking shape [10].

Socio-economic inequality is the existence of unequal opportunities and rewards for various social positions or statuses within the society. It usually contains structured and recurrent patterns of unequal distributions of goods, wealth, opportunities, and even rewards and punishments. There are two main ways to measure social inequality: *inequality of conditions*, and *inequality of opportunities*. *Inequality of conditions* refers to the unequal distribution of income, wealth and material goods. *Inequality of opportunities* refers to the unequal distribution of 'life chances' across individuals. This is reflected in measures such as level of education, health status, and treatment by the criminal justice system. In this review, we will also discuss topics which are manifestations of unequal choices made by groups of individuals, e.g., consensus, religion, bibliometric data etc.

With rapid technological developments and the world getting more and more connected, our social conditions are experiencing changes, rapid in some parts of the world, and much slower elsewhere. The nature of complexity of human social interactions are also changing. The social classes formed out of societal evolution are changing, making it much difficult to predict or even comprehend the near future.

In this review, we concentrate on a handful of issues, which have been of concern to physicists in the same way as they have been drawing attention from social scientists. There have been even deeper and sensitive issues in society, such as gender inequality, racial or ethnic inequality, inequalities in health and education etc. [11] which we do not discuss here. This review is organized as follows. In Sect. 2 we try to discuss the evolutionary origins of socio-economic inequalities. We try to address the question whether statistical physics can be successfully used to address the problems and model situations that lead to such inequalities in Sect. 3. In Sect. 4 we will discuss a number of processes that are relevant in our understanding of the phenomena. The most prominent issue of income and wealth inequality will be discussed in Sect. 5. Urban systems will be addressed in Sect. 6, consensus in Sect. 7, bibliometrics in Sect. 8. We will briefly discuss how networks are important in many contexts in Sect. 9. In Sect. 10 we present the different ways to measure inequalities, and Sect. 11 will discuss some possible ways to deal with socio-economic inequalities. Finally we conclude with discussions in Sect. 12.

2 Evolutionary Perspective to Socio-economic Inequality

The human species has lived as hunter gathers for more than 90 % of its existence. It is widely known that hunter-gatherer societies were egalitarian, as is still evident from the lifestyle of various tribes likes the !Kung people of the Kalahari desert [12]. They traditionally had few possessions, and were semi nomadic in the sense they were moving periodically. They hardly mastered farming and lived as small groups. The mere instinct to survive was driving them to suppress individual interests. They shared what they had, so that all of their group members were healthy and strong, be it food, weapons, property, or territory [13]. When agricultural societies developed, it created elaborate hierarchies, which certainly had more or less stable leadership in course of time. These evolved into clans or groups led by family lines, which eventually blew up as kingdoms. In these relatively complex scenarios, chiefs or kings assumed new strategies for hoarding surplus produce of agriculture or goods, predominantly for survival in times of need, and thus concentrating wealth and power (see Sect. 5.1 for models with savings, also Ref. [14]). There were mechanisms involved that helped in wealth multiplication, along with the advancement of technologies. The transition from egalitarianism to societies with competition and inequality paved way for the development of chiefdoms, states and industrial empires [15].

3 Why Statistical Physics?

Given that many laws of nature are statistical in origin, a well established fact that applies across most areas of modern physics, gives statistical physics a status of a prominent and extremely useful discipline. The subject, being developed in a very general framework, makes it applicable to various areas outside the commonly perceived boundaries of physics, as in biology, computer science and information technology, geosciences etc. It comes as no big surprise that attempts to understand the nature of social phenomena using statistical physics has been going on since some time, with the research activity gaining much more prominence in the recent two decades. What is much more striking is that the usefulness of statistical physics to understand social systems has gone far from being merely 'prominent' to being 'important' [16–21]. Statistical physics treats its basic entities as particles, while in society the basic constituents are human individuals. In spite of this gross difference, human society shows remarkable features which are similar to those widely observed in physical systems. Transitions are observed from order to disorder and emergence of consensus on a specific issue from a situation with multiple options by converging to one or a very few of them. Many measurables in a society show scaling and universality [9]. Dynamical socio-economic systems exhibit self-similarity [22] while going from one phase to another, similar to critical phenomena observed in physical systems. Statistical physics can be used as a tool to effectively understand the regularities at a larger scale, that emerge out of interaction among individual units, the social being.

Statistical physics tells us that systems of many interacting dynamical units collectively exhibit, a behavior which is determined by only a few basic dynamical features of the individual units and of the embedding dimension, but independent of all other details. This feature which is specific to critical phenomena, like in continuous phase transitions, is known as universality [23]. There is enough empirical evidence that a number of social phenomena are characterized by simple emergent behavior out of the interactions of many individuals. In recent years, a growing community of researchers have been analyzing large-scale social dynamics to uncover universal patterns and also trying to propose simple microscopic models to describe them, similar to the minimalistic models used in statistical physics. These studies have revealed interesting patterns and behaviors in social systems, e.g., in elections [24–26], growth in population [27] and economy [28], income and wealth distributions [16], financial markets [29], languages [30], etc. (see Refs. [17, 21] for reviews).

4 Processes Leading to Inequality and Broad Distributions

Can we point out the key processes that make the probability distribution of a quantity broad? Actually, there are several processes known to do so, and can be grossly characterized into a few categories. In this review, we will discuss a few, that are most relevant to the issue of socio-economic inequalities.

4.1 Random Walk

For a random walk of unitary steps, the probability of finding the walker at distance x from origin after N steps is given by

$$P(x) = \frac{2}{\sqrt{\pi N}} \exp\left(-\frac{x^2}{2N}\right). \tag{1}$$

In case the steps are variable, with l being the average length of the path, Eq. (1) gets modified to

$$P(x) = \frac{2}{\sqrt{2\pi N l^2}} \exp\left(-\frac{x^2}{2N l^2}\right), \tag{2}$$

still a Gaussian distribution about the mean displacement.

Let us assume a random walk starting from origin. The time (number of steps) to return to origin t is even and we denote by $2N$, and let the probability of this event be u_t, and f_t be the probability that the first return time is t. One can easily write that the probability u_{2N} that the walker is at position zero after $2N$ steps as

$$u_{2N} = 2^{-2N} \binom{2N}{N} \tag{3}$$

from where it can be shown that the distribution of the first return times is

$$f_{2N} = \frac{\binom{2N}{N}}{(2N-1)2^{2N}}. \tag{4}$$

Now, $\ln f_{2N} = \ln(2N)! - 2\ln N! - 2N\ln 2 - \ln(2N-1)$. Using Sterling's approximation for $N \to \infty$ and simplifying, one can obtain

$$f_{2N} \simeq \sqrt{\frac{2}{N(2N-1)^2}} \sim N^{-3/2}, \tag{5}$$

or equivalently, $f_t \sim t^{-3/2}$, i.e., the distribution of return times has a power law decay with exponent $\frac{3}{2}$.

4.2 Combination of Exponentials

Let a certain quantity y has an exponential distribution $P(y) \sim \exp(ay)$. If the constant $a > 0$, $P(y)$ should have an upper cutoff given by the maximum value of y so that the distribution is normalizable. Now let another quantity x is related to y as $x \propto \exp(by)$, where b is another constant. Then, one can write

$$P(x) = P(y)\frac{dy}{dx} \sim \frac{\exp(ay)}{b\exp(by)} = \frac{x^{\frac{a}{b}}}{bx} = \frac{1}{b}x^{-(1-\frac{a}{b})}. \tag{6}$$

4.3 Self-organized Criticality

In Self-Organized Criticality [31], the steady state of the evolution for a dissipative dynamical system is a fractal. The evolution of the piles of sand inspired this model, known as 'sandpile' models. The sites can support grains of sand until a certain threshold, beyond which the site becomes "critical" and expels the sand to the neighbors. It may happen that many sites can become critical and adding a single grain can create a large scale avalanche. At criticality, the size of the avalanches are power law distributed, and a feedback mechanism ensures that whenever the system is full of sand no more grain can be added until the activity stops, while at times of no activity sand grains need to be added. Thus the amount of sand fluctuates around a mean value, as a result of a dynamical equilibrium.

4.4 Multiplicative Processes

Let us assume that a quantity x evolves in time with a multiplier w.r.t. its state in the previous step. Formally, $x_{t+1} = \varepsilon_t x_t$. Had it evolved from $t = 0$, one can write

$$x_{t+1} = \left[\prod_{i=0,1,\ldots,t} \varepsilon_i \right] x_0, \tag{7}$$

where ε_t is a random multiplier changing with time t. Taking logarithms,

$$\log x_{t+1} = \log x_0 + \sum_{i=0,1,\ldots,t} \log \varepsilon_i. \tag{8}$$

The second term, which is the sum of random variables in a log scale will simply produce a normal distribution independent of the precise nature of the distribution of the ε. Hence, the variable x follows the 'lognormal' distribution

$$P(x) = \frac{1}{x\sqrt{2\pi\sigma^2}} \exp\left\{-\frac{(\log x - \mu)^2}{2\sigma^2}\right\}, \tag{9}$$

where $\mu = \langle \log(x) \rangle$ is the mean value of the logarithmic variable and $\sigma^2 = \langle [\log(x) - \mu]^2 \rangle$ the corresponding variance. However, there is often a confusion in the range where the variance is quite large $\sigma^2 \gg (\log x - \mu)^2$, so that $P(x) \sim 1/x$ in the leading order term.

A small change in the generative mechanism can also produce a power law distribution, which was nicely demonstrated by Champernowne [32] in the context of the power law tail of income distribution (which will be discussed in the following section). Champernowne assumed that there is some minimum income m. For the first range, one considers incomes between m and γm, for some $\gamma > 1$; for the second range, one considers incomes between γm and $\gamma^2 m$. A person is assumed to be in class j for $j \geq 1$ if their income is between $m\gamma^{j-1}$ and $m\gamma^j$. It is assumed that over each time step, the probability of an individual moving from class i to class j, denoted by p_{ij}, depends only on the value of $|j - i|$. The equilibrium distribution of individuals among classes under this assumption gives Pareto law. It is important to note that in multiplicative models, income becomes arbitrarily close to zero due to successive multiplications, while in case of the Champernowne model, there is a minimum income corresponding to the lowest class which is a bounded minimum that acts as a reflective barrier to the multiplicative model. This produces a power law instead of a lognormal distribution [33].

4.5 Preferential Attachment

In one of the most prominent works to discuss preferential attachment, Simon [34] identified that the process can model the emergence of power law distributions, citing examples such as distributions of frequency or words in documents, distributions of numbers of papers published by scientists, distribution of cities by population, distribution of incomes, and distribution of biological genera by number of species. This work was important in the sense that he had discussed small variations which were put into context. Preferential attachment, sometimes referred to as *Matthew effect* [35], is modeled in various ways, and we will discuss a few processes of historical importance, which have found prominence in explaining phenomena in nature and society.

The number of species in a genus, family or other taxonomic group appears to closely follow a power law distribution. Yule [36] proposed a model to explain this in the biological context, but is since then, used to explain several phenomena, including socio-economic systems. In the following, we will discuss the model, following Ref. [37].

Let us assume N be the number of genera, which increases by unity with each time step, when a new species finds a new genus and m other species are added to the already existing genera selected in proportion to the number of species they already have. Let $P(k, N)$ be the fraction of genera having k species, which makes the number of genera $NP(k, N)$. The probability that the next species added to the system adds to a genus i having exactly k_i species is proportional to k_i, and is $\frac{k_i}{\sum_i k_i} = \frac{mk_i}{N(m+1)}$. Now, in the interval between Nth and the $(N+1)$th genera, m other new species are added. The probability that the genus i gains a new species during this time is simply $\frac{mk_i}{N(m+1)}$, while the total number of genera of size k that gain a new species during this same interval is $\frac{mk_i}{N(m+1)} NP(k, N) = \frac{mk_i}{m+1} P(k, N)$. As a result of this evolutionary process, (i) there is a decrease in number of genera with k species at each time step by this number as they become genera with $k + 1$ by gaining a new species; and (ii) along with an increasing term because of species which had $k - 1$ species and now have one extra. One can now write a master equation

$$(N + 1)P(k, N + 1) - NP(k, N) = \frac{m}{m + 1} [(k - 1)P(k - 1, N) - kP(k, N)], \quad (10)$$

while for the genera of size unity

$$(N + 1)P(1, N + 1) - NP(1, N) = 1 - \frac{m}{m + 1}P(1, N). \quad (11)$$

One can easily see that $P(1) = \frac{m+1}{2m+1}$. The stationary solution for $P(k)$ is

$$P(k) = [kP(k - 1) - (k + 1)P(k)] \frac{m}{m + 1} \quad (12)$$

which can be rearranged as

$$P(k) = \frac{k-1}{k+1+\frac{1}{m}}P(k-1) \tag{13}$$

$$= \frac{(k-1)(k-2)\dots 1}{(k+1+\frac{1}{m})(k+\frac{1}{m})\dots(3+\frac{1}{m})} = \left(1+\frac{1}{m}\right)B\left(k, 2+\frac{1}{m}\right),$$

where $B(x, y)$ is the Beta function. For large x and fixed y, $B(x, y) \sim x^{-y}$. Using this, one gets $P(k) \sim k^{-(2+1/m)}$.

In the following, we will discuss in brief two closely related models of preferential attachment, which address the growth of citations to articles. Although the following fit ideally to the Sects. 8 and 9, it is best to discuss it at this juncture in context of their broad applicability.

The first, due to Price [38], was proposed in the context of citations is very similar to the Yule process. It is a model for a directed graph, and uses the fact that citation networks have a broad distribution of incoming citations (in-degree) [39]. The main idea to exploit is that popular papers become popular due to *cumulative advantage* [34]. Let $m = \sum_k kP(k)$ be the mean degree, where $P(k)$ is the degree distribution. Price assumes that the probability that a newly appearing paper cites an older paper is proportional to the number of citations k of that paper, and in fact taken proportional to $k+1$ to take care of the case of first citation. Thus, the probability that a new connection happens on any of the vertices with in-degree k is

$$\frac{(k+1)P(k)}{\sum_k (k+1)P(k)} = \frac{(k+1)P(k)}{m+1}, \tag{14}$$

and the average number of new edges (per vertex added) attached to vertices of in-degree k will be $(k+1)P(k)m/(m+1)$. Given N vertices, let $P(k, N)$ denote the probability distribution of in-degree k. Thus one can write the master equation as

$$(N+1)P(k, N+1) - NP(k, N) = [kP(k-1, N) - (k+1)P(k, N)]\frac{m}{m+1}, \text{ for } k \geq 1, \tag{15}$$

with

$$(N+1)P(0, N+1) - NP(0, N) = 1 - P(0, N)\frac{m}{m+1}, \text{ for } k = 0. \tag{16}$$

The stationary solution for $P(k)$ is

$$P(k) = \begin{cases} [kP(k-1) - (k+1)P(k)]\frac{m}{m+1} & \text{for } k \geq 1, \\ 1 - P(0)\frac{m}{m+1} & \text{for } k = 0. \end{cases} \tag{17}$$

Thus one gets $P(0) = \frac{m+1}{2m+1}$ and $P(k) = \frac{kP(k-1)}{k+2+1/m}$, which leads to

$$P(k) = \frac{k(k-1)\ldots 1}{(k+2+\frac{1}{m})\ldots(3+\frac{1}{m})}P(0) = \left(1+\frac{1}{m}\right)B\left(k+1,2+\frac{1}{m}\right), \quad (18)$$

Using the property of Beta function, $P(k) \sim k^{-(2+1/m)}$, which also comes as a result of the Yule process. The power law exponent in the Price model is always greater than 2.

The second model, known as the Barabási-Albert (BA) model [40], uses *preferential attachment* explicitly in its growth dynamics. The main difference from Price's model is that the graph is treated as undirected, which eliminates the problem with the uncited papers. At each step a new vertex arrives with $m(\geq 1)$ edges. Thus, the probability that a new edge attaches to any of the vertices with in-degree k is given by

$$\frac{kP(k)}{\sum_k kP(k)} = \frac{kP(k)}{2m}. \quad (19)$$

The average number of vertices of degree k who gain an edge upon adding a single new vertex with m edges is simply $mkP(k)/2m = kP(k)/2$. Thus, the master equation is

$$(N+1)P(k, N+1) - NP(k, N) = \frac{1}{2}(k-1)P(k-1, N) - \frac{1}{2}kP(k, N), \text{ for } k > m \quad (20)$$

and

$$(N+1)P(m, N+1) - NP(m, N) = 1 - \frac{1}{2}mP(m, N), \text{ for } k = m. \quad (21)$$

The stationary solution for $P(k)$ is

$$P(k) = \begin{cases} \frac{1}{2}(k-1)P(k-1) - \frac{1}{2}kP(k) & \text{for } k > m, \\ 1 - \frac{1}{2}mP(m) & \text{for } k = m. \end{cases} \quad (22)$$

Thus one gets $P(m) = \frac{2}{m+2}$ and $P(k) = \frac{(k-1)P(k-1)}{k+2}$, combining which one can get

$$P(k) = \frac{(k-1)(k-2)\ldots m}{(k+2)(k+1)\ldots(m+3)}P(m) = \frac{2m(m+1)}{(k+2)(k+1)k}, \quad (23)$$

which for large k gives $P(k) \sim k^{-3}$ [41, 42].

5 Income, Wealth & Energy

> The disposition to admire, and almost to worship, the rich and the powerful, and to despise, or, at least, to neglect persons of poor and mean condition is the great and most universal cause of the corruption of our moral sentiments.
>
> – Adam Smith, Scottish political economist (1723–1790)

The issue of inequality in terms of income and wealth, is a widely debated subject in economics. Economists and philosophers have pondered over the normative aspects of this problem [43–46]. The direct and indirect effects of inequality on the society have also been studied extensively while relatively less emphasis had been put on the origin of the problem. The main non-trivial issues and open questions have been: related to the form of the income and wealth distributed, do they depend upon specific conditions of a country, or if they are universal, and why, if they are so.

More than a century ago, Pareto [47] made extensive studies in Europe and found that wealth distribution follows a power law tail for the richer section of the society, known to be the Pareto law. Independently, Gibrat worked on the same problem and he proposed a "law of proportionate effect" [48]. Much later, Champernowne's systematic approach provided a probabilistic theory to justify Pareto's claim [49]. Subsequent studies revealed that the distributions of income and wealth possess some globally stable and robust features (see, e.g., [16]). In general, the bulk of the distribution of both income and wealth seems to reasonably fit both the log-normal and the Gamma distributions. Economists have a preference for the log-normal distribution [50, 51], while statisticians [52] and rather recently, physicists [14, 53, 54] prefer the Gamma distribution for the probability density or Gibbs/ exponential distribution for the corresponding cumulative distribution. The high end of the distribution, known to be the 'tail', is well described by a power law as was found by Pareto. These observed regularities in the income distribution may thus indicate a "natural" law of economics:

$$P(x) \sim \begin{cases} x^n \exp(-x/T) & \text{for } x < x_c, \\ \frac{\alpha x_c^\nu}{x^{1+\nu}} & \text{for } x \geq x_c, \end{cases} \tag{24}$$

where n and ν are two exponents, and T denotes a scaling factor. The exponent ν is called the Pareto exponent and its value ranges between 1 and 3 [16] (See Ref. [59] for a historical account of Pareto's data and those from recent sources). The crossover point x_c is extracted from the numerical fittings.

Gibrat [48] clarified that Pareto's law is valid exclusively for the high income range, whereas for the middle income range is described by a log-normal probability density (Eq. 9). The factor $\beta = 1/\sqrt{2\sigma^2}$ is known as Gibrat index, which measures the equality of the distribution. Empirically, β is found to lie between 2 and 3 [60].

Figure 1 shows a few examples of income and wealth distributions—incomes as computed from income tax data sources [55] of USA for several years, wealth of top billionaires [56], wealth of individual Bitcoin accounts [57] in terms of Bitcoin units

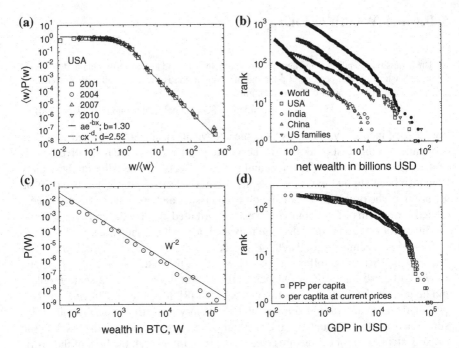

Fig. 1 **a** Probability distribution $P(w)$ of income w rescaled by average income $\langle w \rangle$ for USA for different years. The income is calculated from the IRS tax data [55]. The low and middle income range fits to an exponential $a \exp(-bx)$ while the high income range fits to a power law decay cx^{-d}. **b** The rank plot of the wealthiest individuals, for the whole world, USA, India, China, and also wealthiest families of USA [56]; **c** The probability distribution $P(W)$ of wealth W (in BTC) of Bitcoin accounts [57]; **d** Rank plot of GDP of countries of the world [58]

(BTC), and also GDP of countries [58] in US Dollars, both for purchasing power parity (PPP) per capita as well as per capita at current prices. All of these indicate a power law for the richest.

It has also been seen that the tail of the consumer expenditure also follows power law distribution [61, 62] along with a lognormal bulk, in the same way as income distribution does.

5.1 Modelling Income and Wealth Distributions

Gibrat [48] proposed a law of proportionate effect which explained that a small change in a quantity is independent of the quantity itself. Thus the distribution of a quantity $dz = dx/x$ should be Gaussian, and thus x is lognormal, explaining why small and middle range of income should be lognormal. Champernowne proposed a probabilistic theory to justify Pareto's claim [32, 49], but used a class structure in a multiplicative process setup (discussed in Sect. 4).

In case of wealth distribution, the most popular models are chemical kinetics inspired Lotka-Volterra models [63–65], polymer physics inspired Bouchaud-Mézard model [66] and models inspired by kinetic theory [14, 67–69] (see Ref. [16] for detailed examples). What is quite well understood and established is that two-class structure [54] of the income distribution is a result of very different dynamics of the two classes. The bulk is defined by a process which is at its core nothing but a random kinetic exchange [67, 68] producing a distribution dominated by an exponential function. The dynamics is nothing but as simple as what we know as the kinetic theory of gases [70]. The minimal modifications that one can attempt are to use additive or multiplicative terms.

Inequality creating processes involving uniform retention rates [71, 72] or savings [68] can only produce Gamma-like distributions. In these models, the wealth exchanging entity or *agent* randomly exchanges wealth but retaining a certain fraction (saving propensity) of what they had before each trading process. These models consider each agent the same, assigning each with the same value of the 'saving propensity' (as in Ref. [68]), which could not produce a broad distribution of wealth. What is important to note here is that the richest follow a different dynamic where heterogeneity plays a crucial role. To get the power law distribution of wealth for the richest, one needed to simply assume that each agent is different in terms how much fraction of wealth they save in each trading [69], a very natural ingredient to assume, since it is very likely that agents in a market think very differently from one another. In fact, with this very little modification, on can explain the whole range of wealth distribution [14]. However, these models can show interesting characteristics if the exchange processes and flows be made very asymmetric, e.g., on directed networks [73]. Numerous variants of these models, their results and analyses find possible applications in a variety of trading processes [16, 74].

Statistical physics tools have helped to formulate these microscopic models, which are simple enough yet rich with socio-economic ingredients. Toy models help in understanding the basic mechanism at play, and bring out the crucial elements that shape the emergent distributions of income and wealth. A variety of models, from zero-intelligence variants to the much complex agent based models incorporating game theory have been proposed and found to be successful in interpreting results [16]. Simple modelling has been found to be effective in understanding how entropy maximization produce distributions dominated by exponentials, and also explain the reasons of aggregation at the high range of wealth, including the emergence of the power law Pareto tail.

A rapid technological development of human society since the industrial revolution has been dependent on consumption of fossil fuel (coal, oil, and natural gas), accumulated inside the Earth for billions of years. The standards of living in our modern society are primarily determined by the level of per capita energy consumption. It is now well understood that these fuel reserves will be exhausted in the not-too-distant future. Additionally, the consumption of fossil fuel releases CO_2 into the atmosphere, which is the major greenhouse gas, affecting climate globally—a problem posing great technological and social challenges. The per capita energy consumption around the globe has been found to have a huge variation. This is a challenge at the geo-socio-political level and complicates the situation for arriving at a global consensus on how to deal with the energy issues. This global inequality in energy consumption was characterized [75, 76], and explained using the method of entropy maximization.

5.2 Is Wealth and Income Inequality Natural?

One is often left to wonder if inequalities in wealth and income are natural. It has been shown in terms of models and their dynamics that certain minimal dynamics [77] over a completely random exchange process picture and subsequent entropy maximization produces broad distributions. Piketty [78] recently argued that inequality in wealth distribution is indeed quite natural. He specifically pointed out that before the wars of the early 20th century, the stark skewness of wealth distribution was prevailing as a result of a certain 'natural' mechanism. The two great World Wars followed by the Great Depression helped in dispersion of wealth that, sort of brought the extreme inequality under check and gave rise to a sizable middle class. He argues, analyzing very accurate data, that the world is currently 'recovering' back to its natural state, due to capital ownership driven growth of finance [79] which has been dominant over a labor economy, which is simply a result of which type of institution and policies are adopted by the society. This work raises issues, quite fundamental, concerning both economic theory and the future of capitalism, and points out large increases in the wealth/output ratio. In standard economic theory, such increases are associated with a decrease in the return to capital and an increase in the wages. However, today the return to capital has not been seen to have diminished, while wages have. There is also some prescription proposed—higher capital-gains and inheritance taxes, bigger spending for access to education, tight enforcement of anti-trust laws, reforms in corporate-governance that restrict pay for executives, and finally financial regulations which had been an instrument for banks to exploit the society—all of these might help reduce inequality and increase equality of opportunity. The speculation is that this might be able to restore the shared and quick economic growth that was the characteristic the middle-class societies in the middle of the twentieth century.

6 Urban and Similar Systems

With growing population all over the world, towns and cities have also grown in size (population) [80]. The definitions using precise delimiters can be an issue of debate, whether one is looking at a metropolitan area, a commune, or a central urban agglomeration, but irrespective of that it has been persistently observed that the number of agglomerations of a certain size (or size range) is inversely proportional to the size. Questions has been asked whether this is a result of just a random process, a hierarchical organization, or if they are guided by physical and social constraints like optimization or organization.

6.1 City Size

Auerbach [81] was the first to note that the sizes of the cities follow a broad distribution, which was later bettered by Lotka [82] as $k^{0.93}N_k = const.$, which was subsequently cast as [83] as $kN_k^{\gamma-1} = const.$, where N_k is the population of the city of rank k. Rewriting, one gets $k \propto 1/N_k^{\gamma-1}$, where k is the number of cities with population N_k or more. Thus γ is nothing but the size exponent (Pareto). It was examined again by Zipf [84] by plotting rank-size distribution of quantities, and restated as $N_k \propto k^{-b}$, from where it follows that the two exponents are related as $1/b = \gamma - 1$. $b = 1$ is the value of the exponent, à la Zipf. Rigorous studies [85] showed that the Zipf exponent has deviations from unity, as also found in recent studies of China [86] or former USSR countries [87]. A general view of the broad distribution of city sizes across the world is shown in Fig. 2 (left panel), also showing the non-universal nature of the Zipf exponent.

Several studies have attempted to derive Zipf's law theoretically for city-size distributions, specifically where the exponent of the CDF of size is equal to unity. Gabaix [88] indicated that if cities grow randomly with the same expected growth rate and the same variance (Gibrat's law, see Sect. 5), the limiting distribution will converge to Zipf's law. Gabaix proposed that growth 'shocks' are random and impact utilities both positively and negatively. In a similar approach of shocks, citizens were assumed to experience them, resulting in diffusion and multiplicative processes, producing intermittent spatiotemporal structures [89]. Another study used shocks as a result of migration [90]. In Ref. [88] however, differential population growth

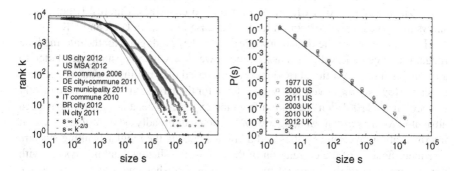

Fig. 2 *Left panel* Plot of size s with rank k of a city for different countries. The two *straight lines* are respectively $s \propto k^{-1}$ and $s \propto k^{-2/3}$, guides to the approximate extremes of the power law behavior of the data sets. Figure adapted from Ref. [92]. *Right panel* Firm size distribution for USA [93] and UK [94]

resulted from migration. Some simple economics arguments showed that expected urban growth rates were identical across city sizes and variations were random normal deviates, and the Zipf law with exponent unity follows. However, there seemed to be a missing assumption which may not produce the power law [91], a random growth model was proposed based on more 'physical' principles, which can generate a power law instead.

6.2 Scaling of Urban Measures

Analysis of available large urban data sets across decades for U.S.A. have concluded that (i) size is the major determinant, (ii) the space required per capita shrinks—the denser the settlement, the more intense is the use of infrastructure, (iii) all socio-economic activities accelerate, indicating higher productivity, and (iv) socio-economic activities diversify and become more interdependent. What comes as a surprise is that as city size increases, several quantities increase by small factor more than linear (*superlinear scaling*) [95, 96]. These relations were tested to be robust across a variety of urban measures, e.g., crime [97]. A theoretical framework was developed [98] that derived the general properties of cities through the optimization of a set of local conditions, and was used to predict the average social, spatial and infrastructural properties of cities in a unified and quantitative way. A set of scaling relations were found that apply to all urban systems, supported by empirical data.

6.3 Firms

Sizes of employment firms in terms of employees are known to follow the Zipf law [99], in a rather robust fashion compared to city sizes. Figure 2 (right panel) shows the typical size distribution features in case of firms in USA and UK.

However, recent studies find that while the Zipf law holds good for the intermediate range, if one considers the firm assets, the power law which is usually attributed to a mechanism of proportional random growth, based on a regime of constant returns to scale. However the power law breaks down for the largest firms, which are mostly financial firms [100]. This deviation from the expected size, à la Zipf law, has been attributed to the *shadow banking system*, which can broadly be described as credit intermediation involving entities and activities outside the regular banking system. The identification of the correlation of the size of the shadow banking system with the performance of the financial markets through crashes and booms sheds some light on the reason for the financial crises that have struck the world in the recent times.

7 Consensus

Consensus is what many people say in chorus but do not believe as individuals.

– Abba Eban, Israeli politician (1915–2002)

Consensus in social systems is a very interesting subject in terms of it dynamics, as well as concerning conditions under which it can happen. Consensus is one of the most important aspects of social dynamics. Life presents many situations where it requires to assess shared decisions. Agreement leads to stronger position, giving it a chance to have an impact on society. The dynamics of agreement and disagreement among a collection of social being is complex. Statistical physicists working on opinion dynamics tend to define the opinion states of a population and the dynamics that determine the transitions between such states. In a typical scenario, one has several opinions (discrete or continuous) and one studies how global consensus (i.e., agreement of opinions) emerges out of individual opinions [101–103]. Apart from the dynamics, the interest lies in the distinct steady state properties: a phase characterized by individuals widely varying in opinions and another phase where the majority of individuals have similar opinions.

7.1 Voting

The most common example of consensus formation in societies are in the process of voting. Elections are among the largest social phenomena, and has been well studied over years. There are many studies concerning turnout rates, detection of election anomalies, polarization and tactical voting, relation between party size and temporal correlations, relation between number of candidates and number of voters, emergence of third parties in bipartisan systems etc. Since the electoral system varies from country to country, there exists debates on the issue of which of the systems are more effective in capturing the view of the society. One of the many interesting questions one can ask is whether the distribution of votes is universal in a particular type of voting system. Recently it was shown that for open list proportional elections, the probability distribution of a performance of a candidate with respect to its own party list follows an universal lognormal distribution [24], later confirmed by an extensive analysis of 53 data sets across 15 countries [25]. While the countries using open list proportional system do show universality, the rest follow different distributions showing little commonality within their patterns. In Fig. 3 (top panel) we show the data for countries which follow the lognormal pattern. The easiest way to see if distributions are similar, is to perform the Kolmogorov-Smirnov (K-S) test, where one can compute a distance measure to express how dissimilar a pair of probability distributions are. Figure 3 (bottom panel) shows these distances. The countries at the left bottom corner, i.e., Italy, Finland, Poland, Denmark, Estonia I (elections held after 2002) are similar and their performance distributions known to be lognormal.

Data from other countries are less similar to this set, and hence the computed K-S distances are much larger.

In *closed list* proportional system, votes are given to lists rather than candidates. Usually, the voting mechanism is complicated, with specific rules adopted by specific countries. What one can measure here is the fraction of votes v given to a list. In Fig. 3 (Bottom Left), we show the rescaled (with average votes $\langle v \rangle$) plots for one election each of Spain, Sri Lanka and Turkey. What is interesting to observe is that the basic nature of the curves are similar, the range of the abscissa being dependent on the size of the constituencies. For a considerable range, the curves seem to follow a power law, approximately $v^{-1.5}$.

The majority rule or the *first-past-the-post* (FPTP) is a simple plurality voting system where the candidate with the maximum votes is declared the winner, and hence referred to as the *winner takes all* system. Usually, several political parties compete in a constituency, each nominating one candidate for the election. Hence, in practice, the list of candidates in a constituency consist of one each from all political parties contesting against one another. A voter casts a single vote in favor of the candidate of his/her choice. Under the FPTP system, the candidate with the maximum vote wins and is elected to the parliament. Figure 3 (Bottom right) shows the distribution $P(v)$ of the fraction of votes v received by a candidate in its constituency [104]. The curves are similar, with a log-normal peak in the bulk, indicative of an effective random multiplicative/branching process, and a second 'leader-board' peak around $v^* \approx 1/2$, indicative of the effective competition in a FPTP system. The gross feature of FPTP elections are similar except that the size and position of the leader-board peak indicated the effective competition in the political scenario.

Mathematical modelling would gain immensely from the information obtained from the empirical results. Having established the regularities for different election scenarios, one of the future challenges will be to develop statistical methodologies which can detect/indicate malpractices or even frauds in elections.

7.2 Religion

A very debatable question is whether religion can be treated as a process of consensus formation? Viewing religious ideologies as opinion states, one can easily visualize the dynamics of religion—as in conversion, growth and decay in population due to change in the level of popularity, paradigm shift in beliefs, etc. Figure 4 shows the rank plot of major religious affiliations. There has already been studies [106] relates to the size distribution of religious groups. The issue of inequality rises from the fact that religious representations are very much inhomogeneous across countries, the usual trend being the prevalence of one dominating religion or religious ideology. The cultural aspects, which are highly correlated with religious ideology shapes the social conditions and level of integration between various groups.

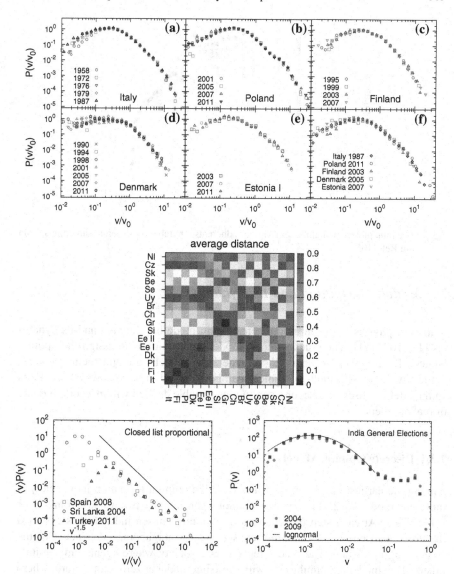

Fig. 3 *Top* Distribution of electoral performance of candidates in proportional elections with open lists. *Middle* Quantitative assessment of the similarity between distributions between countries—The average K-S distance between distributions of electoral performance of candidates. Figures are adapted from Ref. [25]. *Bottom Left* Probability distribution $P(v)$ of fraction of votes v won by lists, rescaled by the average $\langle v \rangle$ for Closed List Proportional voting systems. The *solid line* indicates a power law decay of $v^{-1.5}$. *Bottom Right* Probability distribution $P(v)$ of fraction of votes v won by candidates in parliamentary elections of India, which follow majority rule voting

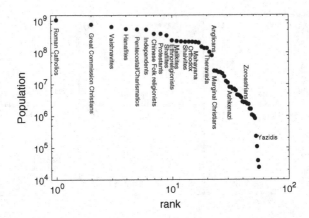

Fig. 4 The rank plot of the number of religious adherents—numbers of different sub groups. Data taken from Ref. [105]

7.3 Modelling Opinion and Its Dynamics

Statistical physics has been used extensively in modelling opinion and its dynamics [17, 107–111]. The crucial step in modelling opinion is to assign the opinion 'states'. In situations of binary choice, the Ising model paradigm seems to work, while situations with multiple choices can be modeled in various ways, e.g. Potts spin model for discrete case, or a continuous variable defined within a certain range of real numbers.

7.3.1 Discrete Opinion Models

Among the earliest models proposed to describe opinion formation in a society is the voter model [112, 113], where opinions can assume two values, 0 and 1 or -1 and $+1$. At each step of dynamics a randomly chosen individual is assigned the opinion of one of its randomly chosen nearest neighbors, independent of the original opinion state. The model represents a society where agents only imitate others. This model has similarities with the Ising model at zero temperature, where the states of the spins evolve depending to the states of the neighbors. But, since a random neighbor is chosen, the probability that any particular opinion is chosen is proportional to the number of neighbors having that opinion state, which is different from the Ising model, in which the probabilities depend exponentially on that number. For dimension $d > 2$, no consensus can be reached for infinite systems, while in finite lattices consensus is asymptotically reached. However, for $d \leq 2$, it is possible to reach a consensus.

The Sznajd model [114] was motivated by the phenomenon of social validation, i.e., agents are influenced by groups rather than individuals. In a one-dimensional lattice, a pair of neighboring spins influence the opinion of their other neighbors in a particular way. In its original version, the rules of dynamic evolution were defined as:

- if $s_i s_{i+1} = 1$, then s_{i-1}, s_{i+2} assumes the opinion of s_i
- if $s_i s_{i+1} = -1$, then s_{i-1} assumes the opinion of s_{i+1}, s_{i+2} the opinion of s_i.

In the majority-rule model [115] a group of size R is randomly chosen and the opinion state which is the majority in that group is assigned to all these R individuals, R being a random variable. One of the 'favored' opinions is adopted when there is a tie. Depending on the initial configuration, the final state is $+1$ or -1 for all opinions. Let the favored opinion be $+1$. It was shown that if one starts initially with p fraction of opinions equal to $+1$, then a phase transition occurs at p_c—for $p > p_c$, the final state is all $+1$. p_c is dependent on the maximum size of a group. It was shown that p_c could be less than $1/2$, which means that an initial minority can win in the end. The bias is possibly the reason for this, and was termed 'minority spreading'. However, if only odd-sized groups are allowed, $p_c = 1/2$. This model framework has been used with minor modifications to explain hierarchical voting, multi-species competition, etc. (see Ref. [116] for details). The problem with constant (odd) R was solved in the mean-field limit [117], where the group can be formed with randomly selected agents. The consensus time was found to be exactly $\ln N$ for N agents. In $d = 1$, where the opinions are not conserved, it was reported that the density of domains decay according to $t^{-1/2}$ so that the time to consensus is $\mathcal{O}(N^2)$. In $d > 1$ there is a broad distribution of the consensus times in higher dimensions and the most probable time to reach consensus shows a power-law dependence on N with an exponent which depends on d. The upper critical dimension is claimed to be greater than 4 and in dimensions greater than one there are always two characteristic timescales present in the system.

7.3.2 Continuous Opinion Models

One of the important models using opinion as a continuous variable is the Deffuant model, uses the idea of 'bounded-confidence' [102]. In a model where a pair of individuals simultaneously change their opinions, it was assumed that two agents interact only if their opinions are close enough, as in people sharing closer points of view would interact more. Let $o_i(t)$ be the opinion of the individual i at time t. An interaction of the agents i and j agents, who are selected randomly, would make the opinions update according to

$$o_i(t+1) = o_i(t) + \mu[o_j(t) - o_i(t)], \qquad (25)$$
$$o_j(t+1) = o_j(t) + \mu[o_i(t) - o_j(t)],$$

where μ is a constant ($0 \le \mu \le 0.5$), known as the 'convergence parameter'. The total opinion remains constant and bounded, i.e., o_i lies in the interval $[0, 1]$. There is no randomness in the original model except the random choice of the agents who interact. Obviously, if $o_j > o_i$, o_i becomes closer to o_j. If δ is the 'confidence level' (agents interact when their opinions differ by a quantity not more than δ), the final distribution of opinions is dependent on the values of δ and μ. The possibilities are: (i) all agents may attain a unique opinion value, which is a case of consensus. It has been reported that the threshold value of δ above which all agents have the same opinion in the end is $1/2$, and is independent of the underlying topology [118]. (ii) Convergence is reached with only two opinions surviving, a case of 'polarization'. (iii) It can happen that the final state has several possible opinions: a state of 'fragmentation'.

Hegselmann and Krause [103] generalized the Deffuant model, where an agent simultaneously interacts with all other agents who have opinions within a prescribed bound. Instead of the convergence parameter, the simple average of the opinions of the appropriate neighbors is adopted. As a result, consensus is enhanced by making the bound larger. In case of a graph, the value of the consensus threshold is seen to depend on the average degree. In these models, starting from continuous values, the final opinions are discretized.

Another class of models employ a kinetic exchange like framework [119]. Including a diffusion term that takes care of the fact that agents can be randomly affected by external factors and subsequently change their opinion. The opinions o_i and o_j evolve following

$$o_i(t+1) = o_i(t) + \gamma \mathscr{P}(|o_i|)[(o_j(t) - o_i(t)) + \eta_i \mathscr{D}(|o_i|)], \qquad (26)$$
$$o_j(t+1) = o_j(t) + \gamma \mathscr{P}(|o_j|)[(o_j(t) - o_i(t)) + \eta_j \mathscr{D}(|o_j|)],$$

where $\gamma \in [0, 1/2]$ represents compromise propensity and η is drawn from a random distribution with zero mean. As in the bounded-confidence models, the opinion of an agent will tend to decrease or increase so as to be closer to the opinion of the other. it is important to note that even if the diffusion term is absent, the total opinion is not conserved unless $P(o)$ is a constant. The functions $\mathscr{P}(.)$ and $\mathscr{D}(.)$ take care of the local relevance of the compromise and diffusion terms respectively. The choice of the functional forms of $\mathscr{P}$ and $\mathscr{D}$ decides the final state.

Another simpler model [120] considers binary interactions, where the opinions evolve as:

$$o_i(t+1) = \lambda[o_i(t) + \varepsilon o_j(t)], \qquad (27)$$
$$o_j(t+1) = \lambda[o_j(t) + \varepsilon' o_i(t)],$$

where $\varepsilon, \varepsilon'$ are drawn randomly from uniform distributions in $[0, 1]$. $o_i(t)$ is the opinion of individual i at time t, and $o_i \in [-1, +1]$, and are also bounded, i.e.,. if o_i exceeds 1 or becomes less than -1, it is set to 1 and -1, respectively. The parameter λ is interpreted as 'conviction', and this models a society in where everyone has the same value of conviction. The model lacks conservation laws. The order parameter is defined as $O = |\sum_i o_i|/N$, the magnitude of the average opinion in a system with N agents. One also measures the fraction of the agents having $o_i = \pm 1$, called the condensation fraction p Numerical simulations indicate that the system goes into either of the two possible phases: for any $\lambda \leq \lambda_c$, $o_i = 0$ $\forall i$, while for $\lambda > \lambda_c$,

$O > 0$ and $O \to 1$ as $\lambda \to 1$ with $\lambda_c \simeq 2/3$, λ_c is the 'critical point' of this phase transition. The critical point is easily confirmed by mean field calculation [121]. Using a fixed point equation $o^*[1 - \lambda(1 + \langle \varepsilon \rangle)] = 0$, the fixed point turns out to be $\lambda_c = 1/(1 + \langle \varepsilon \rangle)$. For a random uniform distributed ε with $\langle \varepsilon \rangle = 1/2$, one can easily see, $\lambda_c = 2/3$.

Several variants including the discrete version [122], and the generalized version with a second parameter 'influence' [123] gave further insights to the class of models. In another model [124], where negative influence was also considered, one could consider both discrete and continuous versions, which gave interesting insight into competing views and phase transitions.

All of the above models, which show a phase transition, can be viewed to correspond to broad distributions of opinions/consensus variables near their critical point. It is much easier to visualize a percolation [125] picture, where the sizes of percolation clusters have a power-law distribution close to the percolation phase transition. The alternative picture is that of self-organized criticality [31], where the activity size is also power law distributed.

8 Bibliometrics

Academic publications (papers, books etc.) form an unique social system consisting of individual publications as entities, containing bibliographic reference to other older publications, and this is commonly termed as *citation*. The number of citations is a measure of the importance of a publication, and serve as a proxy for the popularity and quality of a publication. There has already been a plethora of empirical studies on citation data [21], specifically on citation distributions [126–129] of articles, time evolution of probability distribution of citation [130–132], citations for individuals [133] and even their dynamics [134], and the modelling efforts on the growth and structure of citation networks have produced a huge body literature in network science concerning scale-free networks [39, 40, 135].

The bibliometric tool of citation analysis is becoming increasingly popular for evaluating the performance of individuals, research groups, institutions as well as countries, the outcomes of which are becoming important in case of offering grants and awards, academic promotions and ranking, as well as jobs in academia, industry and otherwise. Since citations serve as a crucial measure for the importance and impact of a research publication, its precise analysis is extremely important. It is quite usual to find that some publications do better than others due to the inherent heterogeneity in the quality of their content, the gross attention on the field of research, the relevance to future work and so on. Thus different publications gather citations in time at different rates and result in a broad distribution of citations. For individual publications, the present consensus is that while most part of the distribution fit to a lognormal [126, 136], the extreme tail fits to a power law close to 3 [128, 137]. This inequality in citations and other associated bibliometric indicators has been a field of interest to scientists in the last few decades.

There has also been studies that address the most popular papers, most cited scientists, and even Nobel prizes, which are awarded for groundbreaking discoveries, which change the face of science for ever. The laureates belong to the elite, and the papers which are identified to be the ones declaring the 'discovery' are timeless classics by their own right. Speaking in terms of *inequality*, these discoveries are indeed rare, and belong to the extreme end of the spectrum of all the associated work done in a field, some leading to the discoveries themselves, others complementing them, naturally true for the Nobel prize winning papers. However, there are limitations to the prize itself, defined by the number of discoveries that can be recognized and number of recipients in a single year in a given discipline. It has been observed that the delay between the discovery and the recognition as a Nobel prize is growing exponentially [138, 139], the most rapidly in Physics, followed by Chemistry and Physiology or Medicine. As a result the age of a laureate at the time of the award is also increasing, seen to be exponential. Comparison with the average life expectancy of individuals concluded that by the end of this century, potential Nobel laureates in physics and chemistry are most likely to expire before being awarded and recognized.

In recent times, big team projects have dominated some of the frontline areas of science, in astrophysics, biology, genetics, quantum information etc. One can easily see that there has been a gross inequality in the number of researchers who coauthor a paper [140].

In what follows, we will limit our discussion to some interesting aspects related to analysis of citations.

8.1 Annual Citation Indices

One of the quantities of major interest is the nature of the tail of the distribution of annual citations (AC) and impact factor (IF). IF are calculated annually for journals which are indexed in the Journal Citation Reports [141]. The precise definition is the following: if papers published in a journal in years $T - 2$ and $T - 1$ are cited $\mathcal{N}(T-2) + \mathcal{N}(T-1)$ times by the indexed journals in the year T, and $N(T-2) + N(T-1)$ are the number of 'citable' articles published in these years, the impact factor in year T is defined as

$$I(T) = \frac{\mathcal{N}(T-2) + \mathcal{N}(T-1)}{N(T-2) + N(T-1)}. \tag{28}$$

The number of annual citations (AC) $n(T)$ to a journal in a given year is

$$n(T) = \sum_{t \leq T} \sum_{i} \mathcal{A}_i(t, T), \tag{29}$$

where $\mathcal{A}_i(t, T)$ is the number of citations received in the year T by the ith paper published in the year $t \leq T$.

Another measure, $r(T)$, the *annual citation rate* (CR) at a particular year T that is defined [142] as annual citations divided by the number of articles published in the same year, i.e.,

$$r(T) = n(T)/N(T). \tag{30}$$

Extensive studies on the historical behavior of the IF ranked distribution [143, 144] have established the behavior of the low ranked (large k) journals and the precise nature of the distribution function. Study of a limited sample of the top 1,000 ranked ($k \leq 1{,}000$) Science journals (SCI) reveal that the distributions remain invariant with time, seen by rescaling the plots by their averages. For small ranks, the citations are almost independent of the ranks implying a cluster of journals with comparable citations that occupy the top positions (Fig. 5a, b). Fitting the curves for $k > 10$ by $f(x) \sim x^{-b_n}$ it was found that $b_n = 0.70(2)$. Similarly, for the IF, the scaled data seemed to fit to the same form with an exponent $b_I = 0.54(1)$ (Fig. 5d, e). These Zipf exponents as they are obtained from the rank plots. For Social Science (SOCSCI), approximate power law fits are possible for the rank plots with Zipf exponents $b_n = 0.70(2)$ and $b_I = 0.40(1)$ respectively (Fig. 5f, h).

The single exponent fitting on the tail does not justify the observed bending of the real data (Fig. 5b, e, h, k), but the curves can be as well be fitted nicely to a function with two exponents $f(x) = k(N + 1 - x)^a/x^b$, where N represents the rank-order data; a and b are two exponents to fit [144]. For the citation data, $b_n = 0.69(2)$ and $a_n = 0.27(2)$ (Fig. 5c). Similarly, for IF data the exponents are $b_I = 0.56(2)$ and $a_I = 0.03(2)$ (Fig. 5f). For SOCSCI data the exponents are $b_n = 0.61(2)$, $a_n = 0.47(1)$, $b_I = 0.39(2)$, and $a_I = 0.06(1)$ (Fig. 5i, l). a_I is close to zero for (Fig. 5f, l) implying that the power law fit is rather accurate, while for (Fig. 5c, i), the two exponents fit appears to be more appropriate.

The probability distribution of both annual citations and impact factors showed monotonic decays and their tails can be fitted to power law forms. The plots showed excellent scaling collapse for different years when the probability distributions are rescaled by their averages. For annual citations, the lower values hint towards a lognormal but the 'tail' of the distribution seems to fit to a power law [145]. The power law exponents are γ_n and γ_I are 2.52(1) and 3.16(1) respectively. The Zipf exponent b as obtained from a rank plot relates to the exponent of the probability distribution γ as $\gamma = 1 + 1/b$ [146]. Using the values of b obtained, the values of γ are found to be 2.42 and 2.85 respectively for AC and IF, quite consistent with the values obtained directly from the distributions (Fig. 6). The same holds true for the SOCSCI data (expected Pareto exponents are 2.42 and 3.50 from the rank plots compared to the best fit values 2.32(2) and 3.13(2)). However, it is apparent that there could be some corrections to the power law scaling. The distribution of the annual citation rate (CR) r is also broad, but non-monotonic, with a peak around half the average value. The distributions fit well to log-Gumbel distributions [142].

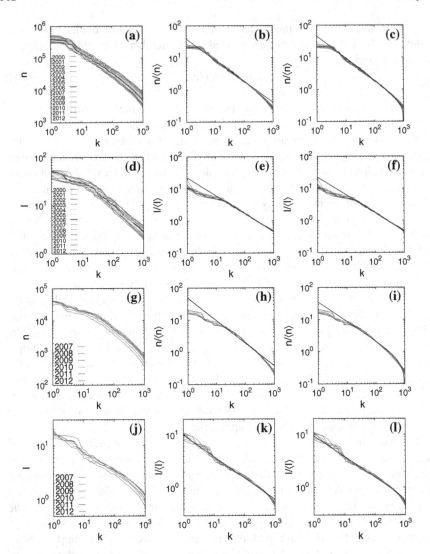

Fig. 5 Plots of annual citation n and impact factor I with rank k: **a** Rank plot of the top 1,000 journals, ranked according to citations, scaling collapse of the same, **b** showing a Zipf law fit: $f(x) = Ax^{-b_n}$, with $b_n = 0.70(2)$ and **c** showing a two-exponent Zipf law fit: $f(x) = k(N+1-x)^{a_n}/x^{b_n}$, with $a_n = 0.27(2)$ and $b_n = 0.69(2)$. **d** rank plot of the top 1,000 journals, ranked according to impact factors, and scaling collapse of the same **e** showing a Zipf law fit: $g(x) = Ax^{-b_I}$, with $b_I = 0.54(1)$ and **f** showing a two-exponent Zipf law fit: $f(x) = k(N+1-x)^{a_I}/x^{b_I}$, with $a_I = 0.03(2)$ and $b_I = 0.56(2)$. The data is from SCI sets. SOCSCI data: **g** Rank plot of the top 1,000 journals, ranked according to citations, scaling collapse of the same, **h** showing a Zipf law fit: $f(x) = Ax^{-b_n}$, with $b_n = 0.7043 \pm 0.001$ and **i** showing a two-exponent Zipf law fit: $f(x) = k(N+1-x)^{a_n}/x^{b_n}$, with $a_n = 0.47(2)$ and $b_n = 0.61(2)$. **j** Rank plot of the top 1,000 journals, ranked according to their impact factors, scaling collapse of the same, **k** showing a Zipf law fit: $f(x) = Ax^{-b_I}$, with $b_I = 0.40(1)$ and **l** showing a two-exponent Zipf law fit: $f(x) = k(N+1-x)^{a_I}/x^{b_I}$, with $a_I = 0.06(2)$ and $b_I = 0.39(2)$. The figure is adapted from Ref. [142]

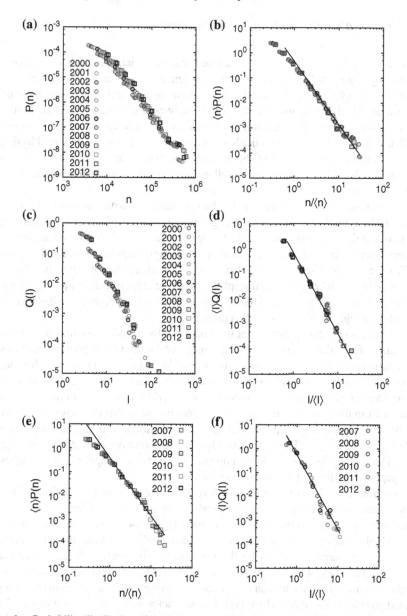

Fig. 6 **a** Probability distribution of annual citations $P(n)$, **b** its scaling collapse, for the top 1,000 journals ranked according to citations. It fits fairly well to a power law $Bx^{-\gamma_n}$; the *straight line* corresponds to $\gamma_n = 2.52$ in the log-log plot; **c** probability distribution of impact factor $Q(I)$, **d** its scaling collapse, for the top 1,000 journals ranked according to impact factor. The *straight line* corresponds to $\gamma_I = 3.16$. The data are for SCI sets. **e** Scaling collapse of probability distribution of annual citations $P(n)$ for the top 1,000 journals ranked according to citations; power law fit with $\gamma_n = 2.42$; **f** Scaling collapse of probability distribution of annual citations $P(n)$ for the top 1,000 journals ranked according to impact factor; power law fit with $\gamma_I = 3.50$. The data are from SOCSCI sets. The figure is adapted from Ref. [142]

8.2 Universality in Citation Distributions

Back in 1957, Shockley [126] first claimed that the scientific publication rate is dictated by lognormal distribution. Later evidences based on analysis of records for highly cited scientists indicate that the citation distribution of individual authors follow a stretched exponential [127]. Another analysis of the data from ISI claims that the tail of the citation distribution has a power law decay with an exponent close to 3 [128]. What followed was a meticulous analysis of 110 years of data from Physical Review, concluding that most of the citation distribution fits remarkably well to a lognormal [136]. However, it is now well agreed that most of the distribution fits to a lognormal but the tail fits to a power law [137].

The distribution of citations of individual publications within a single discipline is seen to be quite broad—some papers get very little citations, a few collect huge citations, but there are many who collect an average number of citations. The average number of citations gathered in a discipline is strongly dependent on the discipline itself. It has been observed [129] that if one can rescale the absolute number of citations c by the average in that discipline $\langle c \rangle$, the relative indicator $c_f = c/\langle c \rangle$ has a functional form independent of the discipline. The rescaled probability distribution fits well to a lognormal for most of its range.

One can also ask the question that what happens for academic institutions, where the quality of scientific output measured in terms of the total number of publications, total citations etc. can vary widely across the world. In the popular notion, there are institutions which are known to 'better' than others, and in fact, rankings go exist among them. One can ask if the nature of the world's best institutions' output is different from the more average ones. It is found that all institutions have a wide distribution of citations to their publications, with a functional form that is independent of their rankings [147]. To see this, one has to use a relative indicator $c_f = c/\langle c \rangle$ again. The scaling function fits to a lognormal for most of its range, while the highest cited papers deviate to fit to a power law tail.

Now, how does the same look at the level of journals? It is quite well known in popular perception that journals present a wide variety in terms of impact factor and annual citations, as also reported in critical studies [142–144]. The natural question to ask is whether the universality that exists across disciplines or academic institutions prevail also for journals. It seems that this is not the case in reality. In fact, it is seen that there are at least two classes of journals, a *General* class, which comprises the most popular standard journals, and an *Elite* class consisting a very small group of highly reputed journals characterized by high impact factor and average citations [147]. The former class is characterized by a lognormal distribution in the bulk, which the latter is characterized by a strong tendency of divergence as $1/c$ at small values of citations, with no clear indication of lognormal bulk. However, for both these classes, the distribution of the highest cited papers decay a power law tail with an exponent close to 3.

9 Networks

Networks have been a subject of intense study in the last 2 decades, and drawn researchers from a variety of disciplines to study the structural, functional, dynamical and various aspects of them, uncovering new patterns and understanding interesting phenomena from physics, biology, computer science and social sciences [148]. The studies of the internet and the web has led to engineering faster routing strategies and developing efficient search engines, biological networks have given insight into the functional elements in bio-chemical processes inside organisms, provided tools for complex experiments, that in social networks has bettered our understanding about spreading of innovations, rumors and even epidemics, leading to devising strategies to make them either more efficient or less efficient according to the case it may be.

Studies of the World Wide Web [149, 150], the internet [151], citation networks [39, 128], email network [152], network of human sexual contacts [153] all show power law tail in the degree distributions for a large range of values, similar to metabolic networks [154] and protein-protein interaction network [155].

Studies on massive and popular social networks like Facebook [156] has shown broad distributions with tails resembling power laws. Experiments done by setting up social networks of communication have shown that the network slowly evolves and the shape of the degree distribution stabilizes with time [157], with characteristics similar to other massive networks, such as clustering and broad degree distribution with power law tail. Since then, almost all possible online social networks have been analyzed, and most of them have been found to have a power law degree distribution (see e.g., Ref. [158]).

10 Measuring Inequality

Socio-economic inequalities can be quantified in several ways. The most common measures are absolute, in terms of indices, e.g., Gini [51], Theil [159] indices. Alternatively one can use a relative measure, by considering the probability distributions of various quantities, whereas the indices can be computed easily from the distributions themselves. The quantity in question in the socio-economic context usually displays a broad distribution, like lognormals, power-laws or their combinations. For instance, the distribution of income is usually an exponential followed by a power law [160] (see Ref. [16] for other examples).

The Lorenz curve [161] is a function representing the cumulative proportion X of (ordered from lowest to highest) individuals in terms of the cumulative proportion of their sizes Y. X can be anything like income or wealth, citation, votes, city population etc. The Gini index (g) is commonly defined as the ratio of (i) the area enclosed between the equality line and the Lorenz curve, and (ii) the area below the equality line. If the area between (i) the Lorenz curve and the equality line is given by A, and that (ii) under the Lorenz curve is given by B (See Fig. 7), the Gini index is simply

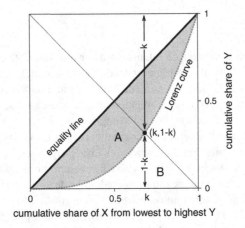

Fig. 7 Schematic representation of Lorenz curve, Gini index g and k index. The *dashed red line* stands for the Lorenz curve and the *black solid line* represents perfect equality. The area enclosed by the equality line and the Lorenz curve is A and that below the Lorenz curve is B. The Gini index is given by $g = A/(A + B)$. The k index is given by the abscissa of the intersection point of the Lorenz curve and $Y = 1 - X$

defined as $g = A/(A + B)$. It is a very common measure to quantify socio-economic inequalities. Ghosh et al. [162] recently introduced a new measure, called 'k index' (where 'k' stands for the extreme socio-economic inequalities in Kolkata), defined as the fraction k such that $(1 - k)$ fraction of people or papers possess k fraction of income/wealth or citations respectively.

When the probability distribution is described by an appropriate parametric function, one can derive, using mathematical techniques, these inequality measures as a function of those parameters analytically. Several empirical evidence show that the distributions can be put into a few types, most of which turn out to be a mixture of two distinct parametric functions with a single crossover point [163].

$$P(m) = F_1(m)\theta(m, m_\times) + F_2(m)\Theta(m - m_\times), \tag{31}$$

where $\theta(m, m_\times) \equiv \Theta(m) - \Theta(m - m_\times)$, $m_\times$ is the crossover point and F_1, F_2 being the two functions. For these, it is possible to compute the general form of the Lorenz curve, Gini index g and k index. It is also easy to check the values obtained from empirical calculations with those from analytical expressions for consistency. By minimizing the gap between the empirical and analytical values of the inequality measures, one can numerically enumerate of the crossover point which is usually determined by eye estimates.

11 How to Deal with Inequality?

So distribution should undo excess, and each man have enough.

[King Lear, Act 4, Scene 1]—William Shakespeare, English playright and writer

(1564–1616)

In this section, we will not discuss socio-political theories such as *socialism* in it different forms, and rather give an objective view of generalized processes that can be adopted to decrease inequalities.

One of the long studied problems that deals with the issue of socio-economic inequality is that of efficient allocation of resources. An inefficient allocation process may lead to unequal distribution of resources which in course of time goes through a reshuffling and multiplicative process, and finally result in a very skew distribution. We will discuss in the following, how statistical physics can help in modelling processes that produce fairly equal distributions.

Most resources are limited in supply, and hence, efficiently allocating them is therefore of great practical importance in many fields. Resources may be either physical (oil, CPUs, chocolates) or immaterial (time, energy, bandwidth), and allocation may happen instantaneously or over a long period. Thus, there is no universal method that solves all allocation problems. In addition, the variety of situations that people or machines (generically called "agents") face, require specific modelling in a first step. However, two features connect all these situations: when resources are scarce, the agents compete for them; when competing repeatedly, the agents learn and become adaptive. Competition in turn has two notable consequences. The agents have a strong incentive to think and act as differently as possible [164], that is, to become heterogeneous. Additionally, competition implies interaction, because the share of resources that an agent obtains depends on the actions of other agents.

While multi-agent modelling in the context of Game theory helps, even simpler modelling using the framework of statistical physics [165] helps in a deeper understanding of the problem using a framework which is very different from those traditionally used by mainstream social scientists. While some models have already been proposed and efficiently utilized in certain fields, there is further need of developing novel and efficient models for a variety of problems.

The optimal allocation of resources is an issue of utmost concern in economic systems. Formalized as the simultaneous maximization of the utility of each member of the economy over the set of achievable allocations, the main issue is that individuals have typically conflicting goals, as the profit of one goes against that of the others. This makes the nature of the problem conceptually different from optimization problems where usually a single objective function has to be maximized. Markets, under some conditions, can solve this problem efficiently—prices adjust in a self-organized manner to reflect the true value of the good.

There have been recent attempts to model and describe resource allocation problems in a collection of individuals within a statistical mechanics approach. The focus is on competitive resource allocation models where the decision process is fully decentralized, that is, communication between the agents is not explicit. Interaction plays a crucial role and give rise to collective phenomena like broad distribution of fluctuations, long memory and even phase transitions. Additionally, the agents usually act very differently from one another, and sometimes have the ability to change actions according to the need their goals. This implies strong heterogeneity and non-equilibrium dynamics, ingredients which are extremely appealing to physicists, who possess the tools and concepts that are able to analyze and possibly solve the dynamics of such systems. The essential difference lies in the behavior of the constituent units, while particles, electrons cannot think and act, social units sometime do, which adds to the complexity.

Let us consider a population of N agents and R resources, which they try to exploit. Generically, R denotes the number of possible choices of the agents, which naturally means $R \geq 2$. For the simplest case $R = 2$, agents must choose which resource to exploit. The El Farol Bar Problem (EFBP) [164]: N customers compete for $L < N$ seats at the bar. At every time step, they must choose whether to go to the bar or to stay at home. The Minority Game (MG) [166] simplifies the EFBP in many respects by taking $L = N/2$. The Kolkata Paise Restaurant problem (KPR) assumes that the number of resources scales with N, in which R restaurants have a capacity to serve only one customer each, so that the agents try to remain alone as often as possible [167].

The KPR [167, 168] problem is similar to various adaptive games (see [169]) but uses one of the simplest scenario to model resource utilization in a toy model paradigm. The simplest version has N agents (customers) simultaneously choosing equal number $R (= N)$ of restaurants, each of which serve only one meal every evening, and hence, showing up in a place with more people gives less chance of getting food. *Utilization* is measured by the fraction of agents f getting a meal or equivalently, by measuring the complimentary quantity: the fraction $(1 - f)$ of meals wasted, because some of the restaurants do not get any customer at all. An entirely random occupancy algorithm sets a benchmark of $f = 1 - 1/e \approx 0.63$. However, a crowd-avoiding algorithm [168] can improve the utilization to around 0.8. If ones varies the ratio of agents to restaurants (N/R) below unity, one can observe a phase transition—from an 'active phase' characterized by a finite fraction of restaurants ρ_a with more than one agent, and an 'absorbed phase', where this quantity ρ_a vanishes [170]. Adapting the same crowd avoiding strategy in a version of the Minority Game where the extra information about the crowd was provided, one could convergence to the steady state achieve in a very small time, $O(\log \log N)$ [171]. In another modification to this problem [172], one could observe a phase transition depending on the amount of information that is shared. The main idea for the above studies was to use iterative learning to find simple algorithms that could achieve a state of maximum utilization in a very small time scale. Although this review touches upon topics seemingly discrete from one another, some of the processes are interlinked, e.g., resource utilization can be seen as a key ingredient to city growth and the broad distribution of city sizes [173].

12 Discussions

Socio-economic inequalities have been around, manifested in several forms, since history. With time, the nature and extent of these have changed, sometimes for good, but mostly for worse. This is traditionally a subject of study of social sciences, and scholars have been looking upon the causes and effects from a sociological perspective, and trying to understand the consequences on the economics. In reality this is not so simple, at times the latter is what is responsible for the effect on the former, making the cause-effect interpretation much more complex.

At the opposite end, had the world been very equal, it would have been difficult to compare the extremes, separate the good from the bad, hardly any leaders people would look up to, lack stable governments if there were almost equal number of political rivals, etc.

Recently, there has been a lot of concern regarding the increase of inequality in income and wealth, as seen from different measurements [78], and has in such, renewed the interest on this subject among the leading social scientists across the globe. However, energy use has been observed to become much more equal with time [76]. In the economics front, how inequalities affect financial markets, firms and their dynamics, and vice versa is an important area to look into. Objects which are directly affected by economy, industrialization and rapidly growing technologies, in terms of social organization of individual entities, as in urban systems, are becoming important areas of study. In climbing up and down the social ladder [174, 175] is something difficult to track, until recent surveys which provide some insight into the dynamics. Many deeper and important issues in society [11] still needs attention in terms of inequality research, available data and its further analysis can bring out hidden patterns which may be used to encounter those situations. The main handicap is the lack of data of good quality, and the abstractness of the issues, which may not always be easily amenable to statistical modelling.

Physicists' interests are mostly concentrated on subjects which are amenable to macroscopic or microscopic modelling, where tools of statistical physics prove useful in explaining the emergence of broad distributions. A huge body of literature has been already developed, containing serious attempts to understand socio-economic phenomena, branded under *Econophysics* and *Sociophysics* [176]. The physics perspective brings new ideas and an alternative outlook to the traditional approach taken by social scientists, and is seen in the increasing collaborations across disciplines [1].

Acknowledgments This mini review is a starting point of a larger material to be written up. Discussions with V.M. Yakovenko were extremely useful in planning the contents. The author thanks S. Biswas, A.S. Chakrabarti and B.K. Chakrabarti for comments, and acknowledges collaborations with F. Becattini, S. Biswas, A.S. Chakrabarti, B.K. Chakrabarti,, A. Chakraborti, S.R. Chakravarty, D. Challet, A.K. Chandra, D. DeMartino, S. Fortunato, A. Ghosh, J-I. Inoue, A. Khaleque, S.S. Manna, M. Marsili, M. Mitra, M. Mitrović, T. Naskar, R.K. Pan, P.D.B. Parolo, P. Sen, S. Sinha, Y-C. Zhang on various projects.

References

1. D. Lazer, A. Pentland, L. Adamic, S. Aral, A.L. Barabási, D. Brewer, N. Christakis, N. Contractor, J. Fowler, M. Gutmann, T. Jebara, G. King, M. Macy, D. Roy, M. Van Alstyne, Science **323**(5915), 721 (2009)
2. P. Ball, *Critical Mass* (Farrar, Straus and Giroux, New York, 2004)
3. L. Bachelier, *Théorie de la Spéculation* (Gauthier-Villars, 1900)
4. E. Majorana, Scientia **36**, 58 (1942)
5. L.P. Kadanoff, Simulation **16**(6), 261 (1971)
6. E.W. Montroll, W.W. Badger, *Introduction to Quantitative Aspects of Social Phenomena* (Gordon & Breach, New York, 1974)
7. P.W. Anderson, K. Arrow, D. Pines, *The Economy as an Evolving Complex System* (Addison-Wesley, Redwood City, 1988)
8. M. Buchanan, To understand finance, embrace complexity. http://www.bloomberg.com/news/2013-03-10/to-understand-finance-embrace-complexity.html (11 March 2013)
9. M. Buchanan, *The Social Atom* (Marshall Cavendish Business, London, 2007)
10. Institute for New Economic Thinking, http://ineteconomics.org/
11. K. Neckerman (ed.), *Social Inequality* (Russell Sage Foundation, New York, 2004)
12. E. Pennisi, Science **344**(6186), 824 (2014)
13. M. Buchanan, Nat. Phys. **10**(7), 471 (2014)
14. A. Chatterjee, B.K. Chakrabarti, Eur. Phys. J. B **60**, 135 (2007)
15. H. Pringle, Science **344**(6186), 822 (2014)
16. B.K. Chakrabarti, A. Chakraborti, S.R. Chakravarty, A. Chatterjee, *Econophysics of Income and Wealth Distributions* (Cambridge University Press, Cambridge, 2013)
17. C. Castellano, S. Fortunato, V. Loreto, Rev. Mod. Phys. **81**, 591 (2009)
18. D. Stauffer, S.M. De Oliveira, P.M.C. de Oliveira, J.S. Martins, *Biology, Sociology, Geology by Computational Physicists* (Elsevier, Amsterdam, 2006)
19. D. Stauffer, J. Stat. Phys. **151**(1–2), 9 (2013)
20. S. Galam, *Sociophysics: A Physicist's Modeling of Psycho-Political Phenomena* (Springer, Berlin, 2012)
21. P. Sen, B.K. Chakrabarti, *Sociophysics: An Introduction* (Oxford University Press, Oxford, 2014)
22. S. Sinha, A. Chatterjee, A. Chakraborti, B.K. Chakrabarti, *Econophysics: An Introduction* (Wiley-VCH, Weinheim, 2010)
23. H.E. Stanley, *Introduction to Phase Transitions and Critical Phenomena* (Oxford University Press, Oxford and New York, 1971)
24. S. Fortunato, C. Castellano, Phys. Rev. Lett. **99**(13), 138701 (2007)
25. A. Chatterjee, M. Mitrović, S. Fortunato, Sci. Rep. **3**, 1049 (2013)
26. M.C. Mantovani, H.V. Ribeiro, M.V. Moro, S. Picoli Jr, R.S. Mendes, Europhys. Lett. **96**(4), 48001 (2011)
27. H.D. Rozenfeld, D. Rybski, J.S. Andrade, M. Batty, H.E. Stanley, H.A. Makse, Proc. Natl. Acad. Sci. **105**(48), 18702 (2008)
28. M.H.R. Stanley, L.A.N. Amaral, S.V. Buldyrev, S. Havlin, H. Leschhorn, P. Maass, M.A. Salinger, H.E. Stanley, Nature **379**(6568), 804 (1996)
29. R.N. Mantegna, H.E. Stanley, *Introduction to Econophysics: Correlations and Complexity in Finance* (Cambridge University Press, Cambridge, 2000)
30. A.M. Petersen, J. Tenenbaum, S. Havlin, H.E. Stanley, Sci. Rep. **2**, 313 (2012)
31. P. Bak, *How Nature Works* (Oxford University Press, Oxford, 1997)
32. D. Champernowne, Econ. J. **63**, 318 (1953)
33. M. Mitzenmacher, Internet Maths. **1**(2), 226 (2004)
34. H.A. Simon, Biometrika **42**(3/4), 425–440 (1955)
35. R.K. Merton, Science **159**(3810), 56 (1968)
36. G.U. Yule, Philos. Trans. R. Soc. Lond. B **213**, 21 (1925)

37. M.E.J. Newman, Contemp. Phys. **46**(5), 323 (2005)
38. D.d.S. Price, J. Am. Soc. Inf. Sci. **27**(5), 292 (1976)
39. D.d.S. Price, Science **149**(3683), 510 (1965)
40. A.L. Barabási, R. Albert, Science **286**(5439), 509 (1999)
41. R. Albert, A.L. Barabási, Rev. Mod. Phys. **74**(1), 47 (2002)
42. S.N. Dorogovtsev, J.F.F. Mendes, Adv. Phys. **51**(4), 1079 (2002)
43. A. Sen, *Poverty and Famines* (Oxford University Press, New Delhi, 1999)
44. M. Foucault, *Le Pouvoir Psychiatrique:Cours au College de France 1973–74* (Gallimard, Paris, 2003)
45. R. Scruton, *Thinkers of the New Left* (Longman, Sussex, 1985)
46. J. Rawls, *A Theory of Justice* (Harvard University Press, Cambridge, 1971)
47. V. Pareto, *Cours d'economie Politique* (Rouge, Lausanne, 1897)
48. R. Gibrat, *Les inégalités économiques* (Sirey, Paris, 1931)
49. D. Champernowne, F. Cowell, *Economic Inequality and Income Distribution* (Cambridge University Press, Cambridge, 1998)
50. E.W. Montroll, M.F. Shlesinger, Proc. Natl. Acad. Sci. **79**, 3380 (1982)
51. C. Gini, Econ. J. **31**(121), 124 (1921)
52. R. Hogg, J. Mckean, A. Craig, *Introduction to Mathematical Statistics* (Pearson Education, Delhi, 2007)
53. A. Chatterjee, S. Yarlagadda, B.K. Chakrabarti (eds.), *Econophysics of Wealth Distributions* (New Economic Windows Series, Springer, Milan, 2005)
54. V. Yakovenko, J.B. Rosser Jr, Rev. Mod. Phys. **81**, 1703 (2009)
55. Internal Revenue Service (IRS) - Individual Tax Tables, http://www.irs.gov/uac/SOI-Tax-Stats---Individual-Statistical-Tables-by-Size-of-Adjusted-Gross-Income Retrieved June 2014
56. Forbes, World's Billionaires list (2014), http://www.forbes.com/billionaires/list/. Retrieved July 2014
57. BitcoinRichList (2014), http://bitcoinrichlist.com. Retrieved March 2014
58. International Monetary Fund (2014), http://www.imf.org/. Retrieved June 2014
59. P. Richmond, S. Hutzler, R. Coelho, P. Repetowicz, in *Econophysics and Sociophysics: Trends and Perspectives*, ed. by B.K. Chakrabarti, A. Chakraborti, A. Chatterjee (Wiley-VCH, Weinheim, 2007), pp. 131–159
60. W. Souma, in *Empirical Science of Financial Fluctuations: The Advent of Econophysics*, ed. by H. Takayasu (Springer, Tokyo, 2002), pp. 343–352
61. T. Mizuno, M. Toriyama, T. Terano, M. Takayasu, Phys. A **387**(15), 3931 (2008)
62. A. Ghosh, K. Gangopadhyay, B. Basu, Phys. A **390**(1), 83 (2011)
63. M. Levy, S. Solomon, Phys. A **242**, 90 (1997)
64. S. Solomon, P. Richmond, Eur. Phys. J. B **27**, 257 (2002)
65. P. Richmond, S. Solomon, Int. J. Mod. Phys. C **12**, 333 (2001)
66. J.P. Bouchaud, M. Mézard, Phys. A **282**, 536 (2000)
67. A. Drăgulescu, V. Yakovenko, Eur. Phys. J. B **17**, 723 (2000)
68. A. Chakraborti, B.K. Chakrabarti, Eur. Phys. J. B **17**, 167 (2000)
69. A. Chatterjee, B.K. Chakrabarti, S.S. Manna, Phys. A **335**, 155 (2004)
70. M.N. Saha, B.N. Srivastava, *A Treatise on Heat* (Indian Press, Allahabad, 1931)
71. J. Angle, Soc. Forces **65**(2), 293 (1986)
72. J. Angle, Phys. A **367**, 388 (2006)
73. A. Chatterjee, Eur. Phys. J. B **67**(4), 593 (2009)
74. L. Pareschi, G. Toscani, *Interacting Multiagent Systems: Kinetic Equations and Monte Carlo Methods* (Oxford University Press, Oxford, 2013)
75. A. Banerjee, V. Yakovenko, New J. Phys. **12**, 075032 (2010)
76. S. Lawrence, Q. Liu, V.M. Yakovenko, Entropy **15**(12), 5565 (2013)
77. A. Chatterjee, S. Sinha, B.K. Chakrabarti, Curr. Sci. **92**(10), 1383 (2007)
78. T. Piketty, *Capital in the Twenty-first Century* (Harvard University Press, Cambridge, 2014)
79. T. Piketty, E. Saez, Science **344**(6186), 838 (2014)

80. M. Batty, Science **319**(5864), 769 (2008)
81. F. Auerbach, Petermanns Geographische Mitteilungen **59**, 74 (2008)
82. A.J. Lotka, *Elements of Physical Biology* (Williams & Wilkins, Baltimore, 1925)
83. H.W. Singer, Econ. J. **46**(182), 254 (1936)
84. G.K. Zipf, *Human Behaviour and the Principle of Least-Effort* (Addison-Wesley, Cambridge, 1949)
85. K.T. Rosen, M. Resnick, J. Urban. Econ. **8**(2), 165 (1980)
86. K. Gangopadhyay, B. Basu, Phys. A **388**(13), 2682 (2009)
87. L. Benguigui, E. Blumenfeld-Lieberthal, Comput. Envir. Urban. Syst. **31**(6), 648 (2007)
88. X. Gabaix, Q. J. Econ. **114**, 739 (1999)
89. D.H. Zanette, S.C. Manrubia, Phys. Rev. Lett. **79**(3), 523 (1997)
90. M. Marsili, Y.C. Zhang, Phys. Rev. Lett. **80**(12), 2741 (1998)
91. A. Blank, S. Solomon, Phys. A **287**(1), 279 (2000)
92. A. Ghosh, A. Chatterjee, A.S. Chakrabarti, B.K. Chakrabarti (2014), arXiv:1403.1822
93. US Census: Business Dynamics Statistics. https://www.census.gov/ces/dataproducts/bds/data.html
94. Office of National Statistics: Data. http://www.ons.gov.uk/ons/datasets-and-tables
95. L.M.A. Bettencourt, J. Lobo, D. Helbing, C. Kühnert, G.B. West, Proc. Natl. Acad. Sci. **104**(17), 7301 (2007)
96. L.M.A. Bettencourt, G.B. West, Nature **467**(7318), 912 (2010)
97. L.G.A. Alves, H.V. Ribeiro, E.K. Lenzi, R.S. Mendes, PloS One **8**(8), e69580 (2013)
98. L.M.A. Bettencourt, Science **340**(6139), 1438 (2013)
99. R.L. Axtell, Science **293**(5536), 1818 (2001)
100. D. Fiaschi, I. Kondor, M. Marsili, V. Volpati, PloS One **9**(4), e94237 (2014)
101. T.M. Liggett, *Stochastic Interacting Systems: Contact, Voter and Exclusion Processes* (Springer, Berlin, 1999)
102. G. Deffuant, D. Neau, F. Amblard, G. Weisbuch, Adv. Complex Syst. **3**(01n04), 87 (2000)
103. R. Hegselmann, U. Krause, J. Art. Soc. Soc. Simul. 5(3) (2002)
104. A. Chatterjee, S. Biswas, P. Sen, in preparation
105. Global statistics for all religions 2001 AD, http://www.bible.ca/global-religion-statistics-world-christian-encyclopedia.htm Retrieved August 2014
106. M. Ausloos, F. Petroni, Europhys. Lett. **77**(3), 38002 (2007)
107. J. Holyst, K. Kacperski, F. Schweitzer, Ann. Rev. Comput. Phys. **9**, 253 (2002)
108. M.S. Miguel, V.M. Eguíluz, R. Toral, K. Klemm, Comput. Sci. Eng **7**(6), 67 (2005)
109. D. Stauffer, in *Encyclopedia of Complexity and Systems Science* (Springer, 2009), pp. 6380–6388
110. P. Sobkowicz, J. Art, J. Artifi. Soc. Soc. Simul. **12**(1), 11 (2009)
111. G. Weisbuch, A. Chakraborti, in *Econophysics and Sociophysics: Trends and Perspectives*, ed. by B.K. Chakrabarti, A. Chakraborti, A. Chatterjee (Wiley-VCH, Weinheim, 2007), pp. 67–94
112. P. Clifford, A. Sudbury, Biometrika **60**(3), 581 (1973)
113. T.M. Liggett, *Interacting Particle Systems* (Springer, Heidelberg, 1985)
114. K. Sznajd-Weron, J. Sznajd, Int. J. Mod. Phys. C **11**(06), 1157 (2000)
115. S. Galam, Eur. Phys. J. B **25**(4), 403 (2002)
116. S. Galam, Int. J. Mod. Phys. C **19**(03), 409 (2008)
117. P.L. Krapivsky, S. Redner, Phys. Rev. Lett. **90**(23), 238701 (2003)
118. S. Fortunato, Int. J. Mod. Phys. C **15**(09), 1301 (2004)
119. G. Toscani, Commun. Math. Sci. **4**(3), 481 (2006)
120. M. Lallouache, A.S. Chakrabarti, A. Chakraborti, B.K. Chakrabarti, Phys. Rev. E **82**(5), 056112 (2010)
121. S. Biswas, A.K. Chandra, A. Chatterjee, B.K. Chakrabarti, J. Phys. Conf. Ser. **297**(1), 012004 (2011)
122. S. Biswas, Phys. Rev. E **84**(5), 056106 (2011)
123. P. Sen, Phys. Rev. E **83**(1), 016108 (2011)

124. S. Biswas, A. Chatterjee, P. Sen, Phys. A **391**(11), 3257 (2012)
125. D. Stauffer, A. Aharony, *Introduction to Percolation Theory* (Taylor and Francis, 1991)
126. W. Shockley, Proc. IRE **45**, 279 (1957)
127. J. Laherrere, D. Sornette, Eur. Phys. J. B **2**(4), 525 (1998)
128. S. Redner, Eur. Phys. J. B **4**(2), 131 (1998)
129. F. Radicchi, S. Fortunato, C. Castellano, Proc. Nal. Acad. Sci. **105**(45), 17268 (2008)
130. R. Rousseau, Scientometrics **30**(1), 213 (1994)
131. L. Egghe, Scientometrics **48**(3), 345 (2000)
132. Q.L. Burrell, Scientometrics **53**(3), 309 (2002)
133. A.M. Petersen, H.E. Stanley, S. Succi, Sci. Rep. **1**, 181 (2011)
134. Y.H. Eom, S. Fortunato, PLoS One **6**(9), e24926 (2011)
135. G. Caldarelli, *Scale-free Networks: Complex Webs in Nature and Technology* (Oxford University Press, UK, 2007)
136. S. Redner, Phys. Today **58**, 49 (2005)
137. G.J. Peterson, S. Pressé, K.A. Dill, Proc. Nal. Acad. Sci. **107**(37), 16023 (2010)
138. S. Fortunato, A. Chatterjee, M. Mitrović, R.K. Pan, P.D.B. Parolo, F. Beccattini, Nature **508**, 186 (2014)
139. F. Becattini, A. Chatterjee, S. Fortunato, M. Mitrović, R.K. Pan, P.D.B. Parolo, Phys. Today (2014), arXiv:1405.7136
140. S. Milojević, Proc. Nal. Acad. Sci. **111**(11), 3984 (2014)
141. Thompson Reuters, J. Cit. Rep. (2013), http://admin-apps.webofknowledge.com/JCR/JCR? RQ=HOME Retrieved March 2013
142. A. Khaleque, A. Chatterjee, P. Sen, (2014). arXiv:1403.1745
143. I.I. Popescu, Glottometrics **6**, 83 (2003)
144. R. Mansilla, E. Köppen, G. Cocho, P. Miramontes, J. Informetri. **1**(2), 155 (2007)
145. M. Golosovsky, S. Solomon, Eur. Phys. J. Spec. Topics **205**(1), 303 (2012)
146. A. Clauset, C.R. Shalizi, M.E.J. Newman, SIAM Rev. **51**(4), 661 (2009)
147. A. Chatterjee, A. Ghosh, B.K. Chakrabarti, (2014). arxiv:1409.8029
148. M.E.J. Newman, A.L. Barabási, D.J. Watts, *The Structure and Dynamics of Networks* (Princeton University Press, 2006)
149. R. Albert, H. Jeong, A.L. Barabási, Nature **401**(6749), 130 (1999)
150. A. Broder, R. Kumar, F. Maghoul, P. Raghavan, S. Rajagopalan, R. Stata, A. Tomkins, J. Wiener, Comput. Netw. **33**(1), 309 (2000)
151. M. Faloutsos, P. Faloutsos, C. Faloutsos, in *ACM SIGCOMM Computer Communication Review*, vol. 29 (ACM, 1999), pp. 251–262
152. H. Ebel, L.I. Mielsch, S. Bornholdt, Phys. Rev. E **66**(3), 035103 (2002)
153. F. Liljeros, C.R. Edling, L.A.N. Amaral, H.E. Stanley, Y. Åberg, Nature **411**(6840), 907 (2001)
154. H. Jeong, B. Tombor, R. Albert, Z.N. Oltvai, A.L. Barabási, Nature **407**(6804), 651 (2000)
155. P. Uetz et al., Nature **403**(6770), 623 (2000)
156. J. Ugander, B. Karrer, L. Backstrom, C. Marlow, (2011) arXiv:1111.4503
157. G. Kossinets, D.J. Watts, Science **311**(5757), 88 (2006)
158. M. Li, S. Guan, C. Wu, X. Gong, K. Li, J. Wu, Z. Di, C.H. Lai, Sci. Rep. **4**, (2014)
159. H. Theil, *Economics and Information Theory* (North-Holland, Amsterdam, 1967)
160. A.A. Drăgulescu, V.M. Yakovenko, Phys. A **299**(1), 213 (2001)
161. M.O. Lorenz, Am. Stat. Assoc. **9**, 209 (1905)
162. A. Ghosh, N. Chattopadhyay, B.K. Chakrabarti, Phys. A **410**, 30 (2014)
163. J.I. Inoue, A. Ghosh, A. Chatterjee, B.K. Chakrabarti, (2014), arXiv:1406.2874
164. W.B. Arthur, Am. Econ. Rev. **84**, 406 (1994)
165. A. Chakraborti, D. Challet, A. Chatterjee, M. Marsili, Y.C. Zhang, B.K. Chakrabarti, Phys. Rep. (2014) (in press), DOI:10.1016/j.physrep.2014.09.006
166. D. Challet, Y.C. Zhang, Phys. A **246**, 407 (1997)
167. A.S. Chakrabarti, B.K. Chakrabarti, A. Chatterjee, M. Mitra, Phys. A **388**, 2420 (2009)
168. A. Ghosh, A. Chatterjee, M. Mitra, B.K. Chakrabarti, New J. Phys. **12**(7), 075033 (2010)

169. D. Challet, M. Marsili, Y.C. Zhang, *Minority Games: Interacting Agents in Financial Markets* (Oxford University Press, Oxford, 2004)
170. A. Ghosh, D. De Martino, A. Chatterjee, M. Marsili, B.K. Chakrabarti, Phys. Rev. E **85**(2), 021116 (2012)
171. D. Dhar, V. Sasidevan, B.K. Chakrabarti, Phys. A **390**(20), 3477 (2011)
172. S. Biswas, A. Ghosh, A. Chatterjee, T. Naskar, B.K. Chakrabarti, Phys. Rev. E **85**(3), 031104 (2012)
173. A. Ghosh, A. Chatterjee, A.S. Chakrabarti, B.K. Chakrabarti, Phys. Rev. E **90**(4), 042815 (2014)
174. J. Mervis, Science **344**(6186), 836 (2014)
175. M. Bardoscia, G. De Luca, G. Livan, M. Marsili, C.J. Tessone, J. Stat. Phys. **151**(3–4), 440 (2013)
176. B.K. Chakrabarti, A. Chakraborti, A. Chatterjee (eds.), *Econophysics and Sociophysics: Trends and Perspectives* (Wiley-VCH, Weinheim, 2007)

Part IV
Discussions and Commentary

Judging the Impact of 'Econophysics' Through Response to Questionnaire

Kishore C. Dash

Abstract Econophysics, after being institutionalized in 1995 has certainly made some impact in the field of social as well as natural science. Many scientists are now interested to work in this field. Some Universities have started courses on Econophysics. Research work is going on in many institutes of repute. Both Physics and Economics based journals are publishing papers on Econophysics. Many books have been written. There is also a lot of media attention on the emergence of this new branch of science. I had asked some questions to the pioneers and people who are instrumental to bring Econophysics up to the level now it has come. Their reactions to the questions have been presented in this article.

1 Introduction

According to Econophysics group of Houston University, Econophysics, also known as the physics of finance, is the study of the dynamical behavior of financial and economic markets. According to University of Maryland group of Econophysicists, Econophysics is the applications of statistical physics to economics and finance. Econophysics: Using statistical physics concepts to better understand economic questions, according to Boston University econophycists [1]. On May 16, 2013, in an interview with Kausik Gangopadhyay, Prof. of Economics, Indian Institute of Management (IIM), Kozhikode, Prof. Gene Stanley (visiting Leiden University at that time) replied to his questions as follows [2] (relevant part of the interview is reproduced here):

> **"KAUSIK: Gene, what is the story behind Econophysics? You are a pioneer in this field and also named econophysics as 'Econophysics'. Could you throw some light on your personal journey towards econophysics?**
> **GENE**: *As a matter of fact, it was in your city of Kolkata that for the first time it was publicly named. This happened in the Statphys conference which was a meeting*

K.C. Dash (✉)
Neelashaila Mahavidyalaya, Rourkela, India
e-mail: kesidash@gmail.com

327

that Bikas Chakrabarti organized in 1995; and when I used the word in this kind of important conference, the word appeared in the proceedings and it stuck. So let's talk very easily about the word. The word is very simple. Then, I'll try to talk about the field. The word tries, as anyone would guess, to coin something that reflects the fact that Econophysics is a little like Biophysics or Geophysics or Astrophysics which tells you that people involved in these fields are interested in Biology or Geology or Astronomy, which they connect from a physics point of view. And those first three spheres are not much different, because Scientists are Scientists they are all empirical. Sometimes all they are doing is that they are introducing a new piece of equipment or something of that sort. But in Economics, it is different shall we say, a different culture, meaning a physicist's culture and a typical economist's culture, are really quite different. I mention this because you'll understand the backdrop of my personal journey better. The personal journey was very simple. My students in Boston University, some of them wanted to study economics and when I would go to Economics Department (which means even if I managed to get this far), I would just ask people "Are you going to give a Ph.D. in physics to this person? To which they replied, "Of course not, it would be in economics". But then in Economics Department, you have to start with undergraduate Economics and by the time you are able to do something, already quite a lot of time are gone. Most importantly, what you do under the guidance of an economist will not be as revolutionary (from a physics perspective) like what you might do with a background of physics. So I decided to give a name to the synergism, like a revolutionary country has to have a name, you know! You need to have some name and this was a nice name, I thought. The other attempts which were made to give name to this subject never really stuck. I am very happy with this word because it makes clear that we are really physicists who are addressing the questions that economists, some economists, care about and some don't.

KAUSIK: Could you please throw some more light on the Kolkata connection of econophysics?

GENE: *That is very simple. There are some very good people there and Bikas Chakrabarti is one of them, someone who as you probably know likes to see change. He likes to make something really happen. So he started to have meetings on econophysics and I think the first one was probably in 1995 (he decided to start it in 1993–1994). Probably the first meeting in my life on this field that I went to was this meeting. In that sense Kolkata is—you can say the nest from which the chicken was born and Bikas gets, deservingly so, a lot of credit for that because it takes a lot of work to have a meeting on a field that does not really exist, so to say! After all who is going to come? If you have a meeting on standard fields like super conductivity there are many people who were happy to come to India to attend that meeting, but econophysics was something different. So he should get a lot of credit for this.*

KAUSIK: Let us talk about something different, something to do with this field's academic credentials, if I may say so. You are an editor of Physica A which is one of the few journals that publish articles on econophysics on a

regular basis. Do you think econophysics has gained enough prominence in academic journals?

GENE: *Yes, it is doing well. I, for one, am not disappointed. It is a fact however, that pure economics journal, such as Econometrica, as also the other top journals, almost never published a paper from physicists. My group has one paper in one of the top four economics journals called the Quarterly journal of Economics. But we have a co-author who is a very accomplished economist. He wrote the paper in the language that's expected in the economics journals. This means a much more mathematical language making very clear what the assumptions are and making very organized definition of each thing, very precise just like mathematicians do. Physicists are much more empirical and they say, you know, the question is how you are going to answer some questions by looking at the data and come out with the findings. But often without a definite ex-ante theory or anything of this sort. So a typical experience is that author (and this applies also to economists by the way) submits some piece of work to a good journal but journal returns it, sometimes not even with a referee report, saying since you don't have a theory yet, you first have a theory and then resubmit it, which we would be happy to consider. But as you know econophysics is a little like a regular physics discipline in the sense that there is simply no theory of many things; for that matter a lot many things in physics have no theory what so ever. One of the most dramatic discoveries in physics, which, happened about 25 years ago, that is high temperature super-conductivity, was made empirically. People who discovered even won Nobel prizes. However, no one ever knows why this works. And the same for regular super conductivity which was discovered here in Netherlands at the beginning of the last century and it was not until roughly 50 years later that there was a theory of regular superconductivity. So we don't have that rule that you cannot publish what seems to be true until you have a theory and that should hold particularly true in economics as well because it is not clear to me (or anyone else) if there is a correct theory of anything that goes on in economics. Something that fits all the day and not just, you know, the central part of a distribution or something of that sort. So I think, our biggest success story is that we are focusing primarily on asking the question of what are the data telling us and we don't look at the subset of the data, we look at every piece of data you can get which in the last 10 years or so has really grown immensely. In economics, there was always a lot of data but economists did not want to look at the whole dataset for whatever reasons.*

KAUSIK: Now to our knowledge, the last major interdisciplinary subject, similar to econophysics, was biophysics which is a well established discipline now. If you consider 1944 as the time of inception of biophysics by publication of an article entitled "what is life" by Erwin Schrödinger, and the eventual ascendance of this subject to glory with the award of Nobel prize, to Francis Crick and James Watson in 1962, there's a span of 17 years. Econophysics as you mentioned started in 1995; again it's coming roughly around the same time span by now. So what's your opinion, the relative success (or say failure) of growth of econophysics in this last 17 years when compared to Biophysics in the similar time span?

GENE: *Well, there is no question that Crick & Watson did their work in 1962; that's a fact. However as to when physicists started to get interested in life, it was long before Schrödinger the physicists were interested in life may be hundreds of years ago. Inception of biophysics was not then 17 years before Watson and Crick—there's no clear beginning. Same is with econophysics, which did not begin in the year that I named it; in fact, people, I mean, physicists have been interested in economics for a long time. Right here in Leiden, a person got his Ph.D. in Physics but was interested in Economics so the focus of his thesis was economics and then went on to win a Nobel prize in economics, quite a few years later. His last name is TINBERGEN.*

KAUSIK: I know, Jan Tinbergen.

GENE: *You know about him? Of course, in economics, everyone knows him. I know that he studied right here in this institute under the famous physicist named Ehrenfest. He studied, if I remember right now, during circa 1925, or so way back, and long before 1995, when I introduced this word. And even [Sir] Isaac Newton is said to have been interested in econophysics.*

KAUSIK: Oh really! There is a field in economics called econometrics which deals with statistical testing of economic theories using available data. So a set of economists is also dedicated to deal primarily on data. How are they different from the econophysicists?

GENE: *Econometrics starts with a theory and then test them with data. If you look into the published articles on econometrics, there is ample evidence of what I said.*

KAUSIK: What are the biggest challenges facing the world where econophysics can hope to make an inroad?

GENE: *I think the main challenge is, knowing the statistical laws which describe fluctuations, fluctuations of finance, fluctuations in economies or anything that is a part of economic systems. And you would think why these laws were not discovered by traditional economists....*

KAUSIK: Of course. What does emerge as the significance of econophysics in business decision making? Or maybe I can put it in another way, can findings of econophysics be taken to the corporate board rooms?

GENE: *See, the trouble is that I do not really know what goes on in corporate board rooms but if it means do they have some practical advantage [by the study of econophysics] then I must say any basic laws have much practical advantage. In other words, without Newton's laws you could not build devices whether it's a satellite or a canon ball or anything else. You have to pay attention to these laws. And so, businesses have to, in fact, pay attention to basic laws of economics and the stylized facts. In economics it is often conventional to not even respect such things as laws because they are just so called stylized facts. We think that businesses have to recognize the fact that these stylized facts are really facts. They are as much a fact as it is that the sun will come up tomorrow, except they are statistical. If we are talking about the sun, then it will be like: Will I see the sun at all? Someday you just see the sun, someday you just see clouds. There are fluctuations, but the main thing is that they really are laws of economics, that are made of lower scale.*

So businesses have to learn to account for the same. In some ways every little thing matters because if the country goes bankrupt then your business is going to be in trouble." This part is reproduced from Ref. [2].

2 Response to Questionnaire

In the discussions above we have come across the story of inception of 'Econophysics' to its possible appearance in the corporate board rooms. This interview was taken in 2013. One year back in 2012, I had taken an attempt to understand the impact of institutionalized econophysics in the circles of scientist pioneering and dealing with this field through a numbers of questions. Although many of the them have not replied, Their reactions, opinions, feedbacks of the people who have replied have been presented below:

2.1 Questions

(1) What is your formal field of research: Physics, Economics, Mathematics, Computer Science, or ... ? Do you identify yourself with the Econophysicsts? What is the fraction of your research time/effort you put today in this kind of interdisciplinary research?

(2) Do you think, the studies made generally under this econophysics banner made any significant contribution either to physics or to economics, or to both? If not yet at any satisfactory level, do you think these studies have any potential to succeed in the near future?

(3) Do you think an 'Econophysics' forum/body should be formed like other fields in physics and international conferences/workshops are conducted under its banner in different countries by rotation, annually or bi-annually?

(4) Some universities have started offering optional graduate courses in econophysics; e.g.,:

http://www.physics.leidenuniv.nl/edu/bachelor/courses_variatie/EF.asp

http://phys.uh.edu/research/econophysics/index.php

http://english.us.edu.pl/econophysics

http://www.sciencedirect.com/science/article/pii/S0378437104009616
and others like

(a) **Trinity College Dublin, the University of Dublin, Ireland,**
(b) **University of Munster, Germany,**
(c) **University of Southern Denmark,**
(d) **University of Ulm, Sweden,**
(e) **University of Fribourg, Switzerland.**
Do you think, your university/institute should also offer such courses to the students? Did you write any textbook/research monograph on econophysics or related fields? If not, do you intend to write one?

(5) Do you think, like many other interdisciplinary fields like biophysics, geophysics etc., 'econophysics' will be recognised as a natural one soon? If so, by whom: physics or economics, or by both? Do you think, econophysics research will be recognised by Nobel Prize ever? If so, in which discipline? Physics or Economics? And when?

2.2 Answers

John Angle The Inequality Process Institute, Pennsylvania, USA

For major part of his career, John has been a statistical consultant to the Economic Research Service (ERS) of the U.S. Department of Agriculture in Washington, DC. He is the founder of The Inequality Process Institute LLC (TIPI), an organization that does both pure and applied research on personal income and wealth. At TIPI, he works full time on research related to the Inequality Process, a particle system model of personal income and wealth distribution, their dynamics, and the dynamics of income and wealth at the micro-level. John is a mathematical sociologist interested in income and wealth distribution phenomena and is not as an econophysicist. He has an outsider's interest in the natural sciences, particularly statistical physics, and envy the excitement of discovery among physicists.

Ans. to Q.1
My Ph.D. is in sociology (University of Michigan, Ann Arbor). I was initially employed as an assistant professor (University of Arizona). Most of my career since has been as a statistical consultant to the Economic Research Service (ERS) of the U.S. Department of Agriculture in Washington, DC. I was employed in the Research Division, National Office, U.S. Internal Revenue Service (IRS) (the income tax people in the U.S.) in 1985 and 1986 as a mathematical statistician. While the task of the Research Division of the IRS was improving compliance with the U.S. tax code, suggestions about new statistical techniques to identify noncompliance were not welcome. Statistical consulting at ERS included teaching econometrics in terms of matrix algebra to agricultural economists who had taken scalar variable econometrics in their youth. I also researched rural income issues at ERS and served on the Federal inter-departmental Confidentiality and Data Access Committee. I took early

retirement from ERS to found The Inequality Process Institute LLC (TIPI), an orga-
nization that does both pure and applied research on personal income and wealth. At
TIPI I work full time on research related to the Inequality Process, a particle system
model of personal income and wealth distribution, their dynamics, and the dynamics
of income and wealth at the micro-level.

My papers on the Inequality Process began appearing in print in 1983. I didn't
refer to the model as a particle system until the late 1980s. It wasn't until 1990
that I published a paper discussing the similarity of the Inequality Process to the
stochastic particle system model of the kinetic theory of gases. So I didn't find the
Inequality Process via statistical physics. The Inequality Process was abstracted
from verbal theory of the inequality of wealth over techno-cultural evolution in eco-
nomic anthropology and sociology. I was delighted to find that some physicists were
interested in this model and similar models. Thanks to Prof. Thomas Lux (Depart-
ment of Economics, University of Kiel, Germany) I learned in 2005 of the work of
Prof. Bikas Chakrabarti and his students as well as that of Prof. Victor Yakovenko
and his students, and that of other physicists.

I'd identify myself as a mathematical sociologist interested in income and wealth
phenomena, not as an econophysicist. I have an outsider's interest in the natural sci-
ences, particularly statistical physics, and envy the excitement of discovery among
physicists. No meeting of the the American Sociological Association has ever been
characterized by a name anything like the "Woodstock of Sociology" (cf. the 1989
meeting of the American Physical Society meeting in which findings about high
temperature super-conducting were presented, called by the New York Times "the
Woodstock of Physics" after a well known rock music concert because of the excite-
ment at the meeting about those findings). Neither has any seminar group of sociol-
ogists ever acquired the fame or historic significance of the Bonzenfreies colloquium.

Ans. to Q.2

Until econophysics models are shown relevant to the empirical phenomena that are
the traditional subject matters of physics, I suspect that econophysics models will be
little more than an avocation, a pleasant excursion into a novel subject area using
familiar mathematical models, for physicists. My opinions about the subject matters
of academic disciplines and their boundaries, i.e., departmental "turf" as U.S. aca-
demics might say (to use American slang for the territory and sustenance of a violent
gang denied to nonmembers) are much less energetic than those of academics. I am
focused on explaining social and economic phenomena related to wealth in either its
stock (assets) or flow (income) form in a pragmatic way that disregards the bound-
aries of academic disciplines, as well as their paradigms, jargon, and xenophobias.
Except of course when it comes to writing up findings for a journal, most of which
are imbued with a discipline's paradigm. I'm not interested in whether the Inequality
Process is classified as economics or mathematical sociology. I think the statistical
signature of the Inequality Process is all over data on income and wealth in the U.S.
I suspect that same will be shown with other countries' data. If it is demonstrated
that the Inequality Process is universal, findings about it aren't going to fade away
as academic fads do. The Inequality Process, if shown to be universal and ubiquitous

in everyone's life, will become a permanent part of science, a new part that is rather difficult to distinguish from statistical laws in the natural sciences, a part, I hope that will eventually, beyond my lifetime, envelop and surpass the discipline of economics rather than vice versa.

Ans to Q.3

Of course. The enlargement of knowledge requires ramification into increasingly specialized subject matter fields and the formation of association of researchers in each new field as it buds off from the older discipline that once laid claim to having an explanation for the new field's subject. A colleague wrote me to announce a new journal she was editing. Its title seemed to me so specialized that my initial impression was to wonder how so specialized a journal could exist. Then I realized that's what most people think about the first journal in a new area of knowledge at its inception, journals that often go on to become the oldest and most prestigious of their fields.

Ans. to Q.4

I'm not in a position to give advice about the development of a curriculum for students of physics. I have discussed with colleagues who are mathematical sociologists syllabi for courses in mathematical sociology. I'd like to see the emergence of mathematical sociology as a viable subfield of sociology. I'm the initial (but not final) organizer of the Mathematical Sociology Section of the American Sociological Association. The book I'm working on is a monograph on the Inequality Process. It's not a textbook.

Ans to Q.5

Writing as a sociologist interested in the sociology of science, I would hypothesize that the likelihood of the emergence of econophysics as a course, or a program within a physics department, or a joint program between departments or a separate department of its own in a university is essentially a question of whether teaching and doing research in it is a financially viable activity. It is a question of demand for the knowledge it produces. So while national science funding agencies, such as in the U.S. the National Science Foundation, might see promise in such an area of knowledge and provide "seed money" for a pioneering program or research institute (e.g., the Santa Fe Institute) in econophysics, the long term and widespread viability of the field depends on how powerfully useful the knowledge it produces is. Practically important findings well funded department or free-standing program or rebranding by an existing department with its name (adoption whole). If one can unify a useful econophysics model with established physical law (about the traditional subject matter of physics) in a novel and surprising way, then, I'd guess, depending on the scope of possibilities such a finding might open, a Nobel Prize in physics might not be out of the question for a Ph.D. in physics. The Nobel Prize in economics is a more remote possibility. Economics is an insular and dogmatic field, in some ways more like a religion than a science, and more xenophobic than other social sciences. A Nobel Prize in economics might be given for rebranding useful statistical law now classified as econophysics as traditional economics, a kind of theological re-interpretation rather than scientific discovery. Otherwise, the main prize that really good work in econophysics can expect to receive from a committee of elite economists will be, I'd guess, the disciplinary analogue of a referral of a Renaissance era schismatic to the Counter-Reformation's Holy Office of the Inquisition, something of negative prize.

Satya R. Chakravarty Indian Statistical Institute, Kolkta, India

Satya is a Professor of Economics at the Indian Statistical Institute, Kolkata. His main areas of interest are Welfare Economics, Public Economics, Mathematical Finance, Industrial Organization and Game Theory. His work spans theoretical, empirical and policy analysis. He has authored six books and co-edited two books "Quantitative Economics: Theory and Practice (with D. Coondoo and R. Mukherjee)", Allied Publishers, New Delhi, 1998 and "Econophysics & Economics of Games, Social Choices and Quantitative Techniques" (with Banasri Basu, Bikas K. Chakrabarti and Kausik Gangopadhyay), Springer-Verlag, Italia, 2009. He received the Mahalanobis Memorial Award of the Indian Econometric Society in 1994. This is the most coveted award for an economist in India. Together with B.K. Chakrabarti, A. Chakraborti, and A. Chatterjee, Satya published the book "Econophysics of Income and Wealth Distributions", Cambridge University Press (2013).

Ans. to Q.1
Economics, a significant part of research time.

Ans. to Q.2
Likely to make significant contributions to economics.

Ans. to Q.3
Yes. conference probably once in two years. I am associated with ecineq (society for the study of economic inequality) and we arrange one conference in every two years.

Ans. to Q.4
I believe some universities/institutes in India should offer such courses.

Ans to Q.5
I see some possibility in economics. My study of the literature tells that an economy can be regarded as a thermodynamic system and the distribution of income among economic agents can be identified with the distribution of energy particles in a gas. There may be other areas of overlap. In view of these I see a high chance.

Mauru Galegati Economics Department, Polytechnic University of Marche, Italy

Mauru obtained his Ph.D. in Economics in 1989 at the University of Ancona. He is Professor of Economics at the Polytechnic University of Marche, Ancona. Prof. Gallegati is on the editorial board of several economic journals, including New Economic Windows (Springer) and Journal of Economic Interaction and Coordination (Springer). His research includes business fluctuations, nonlinear dynamics, models of financial fragility and heterogeneous interacting agents. Mauro Gallegati is well known from his widely cited work with Joseph E. Stiglitz, developing theory of asymmetric information and heterogeneous agents and their applications. He published papers in journals on economics, economic history and history of economic analysis, nonlinear mathematics, applied economics and complexity.

Ans. to Q.1
Economics, no, 25%.

Ans. to Q.2
Yes, yes.

Ans. to Q.3
Yes.

Ans. to Q.4
Yes, no, no.

Ans. to Q.5
No, no.

Aoyama Hideaki Graduate School of Sciences, Kyoto University, Japan

Hideaki is one of Japan's leading theoretical physicists. Though his interests span many areas, his current focus is on econophysics, the new field that hopes to move economics closer to being an exact science. Amongst his many publications are Classical Mechanics (2005), Pareto Firms (2007), Econophysics (2008), all in Japanese, and a new book from the Cambridge University Press, Econophysics and Corporations, -Statistical Life and Death in Complex Business Networks (2012). Prof, Aoyama with others have run four domestic econophysics conferences at Yukawa Institute for Theoretical Physics at Kyoto University, in 2003, 2005, 2007 and 2009. Very recently they have analyzed business cycles in Japan using the indices of industrial production (IIP), an economic indicator which measures the current conditions of production activities over the nation on a monthly basis.

Ans. to Q.1
My field of research is Physics. I identify myself as a Theoretical Physicist. I have been doing research in Particle Physics mostly, and in addition to that, I wrote papers in Linguistics, Economics (Econophysics), and Condensed Matter Physics. Economics and Econophysics work occupied 100% of my research time last year, but it can change as my interest is broad.

Ans. to Q.2
Some works on real economy made impact on Economics. Although there are many who still do not agree, mainstream economics with its utility function and classical Newtonian equilibilium has been shown to be inappropriate. Core of the thinking in Econophysics, the scientific study with emphasis on analysis of actual data and theoretical models with concepts borrowed from statistical physics, is gradually becoming the standard for the next stage of the Economics.

Ans. to Q.3
Yes, to all. In every possible form, it should be discussed, the research results shared, and the young generation of academics should be exposed to its light.

Ans. to Q. 4
I wish we have courses like that. But funding comes very slowly. I and my colleagues published a book on Econophysics from Cambridge Univ. Press and wish to publish another.

Ans. to Q.5
I am not sure if it is good to establish ourselves as "Econophysicists", separating us from Economists. Rather we need to convince them that our view, scientific view, is the view they should adopt, and create next stage of development of Economics, contributing stable and happy global society.

Hagen Kleinert Theoretical Physics, Free University of Berlin, Germany

Hagen is a Professor of Theoretical Physics at the Free University of Berlin, Germany since 1968. For his contributions to particle and solid state physics he was awarded the Max Born prize 2008 with Medal. His contribution to the memorial volume celebrating the 100th birthday of Lev Davidovich Landau earned him the Majorana Prize 2008 with Medal. Kleinert has written 400 papers on main stream physics alongwith the theory of financial markets. He has written several books on theoretical physics, the most notable of which, Path Integrals in Quantum Mechanics, Statistics, polymer Physics, and Financial Markets, has been published in five editions since 1990 and has received enthusiastic reviews.

Ans. to Q. 1
Theoretical Physics, Yes, 10%.

Ans. to Q. 2
These studies have potential to succeed in the near future.

Ans. to Q.3
Yes.

Ans. to Q. 4.
Yes, Yes I wrote book Path Integrals In Quantum Mechanics, Statistics, Polymer Physics, and Financial Markets.
http://www.worldscibooks.com/physics/7305.html

Ans. to Q. 5
Yes, Economics, Yes in Economics, Future.

Reiner Kümmel Universität Würzburg, Germany

Kümmel was awarded Ph.D. degree at the University of Frankfurt. He was DAAD Lecturer and Professor Asistente/Asociado at the Universidad del Valle in Cali, Colombia from 1970 to 1972. During 1974–2004 he was Professor of Theoretical Physics at the University of Wuerzburg. He Retired from service in 2004. Since then he is working as a lecturer in "Thermodynamics and Economics" at the University of Wuerzburg. Kümmel noted for his 2011 book "The Second Law of Economics: Energy, Entropy, and the Origins of Wealth", in which he argues that we need to begin to incorporate energy and entropy thinking into economics. Kümmel's central thesis, which he calls the second law of economics, is that wealth creation by energy conversion is accompanied and limited by polluting emissions that are coupled to entropy production. He argues that we need to begin teaching students about the basics of economic thermodynamics, if we are to avoid, the shrinking of natural resources, environmental degradation, and increasing social tensions.

Ans. to Q.1
Physics. I am a member of the econophysics branch of the German Physical Society. Before my retirement in 2004, I dedicated about one third of my time to econono-physics. Now it is 100%, see my homepage.

Ans. to Q.2
Probably yes. But there is not yet much impact on either field. I believe these studies have any potential to succeed in the near future.

Ans. to Q.3
Yes. Bi-annual international conferences should complement the annual German econophysics conferences. An organization like the "Association on the Study of Peak Oil", or the "International Society of Ecological Economics", would be helpful.

Ans. to Q.4
My university, the University of Würzburg, Germany, offers the course on "Thermodynamik und Ökonomie" (Thermodynamics and Economics), which is also announced on my homepage. The monograph, on which the course is based, is my book "The Second Law of Economics: Energy, Entropy, and the Origins of Wealth",
http://www.springer.com/physics/complexity/book/978-1-4419-9364-9

Ans. to Q.5

I think, like many other interdisciplinary fields like biophysics, geophysics etc., 'econophysics' will be recognized as a natural one soon. It already is by the German Physical Society. Quite a few Nobel prize winners in economics were physicists by training. But economists will never propose an active econophysicist for the prize. They prefer it for themselves, and dislike econophysics anyway. If the NP will ever be awarded to an econonphysicist, it will be after economic breakdowns predicted, and economic recoveries via actions suggested, by econonophysicists.

János Kertesz Institute of Physics, Budapest University of Technology and Economics, Hungary

János is a physicist from University of Budapest, Hungary. He is one of the pioneers of econophysics, complex networks and application of fractal geometry in physical problems. The inaugural meeting on Econophysics was organised 1998 in Budapest by János Kertész and Imre Kondor. János Kertész obtained his Ph.D. from the Institute of Technical Physics (ITP) of the HAS in 1980 and D. Sc from HAS in 1989. He has occupied many positions at ITP and became Professor of Physics at the Budapest University of Technology and Economics (BME) from 1993, also Deputy Director of the Insitute of Physics at BME from 1992 to 2000. From 2001, he was the Director of the Institute of Physics at BME. He is presently the director of the Institute of Physics in Budapest University of Technology and Economics, Budapest, Hungary.

Ans. to Q.1

My formal field: Physics, Econophysics is an important field of research, in my opinion. 30%.

Ans. to Q.2

I think that econophysics has already achieved remarkable results in economics though there has not been a real breakthrough so far; however, I am confident that we are on a good track. (It is perhaps not a cynical comment to say that many of the 'breakthroughs' of financial economics have turned out to be failures, and that the ongoing global crisis is just a painful proof of the lack of fundamental understanding in this area).

Ans. to Q.3

I think that there have already been such forums: APFA, ECCS etc. Econophysics is massively represented in conferences on complex systems. The German Physical Society has a Division for the Physics of Socio-Economic Systems. I think that a similar, world-wide umbrella, which includes social systems too, would make a lot of sense.

Ans. to Q.4

I am in the process of introducing a related project in Budapest. I have not written such a book and will not do so in the near future.

Ans. to Q.5
The goal of econophysics is to understand economics and finance, therefore recognition should also come from that direction. I am optimistic but this will take time. What helps is more respect for the economics literature and groundbreaking results.

Thomas Lux Economics Department, Christian-Albrechts-Universität zu Kiel, Germany

Thomas is an economist, who has important contributions in the field of 'Econophysics'. He is ranked among the 1,000 most often quoted economist within the decade 1990–2000. He has a number of publications in physics journals like Nature Physics, Physica A, European Journal of Physics B, International Journal of Modern Physics, Physical Review E etc.

Ans. to Q.1
My field of education and research is economics and finance. I hold a Ph.D. in Economics, I am currently Professor of Economics at CAU Kiel and have spent my entire career within economics departments. My education in physics is of a more informal nature. As a post-doc I started studying material from statistical physics. My feeling then was that stochastic models for interacting-particle systems could be very useful as a methodological approach for modelling of interacting agents in economics as well. I soon started to use such methods in my own research (since the early 90s). Since my first publications in this vein were received quite positively, from about the mid 90s most of my research output can be attributed to what has later been called 'econophysics'.

Ans. to Q.2
I believe that at least certain branches of the 'econophysics' research have made significant and lasting contributions. In financial economics, for instance, the development of agent-based models of heterogeneous, interacting investors has demonstrated how the stylized facts (economics terminology) or universal statistical features can be explained via the tension between centripetal and centrifugal forces generated by different trading strategies. A large literature has (in my view) convincingly shown that features like fat tails and clustered volatility can be obtained in a robust way as emergent phenomena from microscopic models with a few basic ingredients. In contrast, mainstream finance has always treated the most prevalent features of financial data as 'anomalies'—not a very convincing way to deal with empirical phenomena. There is also quite a broad area of research on topics in quantitative finance (option pricing, risk management etc.) in which research contributed under the banner of 'econophysics' mingles with research conducted by statisticians, mathematicians and others without there being much recognizable difference in scope and methodology. A certain blind spot exists still in macroeconomics, despite some innovative

and interesting work like that by Mauro Gallegatti, Domenico delli Gatti and some others. Macroeconomics would certainly benefit from adapting more of a view of the economy as an adaptive dynamic system with dispersed activity rather than the still dominating representative agent approach.

Ans. to Q.3

There have been quite a number of annual conference series (like Econophysics Colloquium, APFA, the econophysics meetings at Kolkata and others) most of which have been initiated at least ten years ago. Certainly, such focused events are very important for the development of the field. Similarly important are conferences in which 'econophysicists' meet and interact with open-minded economists without a physics background. The annual meetings of the Society for Economics with Interacting heterogeneous Agents and the Society of Computational Economics have usually attracted a large number of colleagues with an interdisciplinary background.

Ans. to Q.4

Certainly, courses providing an overview over the econophysics literature should be very welcome. After all, it has been a very active field of research and, as I see it, has produced some significant insights. However, I don't see a clear 'econophysics' curriculum emerging. Existing courses and textbooks are often pretty idiosyncratic and governed by the particular research interest of the persons in charge. In my view, a major contribution of such programs should be to bring students from economics into contact with the approach and methodology of research in physics which has many important tools to offer and is altogether of a much more practical hands-on nature than the inherited research style in economics (with its leaning towards rigorous mathematical proofs rather than practical data analysis). I myself am offering a course on agent-based modelling (including statistical physics methodology to deal with such systems) in our Ph.D. program and have authored a few surveys on econophysics research.

Ans. to Q.5

Right now it seems quite unthinkable. There is obviously a lot of resistance from both mainstream economics and traditional quarters in physics. While some physicists do not consider research on social and economic phenomena as an appropriate research topic for their colleagues, the attitude of some economists seems to be more that of shyness with strangers who use concepts and terminology they have not encountered before. Often it is also an outrage that some concepts dear to economists like rationality and utility maximization are altogether ignored by econophysicists. And last but not least there is vested interest in the value of one's human capital that might be tied to particular tools and methodological approaches, and gets endangered by new developments. Nevertheless, economics has seen many 'scientific revolutions' and the once heavily rebuked fields of behavioral and experimental economics have meanwhile also been honored with Nobel prizes. So, at the end, solid research will always be recognized sooner or later.

Matteo Marsili The Abdus Salam International Centre for Theoretical Physics, Italy

Matteo is basically a statistical physicist having research interest in Statistical physics, non-equilibrium critical phenomena, disordered systems, probability Interests theory and stochastic processes, complex networks. interdisciplinary applications of statistical physics, including modelling socio-economic phenomena and financial markets, game theory, and biological networks. He has organised many conferences and workshops on socio-economic systems and finance. Together with A. Challet and Y.C. Zhang, Matteo has written an important book on Econophysics of "Minority Games", Oxford University Press (2004).

Ans. to Q.1
*No, I don't think the word 'econophysics' makes any sense. **40% time/effort for this** kind of interdisciplinary research.*

Ans. to Q.2
No; Badly posed question for succeed this field in the near future.

Ans. to Q.3
No, I don't think "econophysics" makes sense. There is no "physics" in "economics". Quantitative training is helpful, but those who study economics should have a proper training in micro-economics, game theory, macro-economics etc. There are already a lot of conferences in "econophysics".

Ans. to Q.4
No. I wrote a book on Minority Games plus a couple of proceedings volumes. I may write another book.

Ans. to Q.5
No, I don't.

Tiziana Di Matteo Financial Mathematics, King's College London, UK

Tiziana is a Reader in Financial Mathematics in King's College, London. She did her Ph.D. in Physics in 1999 from Dipartimento di Fisica-Universitá di Salerno, Italy. She was Associate Professor (Level D), Applied Mathematics at Research School of Physical Sciences, Australian National University, Canberra, Australia. Her research interests are Econophysics, Application of methods from Statistical Physics to Finance, Complex Systems, Science of Networks, etc. In Econophysics, she has publications in refereed journals like New Journal of Physics, Physica A, Journal of Banking & Finance, European Physical Journal B, Quantitative Finance etc. Basically on financial market studies.

Ans. to Q.1
It is Complex Systems and Econophysics. Yes I do, I identify myself as an Econophysicist. 80 %.

Ans. to Q.2
I think they made several contributions and the main contribution so far has been the empirical analysis of real economic and financial data sets that has resulted in the so called "stylized facts" that any model should reproduce.

Ans to Q.3
We have already had some of these including the Genoa forum
http://www.ge.infm.it/~ecph/index.php *the Econophysics forum*
http://www.unifr.ch/econophysics/ *and regarding conferences, meetings and workshops we have had many in the last few years including the Econophysics Colloquium started in Canberra on 2005 and now becoming an annual event* http://www.econophysics-colloquium.org/ *The next edition, the Econophysics Colloquium 2012 will be at ETH in Zurich and the 2013 edition in Korea and the 2014 edition in Japan.*

Ans. to Q.4
I am personally teaching in my Financial Mathematics master courses some materials on Econophysics . I have written some papers and contributed to some volumes in Econophysics.

Ans. to Q.5
I think it is already recognized as a research field mainly within the physics community. When something very relevant to Economics will be discovered, a Nobel prize in Economics to an Econophysicist is certainly possible.

Jürgen Mimkes University of Paderborn, Germany

Jürgen is a Professor at the physics department at the University of Paderborn, Germany. Before that he had taught in many US Universities and Berlin. His main field is solid state physics and thermodynamics. He has been dealing with the subject of the statistics and thermodynamics of social and economic systems since 1990. He calls his present field as "physical economics", especially macro and micro economics, and finance and he has trouble to call his field as econophysics, as this only covers finance. He has many works related to sociological thermodynamics and economic thermodynamics (we can say Econophysics) like "Society as a Many Particle System" (1997), "Chemistry of the Social Bond" (2005) Econophysics and Sociophysics (multi authored, 2006), and many others. Together with P. Richmond and S. Hutzler., Jürgen has published the book "Econophysics and physical economics", Oxford University Press (2013).

Ans. to Q.1
My background is physics and solid state thermodynamics. My present field is "physical economics", especially macro and micro economics, and finance. My time invested in this interdiscipline work is 100%. I have trouble to call my field econophysics, as this only covers finance.

Ans. to Q.2
I think the studies have widened the scope of economics, moving from narrative social science to a more exact "natural science". In addition to it econophysics has shown that physics is far more than a science of dead matter. Physics is the base of natural science including economics and social sciences. In my view the present studies will succeed in the future as the the basis of social and economic science, if econophysics will include macro-economic aspects.

Ans. to Q.3
We have already many international meetings on econophysics, the European COST meetings, the world wide economic WEHIA meetings, the European Complexity meetings, The Indian meetings on Socio-Economics at the SAHA Institute in Kolkata, the annual meeting of members of Socio-Economic Systems by the German Physical Society, meetings in Japan, China, USA and Switzerland, to name a few. These meetings are already linked by members in different meetings, but it would be beneficial to establish international meetings from all these groups.

Ans. to Q.4
I strongly support the idea to install the new field econophysics as an interdisciplinary field in economics and phyics. I am presently writing a textbook phyical economics and econophysics to promote the knowledge of econophysics in different universities.

Ans. to Q.5
I believe that physical economics or econophysics will soon be recognized as a special field in natural science. If the field keeps contributing new ideas a Nobel prize in Economics should be possible. When? People like Nash had to wait many years for this prize!

Bertrand Roehner The Laboratory of Theoretical and High Energy Physics (LPTHE), University of Paris, France

Roehner is a professor of Physics at LPTHE, University of Paris. His Research interests are Speculative trading, price peaks, sharp peak—flat through pattern, stock price bubbles, comparative history, analytical history, market integration, real estate prices, separatism and separatist uprisings, wheat prices, commodity prices, spatial arbitrage, spatial propagation of price fluctuations, social bonds, social cohesion, military occupation, resistance, suicide, apoptosis, intermarriage, mixed couples, white flight. He is the author of several books like "Theory of Markets" (Springer-Verlag), "Hidden collective factors by speculative trading" (Springer), "Patterns of speculation" (Cambridge University Press) etc., and about 60 papers in various scientific journals in the fields of economics, sociology and physics.

Ans to Q.1
My formal field of research is Particle physics. and identify myself as an econophysicist. I am completely involved with this kind of interdisciplinary research.

Ans to Q.2
A: Yes. Particularly because it introduced the idea that models must be able to produce successful predictions instead of just describing past episodes. Describing former episodes without making any NEW testable prediction is just circular reasoning, isn't it? For the second part of the question which is about the future I have no idea.

Ans to Q.3
A: I don't have any clear opinion on this point.

Ans to Q.4
I think, my university should offer econophysics courses to the students. I wrote book on econophysics which were not intended as textbooks but I think that substantial parts of them can be used for that purpose. I have myself used them for teaching a course in econophysics at "Beijing Normal University" in 2011. By the way, this institution can be added to the list given in the questionnaire.

Ans to Q.5
A: No, I don't think so. Economics has a strong political content. In the words used by my colleague Jean-Philippe Bouchaud in one of its papers it is more a religion than a science. So, when econophysicists insist that the tenets and dogma accepted by economists should be tested their wishes are not welcomed by the profession. The Nobel Prize in economics has been awarded to many economists for purely theoretical and untested research something which (as far as I know) has never been done for the Nobel prize in physics. So, one cannot really say that the "Nobel" in economics is a scientific Nobel prize.

Sitabhra Sinha Institute of Mathematical sciences, Chennai, India

Sinha is a Professor in the Physics group of the Institute of Mathematical Sciences (IMSc) at Chennai (formerly known as Madras), and adjunct faculty of the National Institute of Advanced Studies (NIAS), Bangalore. His Ph.D. work was on the nonlinear dynamics of recurrent neural network models done at the Machine Intelligence Unit, Indian Statistical Institute, Calcutta (1994–1998), postdoctoral research is on nonlinear dynamics of spatially extended systems with focus on biological systems at the Department of Physics, Indian Institute of Science, Bangalore (1998–2000 and 2001–2002) and Weill Medical College of Cornell University at New York City (2000–2001). He joined the faculty of IMSc in September 2002. His areas of research fall broadly under complex systems, nonlinear dynamics and theoretical & computational biophysics.

Ans. to Q.1.

My formal field of research is Physics, specifically Complex Systems theory. Yes, I identify myself as beloging to the community of physicists involved in analyzing social & economic phenomena using techniques from statistical mechanics, nonlinear namics and computer simulations. I devote about a third of my research time/effort in exploring econophysics related questions.

Ans. to Q. 2.

Yes, I believe econophysics studies have already made significant contributions to economics—specifically in understanding the behavior of financial systems and also in explaining the ubiquity of inequality distributions. From Bachelier onwards to Mandelbrot and Osborne, there have been many attempts to understand the nature andorigin of fluctuations in economic variables, such as prices and indices. The careful empirical analyses of financial market data pioneered by Stanley, Bouchaud, et al in the 1990s has revolutionized this area of economics. There has not been a critical contribution towards advancing physics as yet—although the advances made in econophysics to understandingthe dynamics of interactions among senient agents may prove very useful in extending the reach of physics to studying biological andsocial systems.

Ans to Q. 3.

Yes. To some extent this has already been done for quite some time now. In India, we have had the annual series of meetings ECONOPHYS-KOLKATA and a few workshops on the subject in IMSc Chennai.

Ans. to Q. 4.

Yes, I believe my Institute should offer an elective course in econophysics as part of the coursework that all physics graduate students have to undergo. Yes, I have written a textbook in Econophysics (Sinha, Chatterjee, Chakraborti and Chakrabarti, "Econophysics: An Introduction" , Wiley-VCH)

Ans. to Q.5.

Yes, econophysics will be recognized soon by both economics and physics. I believe econophysics research will be recognized by Nobel prize in the discipline of Economics in about a decade. I would guess that the work on understanding fluctuation phenomena in financial markets will most likely win the first Nobel for econophysics. While the work on understanding inequality distributions is even more fundamental, I suspect that the real breakthrough discovery in this field is still in the future.

Victor Yakovenko University of Maryland, USA

Victor is one of the pioneers of 'Econophysics'. He is a Professor of Physics, University of Maryland, USA. He had his education from USSR. Did his M.S from Moscow Physical-Technical Institute and Ph.D. in theoretical physics from Landau Institute for Theoretical Physics, Moscow. His research interests are on unconventional superconductors, Andreev bound states in superconductors, strongly correlated materials close to a metal-insulator transition, physical effects of high magnetic fields, the quantum Hall effect, Econophysics: Applications of statistical physics to economics and finance etc. He has written a book entitled "Classical Econophysics" with A. F. Cottrell, P. Cockshott, G. J. Michaelson, and I. P. Wright, Published by Routledge (2009). Many of his articles have attracted media and have been published in news papers and periodicals written by reporters in American Scientist, Australian Financial Review, New Scientist, The Newyork Times Magazine etc. Together with J Barkley Rosser Jr, Victor had written an influential review on Econophysics in Review of Modern Physics (2009).

Ans. to Q.1
About 50 %.

Ans. to Q.2
Even more potential in the future.

Ans. to Q.3
Yes.

Ans. to Q.4
It is not going to happen at my university any time soon. I consider writing a book, but never have time.

Ans. to Q.5
Econophysics is already in the physics PACS classification, but not in the economics classification. There is no Nobel Prize in economics. Alfred Nobel was against it. After his death, the Bank of Sweden established a pseudo-Nobel prize to promote its political agenda. Given these controversies, I don't think a Nobel Prize in econophysics would be awarded any time soon, but it does not matter.

3 Conclusion

I had sent e-mail to about forty two number of esteemed Professors around the globe. I only received fifteen replies. Percentage of response is just 35.7 %, which is not encouraging. Among those replied, it was a mix bag of Physicists (60 %), Economists

(33.5 %), and rest exclusively as Econophysicists (6.5 %). Totally 46 % of them identified themselves as Econophysicists. It is encouraging to note that 31 % of them are
fully dedicated to econophysics research, whereas 33.3 % devote more than 50 % of
their time for econophysics research. Interestingly 80 % of them believe that studies
made under econophysics banner has made significant contributions and 60 % agree
it has contributed to economics and 0 % say it has contributed to physics. 84.6 % are
of opinion to form an Econophysics Forum/body and 80 % want international con-
ference/workshops, conducted under its banner in different countries by rotation. It
is certainly a big news that 80 % are interested for offering courses on econophysics
by their Universities. 53.3 % believe that 'econophysics' will be recognized as biophysics and geophysics soon and even 46.6 % expect econophysics research will be
recognized by Nobel Prize sooner or later that too in Economics and not in Physics.

Although this is a preliminary study and responses that I have received is not
significant in number, I believe that Econophysics certainly has made some impact
among natural as well as social scientists and is stepping forward to establish itself
as a new interdisciplinary field of science. My previous article 'Evolution of Econophysics [3] also established the same fact that econophysics has already shown a
strong presence in different institutes, publishing houses in the form of books and
journals etc. The word 'Econophysics' found in different published articles or papers
is going on increasing day-by-day almost linearly with time [4].

References

1. http://www.eoht.info/page/American+school+of+econophysics
2. K. Gangopadhyaya, Interview with Eugene H. Stanley, IIM Kozhikode Soc. Manag. Rev. 2(2)
   73–78 (SAGE Publications 2013)
3. K.C. Dash, *'Evolution of Econophysics' in Econophysics of Agent-Based Models* (Springer
   International Publishing, 2014), pp. 235–285
4. A. Ghosh, Econophysics research in India in last two decades. IIM Kozhikode Soc. Manag. Rev.
   2(2), 135–146 (SAGE Publications, 2013)

Commentary: Influence of Econophysics Research on Contemporary Researches in Social Sciences

Asim Ghosh

Abstract The term 'econophysics' was introduced about twenty years ago. The impact of the researches under this heading has now percolated both physics and economics. To search for the growth of its impact on physical and social sciences, we make a survey on papers or documents citing the term econophysics. The study indicates that although the field was initially started by physicists, it has now percolated noticeably in different social science disciplines.

1 Introduction

Econophysics is an interdisciplinary research field where physics tools are used to solve problem in economics. The term 'econophysics' was formally coined by H. E. Stanley in a statistical physics meeting held in Kolkata in 1995. Of course, from the beginning last century there were many physicists working on problems of economics. Examples of some such physicists are Daniel Bernoulli, Irving Fisher, Jan Tinbergen (won the first Nobel Prize in economics in 1969) etc.

Many perspective discussions on this recent development of econophysics researches have appeared (see e.g., [1, 2]). Several important books on econophysics researches [3–10] have been written, which are published by well known publishers like Cambridge University Press, Oxford University Press, Springer, Wiley, etc. and important conference proceedings published (e.g., Procs. Econophys-Kolkata series [11–17]). Also many well-known universities have started formal courses on 'econophysics' (see e.g. [2]).

Here we look for the statistics of papers or documents in *Google Scholar* (http://scholar.google.com/schhp) with several keywords:

(a) Number $N(y)$ of all the entries with the keyword 'econophysics' (each year y) anywhere in the document.

(b) We then look for the numbers $N_P(y)$ for all the entries with both 'econophysics' anywhere in the document and the words 'physics' or 'physica' or 'physical'

A. Ghosh (✉)
Condensed Matter Physics Division, Saha Institute of Nuclear Physics, Kolkata, India
e-mail: asim.ghosh@saha.ac.in

© Springer International Publishing Switzerland 2015
F. Abergel et al. (eds.), *Econophysics and Data Driven Modelling of Market Dynamics*,
New Economic Windows, DOI 10.1007/978-3-319-08473-2_14

in 'Publication'. This gives the subset of entry in $N(y)$ which are published in physics journals (the names of which will either have the word 'physics' or 'physical' or 'physica').

(c) We then look for the number $N_E(y)$ for the entries with the word 'econophysics' and the words 'economic' or 'social' or 'financial' in 'Publication'. This $N_E(y)$ gives the subset of entries $N(y)$ which are published in economics or finance journals.

(d) Next, we look for the number of entries $N_W(y)$ with both the words 'econophysics' and 'working paper' anywhere in the document. This word 'working paper' is uniquely used by social scientists (economists in particular) for their unpublished work. This terms is never used by physicists. Hence this number $N_W(y)$ gives clearly number of documents in *Google Scholar* for year y, written by social scientists or economists in particular.

Of course $N(y) \geq N_P(y) + N_E(y) + N_W(y)$ and we look for the relative dominance of N_P, N_E and N_W as a function of time or year y.

2 Result

It is clear from Fig. 1 that since the year 2011, N_W has crossed over N_P, suggesting more entries by social scientists are coming to *Google Scholar* than published in physics journals on econophysics (see inset of Fig. 1).

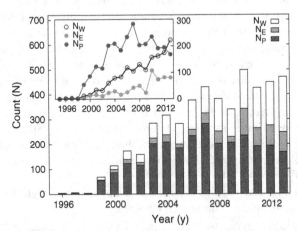

Fig. 1 Histogram shows the count $N(y)$ of articles containing the term 'econophysics' anywhere in the articles over years (y) with specific search options. The relative contribution from different search options are indicated by different shades. The *inset* shows number of articles published in physics journals $(N_P(y))$ or in economics journals $(N_E(y))$. N_P seems to become saturated beyond 2005 while N_E is continuously growing. Note, the number N_W of the documents having both the terms 'econophysics' and 'working paper' anywhere within the document has been increasing very fast over years (y), indicating increasing interest in econophysics by social scientists. All data are taken from *Google Scholar* site

3 Conclusion

Here we have surveyed (from *Google Scholar*) the documents or entries having the term 'econophysics' anywhere in the document to assess growth of econophysics researches in different main stream disciplines. Our study indicates that although the field was initiated by physicists, the field has now percolated in other disciplines. From the year 2011 onwards, N_W has crossed over N_P, suggesting more entries by social scientists are coming to *Google Scholar* than published in physics journals on econophysics researches. Note that all the publications or preprints on econophysics today do not have the word 'econophysics'. Only the books recently published (e.g., Refs. [5–17]) have the word in their title and any reference to such books appear in these entries. Hence the real number of social science papers or documents on econophysics related researches must be much higher. In any case, the growth of the influence econophysics research on social science researches today seems quite perceptible, as our study clearly indicates.

Acknowledgments I am grateful to Prof. Bikas K. Chakrabarti for suggesting the survey and to Dr. Arnab Chatterjee for useful comments.

References

1. M. Buchanan, What has econophysics ever done for us? Nat. Phys. **9**, 317 (2013)
2. A. Ghosh, Econophysics research in India in the last two decades. IIM Kozhikode Soc. Manag. Rev. **2**, 135 (2013). arXiv:1308.2191
3. R.N. Mantegna, H.E. Stanley, *An Introduction to Econophysics: Correlations and Complexity in Finance* (Cambridge University Press, Cambridge, 1999)
4. B.M. Roehner, *Patterns of Speculation: A Study in Observational Econophysics* (Cambridge University Press, Cambridge, 2005)
5. J.L. McCauley, *Dynamics of Markets: Econophysics and Finance* (Cambridge University Press, Cambridge, 2007)
6. S. Sinha, A. Chatterjee, A. Chakraborti, B.K. Chakrabarti, *Econophysics: An Introduction* (Wiley-VCH, 2010)
7. H. Aoyama, Y. Fujiwara, Y. Ikeda, H. Iyetomi, W. Souma, H. Yoshikawa, *Econophysics and Companies: Statistical Life and Death in Complex Business Networks* (Cambridge University Press, Cambridge, 2011)
8. B.K. Chakrabarti, A. Chakraborti, S.R. Chakravarty, A. Chatterjee, *Econophysics of Income & Wealth Distributions* (Cambridge University Press, Cambridge, 2013)
9. P. Richmond, J. Mimkes, S. Hutzler, *Econophysics and Physical Economics* (Oxford University Press, Oxford, 2013)
10. F. Slanina, *Essentials of Econophysics Modelling* (Oxford University Press, Oxford, 2014)
11. A. Chatterjee, S. Yarlagadda, B.K. Chakrabarti (eds.), *Econophysics of Wealth Distributions* (Springer, Milan, 2005)
12. A. Chatterjee, B.K. Chakrabarti (eds.), *Econophysics of Stock and other Markets* (Springer, Milan, 2006)
13. A. Chatterjee, B.K. Chakrabarti (eds.), *Econophysics of Markets and Business Networks* (Springer, Milan, 2007)
14. B. Basu, B.K. Chakrabarti, S.R. Chakravarty, K. Gangopadhyay (eds.), *Econophysics & Economics of Games, Social Choices and QuantitativeTechniques* (Springer, Milan, 2010)

15. F. Abergel, B.K. Chakrabarti, A. Chakraborti, M. Mitra (eds.), *Econophysics of Order-driven Markets* (Springer, Milan, 2011)
16. F. Abergel, B.K. Chakrabarti, A. Chakraborti, A. Ghosh (eds.), *Econophysics of Systemic Risk and Network Dynamics* (Springer, Milan, 2013)
17. F. Abergel, H. Aoyama, B.K. Chakrabarti, A. Chakraborti, A. Ghosh (eds.), *Econophysics of Agent-based Models* (Springer International Publishing, Switzerland, 2014)

Photo

Photograph of a section of the participants in Econophys-Kolkata VIII meeting

© Springer International Publishing Switzerland 2015

F. Abergel et al. (eds.), *Econophysics and Data Driven Modelling of Market Dynamics*,
New Economic Windows, DOI 10.1007/978-3-319-08473-2

Printed in the United States
by Bookmasters